T0338536

DENSITY MATRIX AND TENSOR NETWORK RENORMALIZATION

Renormalization group theory of tensor network states provides a powerful tool for studying quantum many-body problems and a new paradigm for understanding entangled structures of complex systems. In recent decades, the theory has rapidly evolved into a universal framework and language employed by researchers in fields ranging from condensed matter theory to machine learning. This book presents a pedagogical and comprehensive introduction to this field for the first time. After an introductory survey of the major advances in tensor network algorithms and their applications, it introduces, step by step, the tensor network representations of quantum states and the tensor-network renormalization group methods developed over the past three decades. Basic statistical and condensed matter physics models are used to demonstrate how the tensor network renormalization works. An accessible primer for scientists and engineers, this book would also be ideal as a reference text for a graduate course in this area.

TAO XIANG is Professor at the Institute of Physics, Chinese Academy of Sciences (CAS), and specializes in condensed matter theory. He is a CAS academician and a fellow of the World Academy of Sciences, and has received the He-Leung-He-Lee Prize for Scientific and Technological Progress, among other awards.

DENSITY MATRIX AND TENSOR NETWORK RENORMALIZATION

TAO XIANG
Chinese Academy of Sciences

Shaftesbury Road, Cambridge CB2 8EA, United Kingdom

One Liberty Plaza, 20th Floor, New York, NY 10006, USA

477 Williamstown Road, Port Melbourne, VIC 3207, Australia

314–321, 3rd Floor, Plot 3, Splendor Forum, Jasola District Centre, New Delhi – 110025, India

103 Penang Road, #05–06/07, Visioncrest Commercial, Singapore 238467

Cambridge University Press is part of Cambridge University Press & Assessment, a department of the University of Cambridge.

We share the University's mission to contribute to society through the pursuit of education, learning and research at the highest international levels of excellence.

www.cambridge.org
Information on this title: www.cambridge.org/9781009398701

DOI: 10.1017/9781009398671

First published 2023

A catalogue record for this publication is available from the British Library.

A Cataloging-in-Publication data record for this book is available from the Library of Congress.

ISBN 978-1-009-39870-1 Hardback

Contents

Preface

Renormalization group theory of tensor network states has rapidly evolved as a universal framework and language employed by researchers in different fields, from condensed matter theory to statistical physics, quantum field physics, quantum chemistry, quantum information, machine learning, and many other fields. It provides not only a systematical and powerful tool for studying quantum many-body problems but also a new paradigm for understanding entangled structures of complex systems. This field was pioneered by the Nobel laureate K. G. Wilson in the mid-1970s, when he devised a numerical renormalization group method (now named after him) to solve the famous Kondo impurity model. However, it did not enter a fast development track until S. R. White published his seminal work on the density matrix renormalization group (DMRG) in 1992. Since then, the boundaries of the field have greatly expanded, from one to higher dimensions, from real space to an arbitrary nonlocal basis space, from zero temperature to thermodynamics, from pure spin systems to interacting fermion or boson systems, and from static to dynamical or even nonequilibrium evolutions.

The essential property that makes the renormalization group theory of tensor network states appealing is that it allows for an exponential compression of quantum many-body states by modeling their entanglement degrees of freedom. It leads to the establishment of several tensor network methods for accurately and effectively solving complex quantum problems and analytic tools for classifying entangled or topological quantum phases of matter. These methods have laid the foundation for variationally determining the physical properties of complex many-body systems and significantly broadened their spectra of applications.

Over the past three decades, more than ten thousand research articles have been published in this field. However, a systematical introduction to all these methods is still not available. It is difficult for students or junior research scientists to gain a comprehensive picture of this field and to know the subtle differences between various algorithms without reading numerous references written by different authors with different styles. It might be even more difficult for research scientists interested in the progress of tensor network renormalization but not working directly in quantum many-body theory to know where to start. I hope this book will provide a pedagogical and systematic introduction to the renormalization group theory of tensor network states so that it can become a valuable reference for those learning or doing research with this theory.

The book starts with an introductory survey on the major advances in tensor network methods and their applications. After a brief description of the elementary algebra of tensors, it provides an in-depth introduction to the tensor network representations of quantum states and the tensor network renormalization group methods developed in the past three decades, which include but are not limited to the density matrix renormalization group and its variant extensions, matrix product states, projected entangled pair states, variational approaches for optimizing tensor network states, approaches for evaluating low-energy excitation spectra or time-dependent correlation functions, and coarse-graining tensor renormalization group methods. The implementation, including optimization, parallelization, and symmetries, is an essential part of a method and is discussed in the relevant chapter. The performance, including computational cost, stability, and accuracy, is also discussed.

As the relevant literature is enormous, this book is meant to be an introductory text, not a review on any of the subjects discussed. For this reason, the author limits the original citations to those that are essential historically or pedagogically or to those for which the reader may seek further details. Nevertheless, I apologize to many authors whose contributions are not properly represented.

This book intends to benefit senior undergraduate and graduate students, research scientists, and software engineers learning the theory of tensor network renormalization. It would be suitable as a reference for a specialized course. It can also be divided into many small modules and used as a textbook. A prerequisite for reading this book is familiarity with quantum and statistical mechanics, the theory of many-body systems, and linear algebra.

A book like this cannot be written without a great deal of help from others. I would like to express my gratitude to Mingpu Qin, Honghao Tu, and Zongsheng Zhou for their invaluable assistance in the proofreading of the whole text. Moreover, I am indebted to Runze Chi, Ruizheng Huang, Xuan Li, Yang Liu, Zhiyuan Liu, and Haidong Xie for providing the numerical results with the figures used to demonstrate the methods presented in the book. Finally, I want to thank my students, colleagues, and collaborators for making the exploration of tensor network renormalization the pleasure it has been. Although it would be impossible to list them all here, I would particularly like to thank Ruizhen Huang, Wei Li, Haijun Liao, Bruce Normand, Zhiyuan Xie, and Liping Yang for valuable discussions and comments that played a decisive role in shaping the text. Of course, only I am responsible for any errors that may remain in this book.

Unit Used

In this monograph, the reduced Planck constant $\hbar$ and the Boltzmann constant k_B are set to 1 unless otherwise specified.

Notations and Graphical Representations

Without particular specification, the following notations or conventions are used:

- d: dimension of physical basis states at each lattice site
- D: virtual bond dimension or the number of states retained after each DMRG truncation
- D_0: virtual bond dimension before truncation
- σ: spin or local basis index

- The leading eigenvalue of a matrix A is the largest among all eigenvalues in their absolute values. It is also simply called the largest eigenvalue of A without explicitly mentioning the phrase of absolute value. The corresponding eigenvectors are referred to as the leading or largest eigenvectors.
- U or V: without specification, U or V represents a unitary matrix obtained from a singular value decomposition of the matrix introduced in §2.4. It satisfies the equations

$$U^\dagger U = U U^\dagger = I, \qquad (0.1)$$
$$V^\dagger V = V V^\dagger = I. \qquad (0.2)$$

After the column dimension of U (similarly V) is truncated, it becomes a rectangular matrix whose column dimension n is smaller than the row dimension m. In this case, $U_{m \times n}$ becomes an isometric matrix, satisfying the equation

$$U^\dagger U = I_{n \times n}, \qquad U U^\dagger \neq I_{m \times m}. \qquad (0.3)$$

- Q: A matrix obtained from a QR or LQ decomposition. It could be either a unitary matrix if it is a square matrix, satisfying the equation

$$Q^\dagger Q = Q Q^\dagger = I, \qquad (0.4)$$

or an isometric matrix if $Q_{m \times n}$ is a rectangular matrix, satisfying the equation

$$Q^\dagger Q = I_{n \times n} \qquad Q Q^\dagger \neq I_{m \times m} \qquad (m > n), \qquad (0.5)$$

$$QQ^\dagger = I_{m\times m} \qquad Q^\dagger Q \neq I_{n\times n} \qquad (n > m). \tag{0.6}$$

- R: The upper triangular matrix generated by a QR decomposition.
- L: The lower triangular matrix generated by an LQ decomposition.

Graphical Representations

- Vector v: $v_\alpha = $ α

- Matrix A: $A_{\alpha\beta} = $ α —A— β

  The rightward rule: The indices run from left to right (i.e. rightwards).
- Transpose A^T:

$$\alpha -\!\!\fbox{A^T}\!\!- \beta \;\; = \;\; \beta -\!\!\fbox{A}\!\!- \alpha \tag{0.7}$$

- Hermitian conjugate $A^\dagger$:

$$\alpha -\!\!\fbox{$A^\dagger$}\!\!- \beta \;\; = \;\; \beta -\!\!\fbox{A^*}\!\!- \alpha, \tag{0.8}$$

A^* is the complex conjugate of A.
- Unitary matrix U: —U—
- Projection matrices: A pair of matrices

$$A \;=\; -\!\!\fbox{A}\!- \;, \qquad B \;=\; -\!\fbox{B}\!\!- \tag{0.9}$$

form a projection operator $P = AB$, which is idempotent $P^2 = P$, if

$$-\!\fbox{B}\!\!-\!\!\fbox{A}\!- \;=\; I. \tag{0.10}$$

The thicker bond has a higher dimension (D_0) than the thinner one (D) (i.e., $D_0 > D$). This projection is *isometric* if $B = A^\dagger$, otherwise *oblique* if $B \neq A^\dagger$. In the former case, A or B is said to be isometric. In the latter case, A and B form an oblique projection pair.
- Diagonal matrix: A matrix λ, which is not necessary to be a square matrix, is defined as a diagonal matrix if its matrix elements λ_{ij} are nonzero only when $i = j$. Graphically, it is represented as

$$\lambda = -\!\!\diamond\!\!- \;. \tag{0.11}$$

- Vectorized tensor: A matrix or tensor can be converted to a vector by grouping all its legs as one leg. For example, a three-legged tensor can be converted to a three-legged vector as

$$A_{\alpha,\beta}[\sigma] = \quad \alpha -\!\!\left(\!A\!\right)\!\!- \beta \quad = \quad \left(\!A\!\right)\!\!\!\begin{array}{c} -\alpha \\ \sigma \\ -\beta \end{array} \quad = \quad \begin{array}{c} \alpha\ \sigma\ \beta \\ \cup\!\cup\!\cup \\ \left(\!A\!\right) \end{array} \quad = A_{\alpha\sigma\beta}. \quad (0.12)$$

We use the same symbol to represent a tensor and its corresponding multi-leg vector.

Counter-Clockwise Rule

(i) The product of two matrices, A and B, runs from left to right

$$(AB)_{\alpha\beta} = \quad \alpha -\!\!\left(\!A\!\right)\!\!-\!\!\left(\!B\!\right)\!\!- \beta. \quad (0.13)$$

(ii) If B is above or below A, the matrix indices propagate along a counter clockwise path

$$(AB)_{\alpha\beta} = \qquad \begin{array}{c} \beta -\!\!\left(\!B^T\!\right)\!\!\rceil \\ \alpha -\!\!\left(\!A\!\right)\!\!\rfloor \end{array} = \begin{array}{c} \lceil\!\!\left(\!A^T\!\right)\!\!- \alpha \\ \left(\!B\!\right)\!\!- \beta \end{array}. \quad (0.14)$$

In other words,

$$\alpha -\!\!\left(\!A\!\right)\!\!-\!\!\left(\!B^\dagger\!\right)\!\!- \beta \quad = \quad \begin{array}{c} \beta -\!\!\left(\!B^*\!\right)\!\!\rceil \\ \alpha -\!\!\left(\!A\!\right)\!\!\rfloor \end{array}. \quad (0.15)$$

Matrix or Tensor Decompositions

(i) Singular value decomposition (SVD) of matrix: Matrix $A_{m \times n}$ can be diagonalized by two unitary matrices of dimensions m and n, respectively

$$A = U\lambda V, \qquad -\!\!\left(\!A\!\right)\!\!- \quad = \quad -\!\!\left(\!U\!\right)\!\!-\!\!\left\langle\!\right\rangle\!\!-\!\!\left(\!V\!\right)\!\!- , \quad (0.16)$$

where $U_{m \times m}$ and $V_{n \times n}$ are the unitary matrices, λ is an $m \times n$ diagonal matrix whose elements λ_{ij} are zero if $i \neq j$.

If the rank of A, or the total number of nonzero diagonal elements of λ, is r, one can also set λ as a $r \times r$ square matrix. In this case, U becomes an $m \times r$ matrix, which is left isometric if $m > r$. Similarly, V becomes a $r \times n$ matrix, which is right isometric if $n > r$.

(ii) Factorization of tensor: A three or higher-order tensor can be factorized as a product of two tensors by SVD. To do this, the tensor indices are separated into two groups: one is regarded as the row index, and the other is regarded as the column index. This converts the tensor into a matrix that can be factorized by SVD. For example, for a four-indexed tensor A_{ijkl}, we can group (ij) together and (kl) together to convert it into a matrix $A_{ij,kl}$ and factorize it by SVD:

$$i -\!\!\!\!\underset{j}{\overset{l}{(A)}}\!\!\!\!- k \;=\; i -\!\!\underset{j}{\overset{l}{(U)}}\!\!\overset{m}{-}\!\!\diamondsuit\!\overset{n}{-}\!(V)- k \;, \qquad (0.17)$$

where both $U_{ij,m}$ and $V_{n,kl}$ are unitary matrices. λ is the nonnegative singular value matrix. By absorbing one square-root of λ into U and the other square-root of λ in V, we obtain two new tensors:

$$-\!(B)\!- \;=\; -\!(U)\!-\!\!\diamondsuit_{\sqrt{\lambda}} \;, \qquad -\!(C)\!- \;=\; \diamondsuit_{\sqrt{\lambda}}\!-\!(V)\!- . \qquad (0.18)$$

This allows us to factorize A as a product of two tensors:

$$-\!(A)\!- \;\overset{}{\underset{\text{SVD}}{=\!=\!=}}\; -\!(B)\!-\!(C)\!- . \qquad (0.19)$$

Here "SVD" is explicitly added to emphasize that the factorization is obtained with SVD and B and C are defined according to the convention (0.18).

Isometric Tensors

(i) Left canonical tensor: $A[\sigma] = -\!\!\overset{|}{(A)}\!\!- $.

It satisfies the left isometric condition

$$\begin{array}{c} \boxed{A^*}\!\!-\beta \\ \boxed{A}\!\!-\alpha \end{array} \;=\; \left(\sum_\sigma A^\dagger[\sigma]A[\sigma] \right)_{\beta,\alpha} \;=\; \delta_{\alpha,\beta}. \qquad (0.20)$$

(ii) Right canonical tensor: $A[\sigma] = -\!\!\overset{|}{(A}\!\!- $.

It satisfies the right isometric condition

$$\begin{array}{c} \beta -\!\!\boxed{A^*} \\ \alpha -\!\!\boxed{A} \end{array} \;=\; \left(\sum_\sigma A[\sigma]A^\dagger[\sigma] \right)_{\alpha,\beta} \;=\; \delta_{\alpha,\beta}. \qquad (0.21)$$

Transfer Matrix $\mathcal{E}$ of a Uniform MPS

$\Psi(A)$ is an infinite uniform MPS with A the local tensor at each site:

$$\Psi(A) = \quad \cdots \quad -\!\!\!\!\!\!\boxed{A}\!\!-\!\!\boxed{A}\!\!-\ \cdots\ -\!\!\boxed{A}\!\!-\!\!\boxed{A}\!\!-\ \cdots\ . \tag{0.22}$$

- Transfer matrix of $\Psi(A)$ is defined as

$$\mathcal{E}_{\alpha'\alpha,\beta'\beta} \ = \ \sum_{\sigma} A^{*}_{\alpha'\beta'}[\sigma] A_{\alpha\beta}[\sigma] \ = \ \begin{array}{c} \alpha' -\!\boxed{A^{*}}\!- \beta' \\ | \\ \alpha -\!\boxed{A}\!- \beta \end{array} . \tag{0.23}$$

- The dominant left eigenvector $\langle l|$, right eigenvector $|r\rangle$, and the corresponding eigenvalue η of $\mathcal{E}$ satisfy the equations

$$\boxed{l}\!\!\begin{array}{c} \boxed{A^{*}} \\ | \\ \boxed{A} \end{array} \ = \ \eta \ \boxed{l} \ , \quad \begin{array}{c} \boxed{A^{*}} \\ | \\ \boxed{A} \end{array}\!\!\boxed{r} \ = \ \eta \ \boxed{r} \ . \tag{0.24}$$

Abbreviations

AKLT:	Affleck–Kennedy–Lieb–Tasaki
CTMRG:	corner transfer matrix renormalization group
DMRG:	density matrix renormalization group
HOSVD:	higher-order singular value decomposition
HOTRG:	HOSVD-based tensor renormalization group
HOSRG:	HOSVD-based second renormalization group
KT:	Kosterlitz–Thouless
Loop-TNR:	loop tensor network renormalization
MPO:	matrix product operator/operators
MPS:	matrix product state/states
NRG:	numerical renormalization group
PEPS:	projected entangled pair state/states
PESS:	projected entangled simplex state/states
QTMRG:	quantum transfer matrix renormalization group
RG:	renormalization group
SSS:	simplex solid state/states
SVD:	singular value decomposition
TDVP:	time-dependent variational principle
TEBD:	time-evolving block decimation
iTEBD:	infinite TEBD
TMRG:	transfer matrix renormalization group
TNO:	tensor network operator/operators
TNR:	tensor network renormalization
TRG:	tensor renormalization group
VBS:	valence bond solid

1

Introduction

1.1 Quantum Many-Body Problems

According to legend, chess was invented by Grand Vizier Sissa Ben Dahir and given to King Shirham of India. The king was so pleased with the game that he offered to grant Sissa any request within reason. The Grand Vizier asked the king for one grain of wheat to be placed on the first square of the chessboard, two grains on the second square, four on the third, and so on, doubling the amount each time until all 64 squares were occupied. The king, arithmetically unaware, accepted the request. However, all the wheat of his kingdom was not enough to fulfill this offer because the total number of grains the Grand Vizier asked for was

$$1 + 2 + 4 + 8 + \cdots + 2^{63} = 2^{64} - 1 = 18,446,744,073,709,551,615.$$

This amount of wheat is approximately 80 times what would be produced in one harvest at current yields if all of Earth's arable land could be devoted to wheat. This number is significant because the number of grains grows exponentially as the number of chessboard squares increases. It demonstrates the rapid growth of exponential sequences.

This wheat-and-chessboard problem is just what is encountered in the study of quantum many-body theory, in which the Hilbert space grows exponentially with the system size. It limits the application range of many computational methods, such as exact diagonalization. Considering, for example, a system of N interacting electrons, if each electron has d-degrees of freedom, then the total dimension of the Hilbert space of the system is $M = d^N$. To determine the ground state of the system, one needs to minimize the energy within this M-dimensional space. The maximal M that is feasible with the best computer software and hardware currently available is about 10^9. It means that only $N \sim 30$ electrons can be handled, even if $d = 2$ and no extra information (for example, symmetry) is used.

The exponential growth of the Hilbert space with the system size is one of the most severe problems we face in the theoretical study of quantum many-body systems. As vividly pointed out by the Nobel Laureate Walter Kohn [1], it raises an *exponential wall* that hinders the development of quantum many-body theory. In the early stage of quantum mechanics, soon after the successful interpretation of optical spectra of hydrogen atoms and other simple atoms by the Schrödinger equation, many physicists, including particularly Paul Dirac, declared that chemistry had come to an end because its content was entirely contained in the powerful Schrödinger equation. However, as it was realized later, this equation was far too complex to allow a solution for a system with many electrons.

In the search for fundamental interactions and elementary particles, the philosophy of *reductionism* has played an important role. It assumes that the nature of a complex world can be understood by reducing it to the interactions of its parts and that a complex system is nothing but the sum of its parts and can reduce to the account of individual constituents.

From the end of the nineteenth century, with the fast progress in the experimental exploration of the microscopic quantum world, a great many novel phenomena, including superconductivity, helium superfluid, and quantum Hall effect, were discovered. It turns out that the behavior of large and complex aggregates of particles is difficult to understand in terms of the extrapolation of a few particles. Instead, at each level of complexity, entirely new phenomena appear. For example, two hydrogen atoms can form a hydrogen molecule via the valence bond, and many hydrogen molecules can form liquid hydrogen at low temperature and even solid hydrogen under high pressure. On the other hand, if hydrogen molecules mix with oxygen molecules, they will undergo a chemical reaction to form water molecules. Hydrogen atoms behave differently under different circumstances.

Emergence is the way complex systems and phenomena arise out of relatively simple interactions and is central to the theories of integrative levels and complex systems [2]. The exponential wall is an unavoidable problem in the investigation of emergent phenomena. It is a *non-perturbative problem* and cannot be completely solved by conventional quantum field theory, which is established based on perturbation. A detailed description of quantum many-body systems poses formidable difficulties due to the exponentially large dimension of the associated Hilbert space. Certain approximations have to be taken to solve this problem.

In general, two kinds of approximations are used. One is to take a single particle approximation to convert a many-body problem into a single-particle one. The most commonly used methods in this direction include the self-consistent mean-field theory and the first-principles density functional theory. The other is to take a many-body approximation by selecting a finite many-body basis set to represent a target state or physical quantity defined in an intractably large Hilbert space.

It includes, for example, the configuration interaction, coupled-cluster expansion, quantum Monte Carlo methods, and tensor network renormalization group methods to be introduced in this book.

The *renormalization group* (RG) was initially devised in particle physics. It serves as the primary means for constructing the connections between theories at different length scales and becomes a powerful and efficient tool for exploring the systems where perturbation theory fails. Starting from some microscopic Hamiltonian, certain degrees of freedom are iteratively integrated and accounted for by modifying the original Hamiltonian. The new Hamiltonian contains modified couplings by a priori elimination of degrees of freedom. It is tempting to believe that this "renormalized" effective Hamiltonian captures the essential physics of the system on a narrower energy scale.

This method is rooted in modern physics and has played a crucial role in the fundamental theory of microscopic particles and the theory of condensed matter physics, and continuous phase transitions. In quantum field theory, renormalization was first introduced to cancel infinities by redefining parameters at different energy or momentum scales. In condensed matter and statistical physics, a more general RG framework based on scaling analyses was introduced to explain the universality properties of continuous phase transitions. It was even extended to compute wave functions for quantum lattice models directly.

The idea of RG was first anticipated by Stueckelberg and Peterman in 1951 [3] and reformulated in 1953 [4]. It emerges from the renormalization of field variables under a scale transformation. RG is not a group. It just forms a semigroup since the RG transformation is not reversible. It was dubbed as a group probably because in the 1950s, when the RG concept was first proposed, a doctrine of physics was that our world could be understood in terms of symmetries and their implementation through groups. However, the connection between RG transformation and group structure is highly formal, and its group interpretation is almost useless in practical applications.

The modern RG in terms of flow equations was introduced by Gell-Mann and Low [5] in 1954 and reformulated by Callan [6] and Symanzik [7] in 1970. Their essential assumption was that a renormalizable quantum field theory is scale-invariant, which depends on the parameters but not the scale. Thus under the scale transformation, the effective theory makes a self-similar replica of itself, with tiny changes in coupling constants determined by the flow equations.

The application of RG has achieved great success in studying quantum field theory. For example, a renormalization scheme developed by Feynman, Schwinger, and Tomonaga led them to solve the ubiquitous problem of infinities by expressing physical observables in terms of parameters. Their theory was spectacularly successful in quantum electrodynamics. They received the Nobel Prize in 1965.

In the early 1970s, 't Hooft and Veltman showed that Yang-Mills gauge theories are renormalizable, for which they shared the 1999 Nobel Prize. In 1973, Gross, Politzer, and Wilczek found that the beta function describing the renormalization flow equation of the strong interaction is negative and discovered the asymptotic freedom in quantum chromodynamics. They won the Nobel Prize in 2004.

In the second half of the 1960s, ideas to recursively generate flows of coupling constants arose independently in condensed matter physics, which led to a deeper understanding of the physical meaning of RG. In 1966, Kadanoff [8] proposed a *block-spin* idea to define the interactions at large distances as aggregates of components at shorter distances. A block spin transformation consists of the scale and the block spin transformations. This block spin approach, together with many vital contributions of Kenneth Wilson, laid the foundation of the scaling theory of second-order phase transitions and critical phenomena. The success of their approach rests on scale separation. At a critical point, the correlation length diverges and short-range fluctuations can affect long-wavelength behavior quantitatively but not qualitatively.

In the first half of the 1970s, Wilson began to apply the RG techniques to problems not pertaining to a critical point or the computation of ground-state wave functions for quantum systems. His work pioneered the idea of RG and allowed it to be implemented for any physical system, even away from critical points where the scaling invariance breaks. It extended the field of RG from the simple flow equations of coupling parameters to the whole Hilbert space and opened the field of the *numerical renormalization group* (NRG) [9]. He demonstrated the potential of this powerful method by constructing a successive RG solution of the famous Kondo impurity model, which describes a single magnetic impurity in a nonmagnetic metal. Wilson was awarded the Nobel prize partly for this contribution in 1982.

The success of Wilson's method in the single magnetic impurity problem relies on two peculiar features of the Kondo model. First, the width of the Kondo resonance sets an energy scale such that the contributions of energy levels far from the resonance can be integrated out. Second, the Kondo Hamiltonian can always be mapped onto an effective one-dimensional model since the Kondo interaction couples only with the s-wave part of the electron wave function. It dramatically lowers the barrier to solving this problem.

There are two ways in which NRG differs from the conventional (analytic) RG: First, in the conventional RG of quantum field theory, it is the charge, mass, interacting coupling constants, and a few other physical parameters renormalized during the reduction of energy scales. In contrast, in NRG, it is the wave function of a quantum state or the partition function of a Hamiltonian that is calculated using the RG transformation. A wave function or partition function contains the information needed to evaluate all physical observables. Second, in the conventional RG calculation, the scaling invariance is assumed to keep the formula of interacting potentials unchanged. This assumption is no longer needed in the implementation of NRG.

1.2   From NRG to DMRG

Shortly after Wilson's dramatic success in solving the Kondo problem, there was considerable excitement about the possibility of applying the real-space NRG together with the Kadanoff blocking spin scheme [8] to a variety of quantum lattice problems with lattice sites replacing energy levels [10, 11, 12, 13]. The basic idea was to take Kadanoff's block spins as building cells and aggregate every two cells into a new block at each RG transformation. It was hoped that the ground-state properties of other many-body systems, for example, the Hubbard model [14], could be solved by this approach. However, the performance of Wilson's NRG was poor in treating these many-body models. For example, the error of the ground-state energy obtained with this method for the one-dimensional Hubbard model with 16 sites by keeping about 1,000 basis states is about 5–10% [15].

The essence of Wilson's NRG is to select a small set of basis states to represent the ground state or other targeted states of a Hamiltonian through successive local basis transformations. In the NRG calculations, the basis space is not fixed, unlike in the exact diagonalization. Instead, it changes dynamically at each step of RG iterations. The truncation error is determined by the criterion used in the basis space truncation. Of course, an optimal RG scheme should use a criterion that minimizes the error in the basis truncation. In addition, the number of basis states retained is a key parameter that can affect the truncation error – the more basis states retained, the smaller the truncation error.

A basic assumption of Wilson's NRG is that the ground state of a large system is determined by the low-lying excitations of the building blocks, whose weight is determined by the *thermal density matrix*

$$\rho = e^{-\beta H} \tag{1.1}$$

up to a normalization constant. Here, $\beta = 1/k_B T$, k_B is the Boltzmann constant, and T is temperature. Using the eigenvalues and eigenstates of the Hamiltonian

$$H|\Psi_n\rangle = E_n|\Psi_n\rangle, \tag{1.2}$$

the density matrix can also be expressed as

$$\rho = \sum_n e^{-\beta E_n}|\Psi_n\rangle\langle\Psi_n|. \tag{1.3}$$

The weight of the eigenstate $|\Psi_n\rangle$ in the partition function is proportional to $\exp(-\beta E_n)$. Thus, the lower the energy, the larger the weight. This criterion is not optimal because some highly excited states in a small system may lower their energies and become increasingly important with the increase of the system size. However, in Wilson's NRG, energy is the only parameter used for judging whether a state is retained or discarded.

The block spin NRG starts by considering a block B where the Hamiltonian H_B and the operators at the two ends of the blocks are defined. Then a new block of the same size is added to the system, which augments the Hamiltonian to H_{BB}. After diagonalizing this Hamiltonian, the enlarged block is replaced by a new effective block B_{new}, formed by D lowest eigenstates of H_{BB}, and the iteration continues, where D is the number of basis states retained. A schematic representation of these NRG iteration steps is

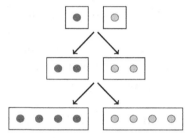

At each step, a block (left) is added with a duplicated block (right) to form an augmented block after the basis truncation.

This block spin NRG scheme was applied to the one-dimensional Hubbard model by Bray and Chui [10] in 1979, and later to the one-dimensional Heisenberg model by Pan and Chen [11], and by Kovarik [12]. Unlike the Kondo impurity model, their results were not that encouraging. The reason for this lies in the physical difference between the Kondo problem and the translationally invariant quantum lattice models. The most significant difference between the Kondo system and a one-dimensional lattice model is that the couplings between adjacent sites decrease exponentially in the Kondo system under Wilson's logarithmic discretization scheme. In contrast, it remains constant in the Hubbard or Heisenberg model. This exponential decrease is the key to the success of the method for the Kondo impurity systems. However, it is flawed in treating quantum lattice problems in a real-space blocking form.

The failure of Wilson's block spin NRG, as pointed out by White and Noack [16], results from the boundary or interface effect between blocks. This can be understood by considering a toy model – a single particle on a tight-binding chain. For this model, the ground state is a standing wave that vanishes at the two ends of each block. As an example, Fig. 1.1 shows how the ground-state wave function

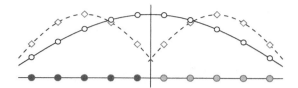

Figure 1.1 Ground-state wave functions of a particle hopping on a tight-binding chain. Two five-site blocks form a large block of ten sites. The ground state is a standing wave on each block. The wave function from the ground state of the two five-site blocks (diamonds) vanishes at the interface where the ground-state wave function of the ten-site block (circles) shows a maximum.

changes when a ten-site block is formed by combining two five-site blocks. The ground state is a standing wave on each block. The wave function takes a minimum at the interface of two small blocks. By combining two blocks into a larger one, the lowest-lying states of each block have nodes at the compound block center. On the contrary, the ground state of the compound block shows a maximum there. It suggests that the basis states of a larger block cannot be accurately approximated by a restricted number of states of smaller blocks without adequately considering the interface effect.

In 1992, Xiang and Gehring showed that Wilson's NRG could be improved by adding just one spin, instead of a block spin, at each iteration [13]. Their approach reduces both the truncation error and the interface effect and improves the accuracy of the ground-state energy of the one-dimensional Heisenberg model by a few orders of magnitude [13, 17]. By keeping about 200 states, for example, they obtained an error of about 0.5% in the ground-state energy. Their approach is similar to that used by Wilson for treating a single-impurity Kondo problem. The difference is that in the Kondo impurity problem, there are distinct energy scales to separate each lattice point in the effective Hamiltonian, and it is natural to add one site at each step of RG transformation.

This simple approach has two advantages compared to the block spin NRG. First, as shown in Fig. 1.2, this approach reduces the boundary error encountered in the double-blocking NRG scheme. Second, in the conventional NRG algorithm, the total number of basis states is squared by combining two blocks into a larger block. Namely, if the number of states at each block is D, then the total number of states is D^2. After truncation, only D of them are retained. However, in Xiang and Gehring's approach [13], the total number of states is dD with d the number of basis states at each lattice site. After truncation, the number of states retained is still D, but its percentage in the total number of states is much higher than in the former case since $d \ll D$. It significantly reduces the truncation error. Furthermore, because the total number of states in this improved NRG scheme is just dD, a larger D can be used in the RG calculation, which can also improve the accuracy.

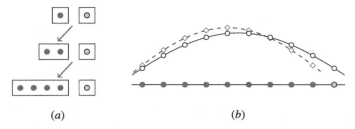

Figure 1.2 (a) Illustration of the one-site growing scheme. After obtaining the new block from the previous step, a new site is added. The size of the block grows by one at each iteration step. (b) The lowest-energy states for a particle on a tight-binding chain. A ten-site block is obtained by adding the rightmost site to a nine-site block.

In 1992, White invented the *density matrix renormalization group* (DMRG) [18]. This iterative and variational method relies on an exact diagonalization of the Hamiltonian. The premise is to obtain a wave function in a reduced Hilbert space that approximates the actual ground state, minimizing the loss of information. It was originally developed for studying one-dimensional quantum lattice models in real space at zero temperature. But its core idea to construct the RG flow in terms of *reduced density matrices* works very generally. It yields an optimal scheme to truncate Hilbert space and a powerful approach for evaluating static, dynamical, and thermodynamic properties of low-dimensional quantum many-body systems without introducing any external bias. Its applicability has now been extended successfully to statistical mechanics, quantum chemistry, nuclear and high-energy physics, quantum information, machine learning, and other fields.

The idea of making a systematic approximation by truncating the basis space is not new. It is also the idea used in NRG and Monte Carlo simulations. Unfortunately, this truncation is often not very effective, especially in strongly correlated electronic systems. In DMRG, however, the basis states are first rotated to reduce the truncation error so that the ground state could be accurately represented just by a small set of basis states.

The DMRG technique splits a system into two blocks, called left (L) and right (R) blocks, and two sites that are often placed between these two blocks. These two blocks need not be of equal size. A set of representative states is retained for each block during the warmup. This system of left block plus two sites plus right block, abbreviated as ($L \bullet \bullet R$), is defined as a superblock. We call the left block plus the left-added site as the *system* block, denoted as $S = L\bullet$, and the right-added site plus the right block as the *environment* block, denoted as $E = \bullet R$. At each step of the iteration, the ground-state wave function of the superblock is calculated by diagonalizing the Hamiltonian. The reduced density matrix for the system block is then evaluated from the density matrix of the ground state by tracing out all the degrees of freedom in the environment. The eigenvalues of the reduced density matrix determine the probabilities of the corresponding eigenvectors in the ground state. These eigenvectors form a new set of basis states of the system block, which are now truncated according to their probabilities. In this scheme, the system block is augmented by just adding one site at each iteration, similar to the conventional NRG scheme of Xiang and Gehring [13].

Recently, DMRG has become one of the most reliable and versatile methods developed in modern computational physics. In particular, the real-space DMRG is one of the most accurate and efficient methods for studying quantum lattice models with short-range interactions in one dimension. It can treat large systems with controllable precision. In particular, in many one-dimensional systems, the error can be reduced to the level at which the DMRG results can be regarded as *quasi-exact* if the ground state has a finite excitation gap. A striking illustration of the precision of this method was first given by White and Huse [19]. They showed, by taking the $S = 1$ antiferromagnetic Heisenberg spin chain as an example, that precision of 10^{-12} for the ground-state energy, $E_0 = -1.401484038971(4)$ in the unit of the exchange constant, could be achieved just using modest computational resources.

1.3 From DMRG to Tensor Network Algorithms

Since the invention of DMRG, it has been successfully applied to study ground-state properties of one-dimensional and quasi-one-dimensional quantum lattice systems. Meanwhile, several methods have been developed to extend DMRG from zero temperature to finite temperatures, from local to nonlocal basis space, and from the calculation of static quantities to time-dependent or dynamical correlation functions. Such progress has extended the application scope of DMRG and stimulated the development of a new breed of algorithms based on the so-called *tensor network states*. These algorithms treat quantum states as products of interconnected tensors and offer new tools for probing quantum many-body systems.

Matrix Product State (MPS)

In 1995, Östlund and Rommer [20] pointed out that the wave function generated by DMRG is an MPS, and the success of DMRG is related to the fact that it is a variational method within the MPS space. They also pointed out that an MPS can be viewed as a variational wave function that can be optimized without invoking DMRG [20, 21]. Their work sheds light on the understanding of DMRG. It offers not just a systematical way to parameterize quantum many-body states but also a new route to extend the DMRG algorithm, especially to higher dimensions.

An MPS provides a highly versatile parametrization for the ground states of local Hamiltonians [22, 23]. It is a collection of so-called local tensors defined on each lattice site. A local tensor contains a dangling physical bond, whose dimension is just the total number of basis states at each lattice site, and two virtual bonds, whose dimensions equal the number of states retained at the DMRG calculation. The local tensors are connected by virtual bonds. The number of parameters required to specify these tensors grows linearly with the system size, significantly smaller than the exponentially large dimension of the Hilbert space.

The matrix product representation of quantum states is not new. It has been introduced under various names over the past five decades. The concept of MPS was first introduced in classical statistical physics by Baxter [24], and later on in the context of quantum Markov chains [25]. In 1987, Affleck, Kennedy, Lieb, and Tasaki (AKLT) proposed an extended Heisenberg model and showed that its ground state is a valence bond solid (VBS) state [26, 27]. Their work provided crucial insight into the physics of Haldane's conjecture that an integer spin antiferromagnetic Heisenberg chain has a gap in the excitation spectrum [28, 29, 30]. In 1991, Klümper *et al.* showed that the one-dimensional VBS state can be parameterized as a translationally invariant MPS [31, 32]. These works stimulated extensive studies of the translationally invariant subclass of MPS in the community of mathematical physics under the name of *finitely correlated states* [33].

The structure of MPS is simple and conceptually useful. However, in the first ten years of DMRG, MPS was not frequently used because the standard DMRG calculation does not benefit much from this kind of representation. The potential of MPS was released when this kind of wave function was explored using the language of quantum information. In particular, it was shown that the expression power of MPS (or DMRG) could be quantified by the *entanglement entropy*, a concept first introduced in quantum information theory [34, 35]. DMRG correctly characterizes the entanglement structure of the ground state governed by the *area law of entanglement entropy* in one dimension [36, 37]. This is the reason why it is so successful.

In 2004, Vidal introduced the *time-evolving block decimation* (TEBD) method to evaluate the time evolution of a quantum state represented by an MPS [38] without invoking direct renormalization of the Hamiltonian and other physical operators. Verstraete and Cirac also showed that one could variationally calculate the ground-state energy for a periodic system as accurately as for an open boundary system if the ground-state wave function is represented by a periodic MPS [39]. Their works revealed a deep connection between DMRG and quantum information, which has allowed the extension of DMRG to fields quite far from its origin.

From the perspective of MPS, DMRG can be regarded as an algorithm that locally updates the matrix elements of an MPS at two lattice sites at a time. By exploiting the translational invariance, it is possible to develop algorithms for infinite-size systems with MPS. This means it is feasible to explore the physical properties of a given system directly in the thermodynamic limit without relying on finite-size extrapolations.

In the framework of MPS, the low-lying excitations can also be evaluated under the *single-mode approximation* that was first introduced by Bijl *et al.* [40] and by Feynman [41, 42] in the 1950s. In 1995, Östlund and Rommer proposed a variational ansatz for a single-particle excitation by representing it as a momentum superposition of a locally excited state from the ground state in the framework of MPS [20]. The local excitation is implemented through a bond matrix that is variationally optimized. In 2009, instead of using a bond matrix, Chung *et al.* suggested using some relevant local operators on the lattice sites to generate local excitations [43]. Pirvu *et al.* [44], on the other hand, generated the local excitation by replacing one of the local tensors in the ground-state MPS with a perturbed local tensor that is variationally optimized. Haegeman *et al.* [45] first explored the extension of the single-mode approximation with MPS in the thermodynamic limit.

As mentioned, MPS faithfully represents a quantum state that satisfies the area law of entanglement entropy in one dimension [36, 37]. However, in a critical system where the correlation length diverges, the entanglement entropy grows logarithmically with the system size [46, 47, 48]. This *logarithmic correction* to the entanglement entropy does not affect the calculation of short-range correlation functions much. However, as it diverges with the system size, it can strongly affect the calculation of long-range correlation functions. An MPS cannot describe this logarithmic correction with a finite bond dimension because the bond dimension has to grow polynomially with system size. Nevertheless, it could be described by an infinite-dimensional MPS [49, 50]

In 2007, Vidal proposed a new kind of tensor network state, called *multiscale entanglement renormalization ansatz* (MERA), to represent a critical state effectively [51]. A MERA yields a tensorial representation of a critical system with fixed bond dimensions independent of system size. It consists of a network of isometric tensors that are connected by another class of unitary tensors, called *disentanglers*. These tensors are introduced to mix up different length scales and to locally minimize the entanglement between neighboring sites on the same length scale. Physically, a MERA could be understood as a network on which quantum entanglement propagates. It opens up exciting perspectives for studying one-dimensional quantum or two-dimensional classical systems. The price to pay is an increase in the overall computational cost.

MPS was originally defined in one-dimensional lattice systems. In 2010, Verstraete and Cirac extended it into the continuous limit and introduced *continuous MPS* [52]. Their work allows, for the first time, variational tensor network algorithms to be applied to quantum field theories in one spatial dimension without lattice discretization. In 2013, Haegeman *et al.* proposed a continuous MERA theory for constructing RG flows of quantum field theories in real space [53]. In 2015, Jennings *et al.* [54, 55] introduced a higher-dimensional generalization of continuous MPS in the framework of path-integral representation.

Nonlocal Basis Space DMRG

In 1996, Xiang extended DMRG from real space to momentum space [56]. In momentum space, a "lattice" is simply a set of properly ordered momentum points allowed by periodic boundary conditions, with each momentum point representing a lattice site. An advantage of implementing DMRG in momentum space is that it enables simple manipulation of the kinetic energy term and treats the momentum as a good quantum number to block diagonalize the Hamiltonian. However, the interaction is highly "nonlocal" in the momentum space. It contains a sum of N^3 terms of four fermion or boson operators, which link two or more momentum points if considering the momentum conservation, where N is the lattice size. To compute and store the matrix elements of these N^3 operators is computationally costly. Xiang solved this problem by introducing a *regrouping method* to reduce the number of operators whose matrix elements need to be computed and stored [56]. This regrouping method works in arbitrary basis space. It paved the way for the application of DMRG in quantum chemistry [57], nuclear physics [58], and quantum Hall effects [59, 60, 61].

The DMRG algorithm was first applied to a molecular system of cyclic polyene chain modeled by an extended Hubbard [62, 63] or the Pariser–Parr–Pople Hamiltonian [64, 65]. It was followed by the first DMRG calculation with the full electronic Hamiltonian of 25 Hartree–Fock orbitals for a water molecule by White and Martin in 1999 [57]. A more accurate DMRG calculation for water molecules outperforming the best coupled-cluster result was first made by Chan and Head-Gordon using 41 orbitals in 2003 [66]. Since then, DMRG has been broadly used to benchmark other molecules. It has also been shown that DMRG is an efficient method for optimizing single-particle basis states from a large reservoir of Hartree–Fock orbitals [67].

Extension to Finite Temperatures

Extension of DMRG to finite temperature was first made by Nishino in 1995 [68] when he realized that the DMRG idea could also be used to diagonalize transfer matrices of two-dimensional classical statistical models. This led to the method of *transfer matrix renormalization group* (TMRG). In 1996, Nishino and Okunishi combined the idea of *corner transfer matrix*, first introduced by Baxter [69], with DMRG and developed the *corner transfer matrix renormalization group* (CTMRG) method [70]. A variant of this CTMRG, called the *directional corner transfer matrix* method, was introduced by Orus and Vidal in 2009 [71]. This method treats the four directions of the lattice independently and is more flexible in application. It is commonly used, for example, in the contraction of two-dimensional tensor network states.

In 1996, Bursill, Xiang, and Gehring extended DMRG to finite temperatures for one-dimensional quantum systems [72] by introducing a quantum transfer matrix representation for the partition function using the *Trotter–Suzuki decomposition* [73, 74]. In this case, the lattice size is infinite, and the finiteness is at the level of the Trotter approximation along the imaginary time (i.e. inverse temperature) direction. As a quantum transfer matrix is generally non-symmetric and its left and right eigenvectors may not be a conjugate pair, a crucial step in establishing an efficient *quantum transfer matrix renormalization group* (QTMRG) was the introduction of *nonsymmetric density matrix*. This was done by Wang and Xiang [75] and independently by Shibata [76]. However, the nonsymmetric density matrix is not always positive definite, although it should be physically. The accumulated truncation error may ruin the positivity of this density matrix and lead to numerical instability, which dramatically reduces the accuracy of results at low temperatures. To avoid this instability, Huang introduced a *biorthonormalization scheme* to perform the QTMRG calculation [77]. In this scheme, a pair of basis sets obtained from the *singular value decomposition* (SVD) of the density matrix are biorthonormalized and used to construct the renormalized transfer matrix. It reduces the numerical instability and allows much lower temperatures to be reliably accessed by QTMRG.

Thermodynamic properties can also be calculated on a finite lattice system using TEBD based on the *purification* of the thermal density matrix [78]. The purification enlarges the Hilbert space with auxiliary sites, called ancillas. An ancilla contains the same number of basis states as a physical site. It pairs with a physical site to form a maximally entangled state at infinite high temperatures. The whole system resembles a ladder with the ancilla lattice appearing geometrically as another chain parallel to the original one. This ancilla approach is convenient to implement within the traditional framework of DMRG [79].

Time Evolution and Dynamical Correlation Functions

Dynamical correlation functions, such as optical conductivity or single-particle spectral function, can be measured experimentally. However, the calculation of these quantities is rather challenging. During the last three decades, several approaches have been proposed to calculate spectral functions using DMRG or related methods.

In 1987, Gagliano and Baliseiro [80] proposed a continued-fraction method (also called the Lanczos vector method) to evaluate dynamical correlation functions based on the Lanczos diagonalization. Their method was first adopted in the DMRG calculation by Hallberg [81] in 1995. While this method requires only modest numerical resources, reliable results that can be obtained are only limited to low frequencies.

The continued-fraction method uses sequential basis states generated by the Lanczos iteration to represent the whole Hilbert space of the system. It is not accurate nor efficient for describing the correlation at a particular frequency. To remedy this deficiency, the *correction vector method*, initially proposed by Soos and Ramasesha [82], was introduced in the DMRG calculation, first by Ramasesha *et al.* [83] in 1997, and later refined by Kühner and White [84] in 1999. The critical idea is to take Green's function at a particular frequency as a target state and calculate it by solving a set of large but sparse linear equations. This method can generate extremely accurate results at a given frequency, but the computational cost is very high. In 2002, Jeckelmann [85] improved the correction vector method and showed that it is much more efficient and accurate to determine the correction vector by minimizing a cost function. In 2011, Dargel *et al.* proposed an *adaptive Lanczos vector method* [86] to improve efficiency in the calculation of dynamical correlation functions. In 2016, Nocera and Alvarez [87] proposed a Krylov-space approach to replace the *conjugate gradient method* in the calculation of correction vectors.

It is more convenient to investigate dynamical spectral functions using either a Chebyshev expansion [88, 89] or a Lanczos expansion [90] in the framework of MPS. The key idea is to use MPS to represent the Chebyshev or Lanczos vectors generated in the corresponding expansion. However, as each vector obtained at each step of expansion is represented by an MPS, the recurrence or orthogonal relation between different vectors is satisfied only approximately. Thus, a reorthogonalization of these vectors is desired to improve the accuracy of dynamical correlation functions [89, 90]. This kind of method offers a balanced scheme between cost and accuracy. It yields results with accuracies comparable to those of the correction-vector DMRG but at dramatically reduced computational cost. Furthermore, one can also use the orthogonalized Chebyshev or Lanczos vectors to represent the Hamiltonian. Diagonalizing this Hamiltonian can also accurately determine the spectral weight at each energy eigenvalue, offering a simple but accurate approach to performing the finite-size scaling in the entire energy range [89].

Dynamic quantities can also be calculated from the Fourier transform of the time-dependent Green's function. However, to obtain good frequency resolution, one has to calculate the correlation functions over a long time interval, which is limited by either a loss of accuracy due to the approximation used or by finite-size effects such as reflections from the two open ends. In 2002, Cazalilla and Marston evaluated the time-dependent correlation functions for a one-dimensional system under an applied bias using DMRG [91]. In 2003, Luo *et al.* [92] proposed a *pace-keeping approach* to optimize the basis states that are retained with the time evolution. TEBD, on the other hand, provides a method to efficiently simulate the time evolution of one-dimensional quantum systems with short-range interactions using MPS [38]. Based on this approach, an *adaptive time-dependent DMRG* was proposed by Daley *et al.* [93] and by White and Feiguin [94].

TEBD is implemented by taking the Trotter–Suzuki decomposition. At each time of evolution, an MPS with a larger bond dimension than the original one is obtained. To proceed, one has to truncate the MPS by discarding less important parameters. In order to reduce both the truncation and the Trotter errors, Haegeman *et al.* introduced a *time-dependent variational principle* to optimize an MPS that preserves all symmetries of the system [95].

Investigation of long-time dynamics, however, remains challenging due to the linear growth of the entanglement entropy with the evolving time [96]. To catch up with the growing speed of the entanglement entropy, the bond dimension of MPS has to grow exponentially with time. In 2009, Bañuls *et al.* [97] proposed a *folding scheme* to reduce the entanglement of transfer matrices defined in the time direction. It yields an accurate approach for evaluating the long-time dynamics of a quantum state in the thermodynamic limit. A QTMRG extension of this method to finite temperatures was proposed by Huang *et al.* [98]. DMRG has also been extended to simulate time evolution at finite temperatures by purifying thermal density matrices with ancillas [99, 100].

Two- or Higher-Dimensional Quantum Systems

Investigation into physical properties of two-dimensional quantum systems with strong correlations, including high-temperature superconductivity, quantum Hall effects, frustrated antiferromagnets, and quantum spin liquids, has been at the heart of condensed matter physics in the past four decades. After the successful application of DMRG in one dimension, it is natural to extend this method to two or more dimensions. Liang and Pang made the first attempt to apply DMRG to two dimensions [101]. Unlike in one dimension, they found that DMRG is only moderately successful in two dimensions because computational resources need to increase exponentially with the width of the lattice, implying that DMRG can reliably access only narrow two-dimensional lattices. From an analytical study for the spectra of a two-dimensional system of coupled oscillators, Chung and Peschel also found that the eigenvalues of density matrix decay much slower than in one dimension [102].

A better understanding of this exponential breakdown of DMRG in two dimensions is revealed by the entanglement area law, which provides a systematic way to quantify quantum correlations. The entanglement area law is believed to hold in both one and two dimensions if the ground state is gapped so that the correlation length is finite. According to the area law, the entanglement entropy of the ground state between two subsystems scales with the size of the interface. In one dimension, the interface contains just one site, independent of the lattice size. In two dimensions, however, the interface grows with the lattice size.

To capture the entanglement feature of the ground state correctly, the minimal number of many-body states should grow exponentially with the entanglement entropy. It puts an upper bound on the system sizes that can be accurately simulated. Nevertheless, DMRG is still one of the most promising methods for studying two-dimensional quantum lattice models because other methods, such as the exact diagonalization and the quantum Monte Carlo, all have their limitations. DMRG has been successfully applied to study, for example, the t-J model [103], Hubbard model [104], frustrated quantum spin models [105, 106, 107], and other quantum lattice models in two dimensions.

A simple extension of DMRG to two dimensions would be to replace every site added between the left and right blocks with a column of sites. However, the extra degrees of freedom added to the system would make the size of the Hilbert space prohibitively large. A practically feasible extension of DMRG to two dimensions is to add a single site to a block at a time. To do this, one needs to map a two-dimensional lattice onto a one-dimensional one or select a path to order all lattice points at the price of breaking the lattice symmetry and introducing long-range interactions.

A typical mapping is to fold a two-dimensional lattice, generally a multichain system, into a one-dimensional zipper by ordering the lattice sites in a snake path extending along the chain direction. Alternatively, one can zip a two-dimensional lattice along the diagonal direction [34]. This diagonal map would allow one to build up a $L \times L$ lattice using the blocks of an $(L-1) \times (L-1)$ lattice.

In a two-dimensional DMRG calculation for a multichain system, open or cylindrical boundary conditions are generally assumed instead of fully periodic boundary conditions. This is to avoid squaring the number of basis states required for a given accuracy. By cylindrical boundary conditions, we mean open boundary conditions along the chain direction and periodic boundary conditions along the direction perpendicular.

As mentioned before, the wave function generated by DMRG is an MPS. In two dimensions, the entanglement entropy scales with the linear size of a lattice in a gapped system. To capture this effect, the bond dimension of MPS has to increase exponentially with the cylinder's circumference. To resolve this exponentially growing problem, one needs to find a scalable representation of the ground-state wave function that captures the entanglement area law.

Niggemann *et al.* made the first tensor network extension of MPS to two dimensions in the study of an extended spin-3/2 Heisenberg model on a honeycomb lattice [108]. They called this kind of tensor network state a *vertex-state model* to emphasize its resemblance to a standard vertex model. Similar to an MPS, they defined a local tensor at each lattice site that maps a set of bond variables onto a physical state. Later, Sierra and Martin-Delgado made a similar ansatz for the tensor network wave function [109]. The use of a tensor network state as a variational wave function for the three-dimensional classical lattice model was also suggested by Nishino [110].

In 2004, the idea of two- or higher-dimensional tensor network states was discussed by Verstraete and Cirac from a perspective of quantum information theory [111]. They reinvented the tensor network wave function first suggested by Niggemann *et al.* [108], and called it a *projected entangled pair state* (PEPS), since it can be understood in terms of pairs of maximally entangled virtual basis states defined on the bond linking two sites. Their work has attracted significant attention because it reveals more clearly the physical picture embedded in the tensor network representation of quantum states. Furthermore, they pointed out that PEPS satisfies the entanglement area law since the number of entangled bonds between any two blocks grows linearly with the size of their interface [112]. In other words, PEPS presents a faithful representation of a ground state that is governed by the entanglement area law.

PEPS accurately describes the entanglement between any two neighboring sites in the ground state. It is, for example, an exact representation of the ground state of the two-dimensional AKLT model [27], which is a VBS. In principle, it can be used to accurately describe an arbitrary quantum state that satisfies the entanglement area law if there is no limitation on the bond dimensions of local tensors. However, in practical calculations, the bond dimension of PEPS that can be reliably handled is generally not much more than ten without imposing any symmetry. In this case, PEPS is no longer an efficient representation of a quantum state in which the entanglements among three or more particles become important. An example of this kind of quantum state is the so-called *simplex solid state* (SSS), first introduced by Arovas [113].

A simplex is a building block of a lattice. For example, a four-site square is a simplex of a square lattice, and a three-site triangle is a simplex of a Kagome lattice.

Unlike VBS, SSS emphasizes the group entanglement within a simplex. The wave function that characterizes the SSS-type state is given by the so-called *projected entangled simplex state* (PESS) first introduced by Xie *et al.* in 2014 [114]. PESS presents a natural extension of PEPS to a system in which many-particle or cluster entanglement becomes important. It also satisfies the entanglement area law. In the limit that a simplex contains two sites, it reduces to PEPS.

Either PEPS or PESS can accurately describe a quantum state with short-range entanglement and keep the locality of interactions. The minimal bond dimension that is needed for accurately describing a quantum state does not depend on the system size in a gapped system. This is an advantage of using this kind of ansatz. Furthermore, it can be directly used in a thermodynamic limit like an MPS.

Unlike an MPS with open boundary conditions, we cannot choose a gauge to simultaneously orthogonalize all the bond indices of PEPS so that all local tensors are canonicalized. It also happens for MPS with periodic boundary conditions or, more generally, if there is a loop in a tensor network state. A loop in a tensor network state means that one cannot split the network into two parts by just cutting one bond.

The biggest challenge in developing a two-dimensional tensor network algorithm is the optimization of local tensors, which give the best approximation to the ground state for a given Hamiltonian. Two kinds of approaches can be used to determine the local tensors of PEPS (similarly to PESS). One is to determine the local tensors by variationally minimizing the ground-state energy [111]. The other is the so-called update approach in which a PEPS is determined by performing an imaginary time evolution, that is, by applying the projection operator, or the density-matrix operator, $\exp(-\beta H)$ onto an arbitrary initialized PEPS which is not orthogonal to the true ground state. This projection is made by dividing β into many small pieces so that the projection operator can be readily evaluated using the Trotter–Suzuki decomposition formula [73, 115]. At each projection, an approximation has to be taken to truncate the bond indices of PEPS. The PEPS obtained with this approach would converge to the ground state in the zero-temperature limit, $\beta \to \infty$.

There are two kinds of update approaches: the simple update [116] and the full update [117]. The simple update is essentially an entanglement mean-field approach. It approximates the environment tensors by entanglement spectra defined on the bonds separating the system and environment sites. In other words, these bond entanglement spectra serve as an effective mean field acting on the system tensors whose elements are updated. This simple update approach becomes exact on the Bethe lattice or any other kind of lattice without loops [118].

The simple update is a local update approach. It underestimates the long-range correlations but works very efficiently and allows a PEPS or PESS with a bond dimension as large as 100 or more to be approximately determined. With this approach, a system block with one or a few tensors is updated at one time. A PEPS determined by the simple update can be directly used to evaluate physical observables. It can also be used as an initial state in the full update calculation. One can also update a relatively large cluster of tensors at one time. In this case, the simple update is referred to as a cluster update approach [119]. The entanglement mean-field approximation used in the simple update can also be used in the variational minimization of PEPS or PESS [120]. It allows us to efficiently determine PEPS or PESS with a relatively large bond dimension.

The full update evaluates the environment tensor by directly contracting all the local tensors in the environment block [117] using, for example, a boundary MPS or CTMRG method. It is more accurate than the simple update but has a much higher computational cost. The bond dimension this approach can handle is generally less than ten if the tensors are not block diagonalized by imposing symmetries.

Both PEPS and PESS are variational wave functions and satisfy the variational principle. Hence, the energy calculated with either PEPS or PESS should be an upper bound of the true ground-state energy. However, as it is difficult to rigorously contract a PEPS or PESS, certain approximations have to be introduced in computing observables. These approximations may violate the variational principle and ruin the condition for the calculated energy being a rigorous upper bound.

The PEPS or PESS can also represent a low-lying excitation in a translation-invariant system under the single-mode approximation. This approach was first explored by Vanderstraeten *et al.* in 2015 [121]. The single-particle excitation is assumed to be a momentum superposition of a locally excited state by replacing one of the local tensors in a PEPS ground state with a perturbed local tensor that is variationally optimized. Dynamical spectral functions can be evaluated using this kind of tensor network states [122, 123].

Besides PEPS and PESS, a number of other tensor network states have also been introduced in two or more dimensions. This includes but is not limited to the tree tensor network state [124], string-bond state [125], entangled-plaquette state [126], and branching MERA [127].

Coarse-Graining Tensor Renormalization

Before the tensor network wave function was introduced in quantum systems, tensor network representations for the partition functions of classical lattice models were already used in statistical physics. In fact, as discussed in Chapter 3, all classical lattice models with local interactions can be represented as *tensor network models*. A tensor network model is a special kind of tensor network state that does not have any dangling physical bonds. Moreover, this representation is rigorous, unlike a quantum TNS whose local tensors have to be variationally determined. The partition function is determined by contracting all local tensors.

Similarly, in the determination of a PEPS, either by the variational minimization or by the full update, or in the evaluation of its observables, one needs to calculate a scalar product of two PEPS. This scalar product of two PEPS forms a double-layer tensor network model. The physical bonds connect the two layers of local tensors. By contracting out these physical bonds, the double-layer tensor network model becomes a single-layer one.

To exactly solve these tensor network models by contracting out all local tensors is, unfortunately, an exponentially hard problem. The computational cost scales exponentially with the number of sites no matter how the local tensors are contracted. To resolve this problem, we have to rely on some approximate schemes.

A two-dimensional tensor network model can be approximately solved, for example, using TMRG or CTMRG. It can also be solved by employing a modified TEBD method, which is referred to as a boundary MPS method in the literature. These methods are efficient and accurate in performing this kind of calculation, and their costs scale linearly with system size.

The scheme of coarse-graining tensor renormalization has also been developed to solve two-dimensional tensor network models. The idea is to perform a scaling transformation by coarse-graining local tensors until some fixed-point tensors are reached, similar to the Kadanoff block spin scheme [8] in classical statistical models. In 2007, Levin and Nave introduced the first coarse-graining renormalization group method of tensor network models [128]. They coined it *tensor renormalization group* (TRG). This method truncates the basis space according to the singular-value spectra of local tensors. It provides a local optimization of the truncation space. However, it does not consider the renormalization effect of the environment tensors on the singular-value spectra, which is key to the success of DMRG. To solve this problem, Xie *et al.* introduced a *second renormalization group* (SRG) method [129, 130] to account for the environment contribution. The SRG improves the accuracy of TRG significantly. For example, for the two-dimensional Ising model, the accuracy is improved by more than two orders of magnitude at the critical point and more than five orders of magnitude far away from the critical point by keeping 24 states.

The difference between TRG and SRG is similar to the difference between Wilson's NRG and DMRG. In Wilson's NRG, a block Hamiltonian is optimized without considering the interaction between its building cells. On the contrary, in DMRG, the truncated basis states of the system are fully optimized by considering the interplay between the system and the environment through the reduced density matrix, which measures the entanglement spectra between these two blocks.

The TRG algorithm can also be improved by just taking *higher-order singular value decomposition* (HOSVD) or more generally *Tucker decomposition* for local tensors [131]. This *HOSVD-based tensor renormalization group* (HOTRG) method determines the entanglement spectra on all the bonds of a local tensor at the same time. It presents a better scheme to truncate the basis states than TRG. HOTRG can also incorporate SRG to improve the accuracy of results further. This HOSVD-based SRG method is called HOSRG [131].

However, TRG and HOTRG do not remove all short-range entanglement embedded in each loop that is gauge invariant during the coarse-graining process. As a result, the effective tensor network at a given length scale still contains irrelevant correlations belonging to shorter length scales. The accumulation of these short-range correlations over successive coarse-graining transformations would ruin the scaling invariance at the critical point, leading to a large truncation error around the critical point. Nevertheless, universal information, such as critical exponents, can still be obtained from the fixed-point tensors.

To remove short-range entanglement, Gu and Wen proposed a tensor entanglement filtering renormalization approach and pointed out that it is crucial to optimize the tensor configurations that contain a loop [132]. However, this approach is computationally inefficient. A better approach for removing short-range entanglement or correlations at each coarse-graining step was introduced by Evenbly and Vidal based on the idea of MERA [133]. This approach, which is referred to as *tensor network renormalization* (TNR), is based on the insertion of optimized unitary disentanglers and isometric tensors into the tensor network. To a large extent, TNR recovers scale invariance at criticality.

In 2017, Yang *et al.* proposed a loop tensor network renormalization (Loop-TNR) approach to remove short-range entanglement [134]. In this approach, the short-range entanglement within a loop is removed by optimizing the rewiring tensors at each coarse-graining step. Like TNR, Loop-TNR improves the RG flow around a critical point. It produces accurately critical exponents, including central charge and scaling dimensions, with a lower computational cost than TNR.

The methods introduced here each have advantages and disadvantages. Some are easier to implement, and some are more stable in catching critical behaviors. In general, the cost of DMRG- or MPS-based methods is lower than that of coarse-graining methods. But the coarse-graining methods scale logarithmically with system size and allow the scaling exponents to be directly computed.

Among these methods, only HOTRG and HOSRG are applicable to three-dimensional tensor network models. These two methods can be used to directly contract three-dimensional classical lattice models or equivalent (2+1)-dimensional quantum lattice models. A HOTRG calculation for the three-dimensional Ising model, for example, produces a very accurate estimation for the critical temperature [135] as well as the temperature dependence of the specific heat and magnetic susceptibility [131].

Other methods are either not applicable or difficult to extend to study of three-dimensional tensor network models. In order to use these methods, we have to make a dimension reduction to contract a three-dimensional tensor network model using a boundary PEPS, whose local tensors are determined using a variational optimization approach. The computational cost for determining this PEPS and its expectation values is generally higher than HOTRG.

1.4 Applications

Density-matrix and tensor network renormalization provides a state-of-the-art method for the classical simulation of quantum systems. It was initially introduced to investigate classical statistical and quantum lattice models in condensed matter physics, statistical mechanics, and quantum information, which significantly deepened our understanding of so-called correlated quantum phenomena. Later they were extended to new fields and generated significant impact, for example, in quantum chemistry, cold atoms, quantum computing, artificial intelligence, and, more generally, in the study of complex systems with a large number of degrees of freedom or variables. During the past thirty years, more than ten thousand scientific papers have been published in the development and applications of density-matrix and tensor network renormalization.[1] It is beyond my ability to comprehensively review the vast literature of papers published in this field. A survey of DMRG with its application up to late 1998 was given in a collection of lectures and articles [136]. To gain a more comprehensive picture of the DMRG or other tensor network-related methods and their applications, we refer the interested readers to the review articles [137, 138, 139, 140, 141, 142, 143, 144, 145, 146, 147, 148, 149, 150, 151] and references therein. A comparison between the early real-space renormalization group methods and the newly developed tensor-network-based renormalization group methods can be found in Ref. [152].

One Dimensional Quantum Systems

As an algorithm that is simple to implement, DMRG, together with its finite temperature and finite frequency extensions, has now become the most effective and accurate numerical method for studying not just the ground state but also thermodynamic and dynamic properties in one dimension. It has been successfully applied

---

[1] See the DMRG Home Page, http://quattro.phys.sci.kobe-u.ac.jp/dmrg.html, run by Tomotoshi Nishino.

to solve nearly all fundamentally interesting models with short-range interactions in one dimension (see Refs. [136, 137, 138, 139, 141]), including, but not limited to, the following quantum systems:

(i) Quantum spin models

This includes the antiferromagnetic Heisenberg models [19, 20, 153, 154, 155, 156, 157, 158] and their experimentally relevant generalizations by adding anisotropy [159, 160], dimerization [161], frustration [162, 163, 164], defects [165, 166, 167], randomness [168], or other interactions [163, 169, 170] to the Hamiltonians. A number of interacting spin models with higher symmetries, such as the SU(N) symmetry ($N > 2$) [171, 172], have also been studied.

The quantum spin models are physically interesting for three reasons. First, as the charge degree of freedom is frozen at each lattice site, this kind of model is relatively simple to study compared to the interacting fermion models, which serves as a playground for testing various numerical methods. Second, in 1983, Haldane predicted that the one-dimensional Heisenberg model with integer spin has an excitation gap and a finite correlation length [28, 29]. Since then, much experimental and theoretical effort has gone into understanding the difference between half-integer and integer spin chains. Third, there are a lot of quasi-one-dimensional compounds whose physical properties can be adequately described within the framework of interacting spin chains governed by quantum spin models.

The spin-1 Heisenberg model was the first quantum lattice model used by White to demonstrate the efficiency and accuracy of DMRG [18]. A benchmark calculation of DMRG was made by White and Huse [19] to evaluate the spin excitation gap and the correlation length for this model with an accuracy that was difficult to achieve with other numerical methods. The results confirmed the valence-bond-solid state picture proposed by AKLT [26, 27].

(ii) Interacting fermion models

This includes the Hubbard model [14, 173, 174, 175], extended Hubbard model [176, 177], multiband Hubbard model [178], SU(N) Hubbard model [179], t-J model [180, 181], periodic Anderson [182, 183], and Kondo lattice [184, 185] models, with randomness [186, 187], or other interactions [188]. The study of interacting fermion models is somewhat more complicated because of the larger number of degrees of freedom at each lattice site. However, DMRG is not bothered by the negative-sign problem that hampers quantum Monte Carlo simulations.

The Hubbard model [14], named after J. Hubbard, is a basic model for describing the Mott metal-insulator transition, charge-spin separation, quantum magnetism, charge density waves, and other fundamental properties of interacting electrons. It reduces to the t-J model in the strong coupling limit. The Hubbard [189] and t-J models [190] are also the two simplest models used for understanding the mechanism of high-temperature superconductivity discovered in copper oxides. The Kondo lattice model is equivalent to the periodic Anderson lattice model in the strong coupling limit [191]. They are the two basic models for understanding the physical properties of heavy fermion systems in which itinerant electrons interact strongly with localized spins.

(iii) Systems with bosons

Bosonic systems are generally more challenging to treat than fermionic ones in DMRG because a single bosonic mode has an infinite state space. The basis space of a bosonic mode must be truncated before carrying out DMRG calculations [192, 193, 194, 195]. A commonly adopted approximation is to restrict the number of bosons in a finite interval around its average at each lattice site. This approximation is valid when the fluctuation in the number of bosons at a site is small.

There are three kinds of bosonic models that have been investigated with DMRG. The first includes the systems that contain just bosons [196, 197]. A typical example is the Bose–Hubbard model [196]. This model mimics the dynamic properties of ultracool atoms in optical lattices generated by laser beams [198]. Both the doping of particle numbers and the interaction strength can be readily controlled in an optical lattice of ultracool atoms. The phase diagram of this model, particularly the Mott insulator to superfluid condensate transition, has been thoroughly investigated with DMRG in conjunction with experimental measurements [199, 200]. The second model includes the systems of bosons coupled with fermions, such as the Su–Schrieffer–Heeger model or other electron–phonon coupled systems [193]. The third includes the interacting systems of bosons coupled with spins, such as the spin-Peierls system [195, 201] and the multiconnected Jaynes–Cummings lattice model [202]. A multiconnected Jaynes–Cummings lattice can be realized by coupling qubits, described by spins, with some cavity photon modes.

(iv) Quantum field theory and quantum gravity

The application of tensor network methods to the lattice gauge theory is a forefront field undergoing rapid development [148]. In particular, DMRG and other tensor network methods offer a powerful tool for defeating the notorious sign problem encountered in the simulation of lattice quantum chromodynamics [203, 204] and for exploring physical properties of the (1+1)-dimensional Abelian Schwinger model [205, 206, 207], $\lambda\phi^4$ scalar field theory [208, 209], Z_2 lattice gauge theory [210], SU(2) [211] and SU(3) [212] gauge theory, O(N) ($N = 2, 3, 4$) nonlinear sigma models [213, 214, 215, 216],

Gross–Neveu–Wilson model [217], lattice Abelian Higgs model [218], and massive Thirring model [219].

Continuous MPS provides a powerful variational ansatz for the ground state of strongly interacting quantum field theories in one dimension [52]. This ansatz, formulated directly in the continuum, enables us to parameterize the ground-state wave function without resorting to a lattice discretization scheme. It was first illustrated with the Lieb–Liniger model [52, 220], which is a basic model of quantum field theory describing one-dimensional bosons interacting through a repulsive contact potential. This approach was then extended to the study of excited states of the Lieb–Liniger model [221], free massive Dirac fermions and the Gross–Neveu model [222], coupled Lieb–Liniger models [223], and systems of two species of fermions [224]. It was also generalized for studying thermodynamic properties of one-dimensional quantum lattice models in the continuous imaginary time representation of quantum transfer matrices [225].

Tensor networks are relevant to quantum gravity. This connection was first pointed out by Swingle [226]. It was proposed that MERA [51] is linked to the geometry of space through a conjectured relationship or correspondence between the anti-de Sitter space in theories of quantum gravity and conformal field theories. Particularly, it was assumed that MERA could be understood as a lattice realization of an anti-de Sitter space with some geometry, in which the curvature is somehow linked to entanglement. The correspondence implies that space-time geometry may emerge from the underlying entanglement structure in a complex quantum state. This intriguing connection has attracted great interest in the community of superstrings. However, a report also claims that MERA is actually a lightcone geometry rather than an anti-de Sitter space [227].

Two- or Higher-Dimensional Quantum Lattice Models

A thorough investigation of two-dimensional quantum systems is essential to the microscopic understanding of high-temperature superconductivity, frustrated magnetism, quantum spin liquids, and many other novel quantum phenomena discovered in quasi-two-dimensional materials. However, the entanglement entropy grows much faster in two dimensions than in one dimension. As a result, it leads to a dramatic increase in the demand for computational resources compared to one-dimensional calculations. Yet geometrical constraints in two dimensions are much more relaxed than in one dimension, allowing exotic elementary excitations, such as anyons, and competing quantum fluctuations to exist.

Two-dimensional quantum models have been investigated with both DMRG and the tensor network methods based on the PEPS representation of wave functions

[138, 140, 141, 142, 143, 144, 149, 150]. DMRG calculations start by unzipping a two-dimensional lattice into a one-dimensional one [34, 101]. Due to the constraint imposed by the entanglement entropy, DMRG is limited to the study of two-dimensional lattices with a relatively small width. PEPS maintains the lattice structure and distributes the entanglement more evenly on the whole lattice. It reduces the dimension of the virtual basis state on each bond and allows even an infinite lattice system to be studied.

The first applications of both DMRG- and the PEPS-related RG methods were in the area of magnetism [228]. The results obtained from the calculation of the two-dimensional Heisenberg model were encouraging, demonstrating the potential of these methods in solving strongly correlated problems in two dimensions [34, 105, 106, 107, 116, 229, 230] as well as in three dimensions [231]. For example, both the ground-state energy and the magnetization of the square lattice Heisenberg model obtained with either DMRG [34, 229] or PEPS [230] are consistent with the quantum Monte Carlo results [232].

The tensor network RG study of the frustrated Heisenberg models on the triangular [229], kagome [105, 106, 107], honeycomb [233], and the Shastry–Sutherland lattice [234], as well as models frustrated by further-neighbor [235, 236, 237] or multispin interactions [238, 239], has yielded fruitful results, particularly in search of quantum spin liquids. It has settled several problems that have been long debated [105, 106, 107]. A quantum spin liquid is difficult to identify because it is topologically nontrivial and may possess long-range entanglement. Furthermore, it does not break any symmetry and cannot be characterized by a local order parameter.

Another set of systems extensively simulated by DMRG or PEPS is the interacting fermionic systems described by the t-J model [240, 241, 242], the Hubbard model [243, 244], and the Kondo–Heisenberg lattice model [245]. A common feature revealed is the formation of the so-called stripe phase in lightly hole-doped systems [240, 243, 244]. A stripe is a charge-density wave state separated by magnetically ordered states. It was first discovered in cuprate superconducting materials by neutron scattering spectroscopy [246]. Superconducting pairing states have also been studied for these models by DMRG [245]. It seems that the long-range pairing order is absent in the thermodynamic limit of the Hubbard model if there is only nearest-neighbor hopping.

The DMRG and PEPS have also been applied to interacting boson systems [231, 247, 248]. Particular attention has been devoted to the study of so-called supersolid [249] in the extended Bose–Hubbard model and Bose metal [250]. Supersolid is a phase with simultaneous charge-density wave and superfluid order. On the other hand, Bose metal is not a phase characterized by an order parameter but by a pattern of correlations associated with a surface of gapless modes.

Quantum models defined on a Bethe or, more generally, a tree lattice have also been extensively studied with tensor network methods. It is natural to represent a quantum state as a *tree tensor network* on a tree lattice. The lack of loops implies that a tree tensor network state can be rigorously canonicalized by minimizing the ground-state energy with DMRG or by taking an imaginary time evolution with a proper local tensor update scheme like simple update [116]. The simple update is an entanglement mean-field approach for canonicalizing a tensor network state, which works particularly well on a tree lattice [118]. It leads to a thorough understanding of magnetic orders and other physical properties of the spin [251, 252, 253, 254], and interacting fermion [255, 256] models on the Cayley or Bethe trees. However, the correlation lengths are always finite, even at a critical point [118]. This implies that the entanglement entropy is upper bounded, not like in a regular lattice, which can be used to evaluate accurately low-temperature thermodynamic quantities just by taking imaginary time evolution [257].

Quantum Systems with Nonlocal and Off-Diagonal Interactions

In an arbitrary basis space, one can order all the single-particle basis states to form an effective "one-dimensional lattice" and carry out the DMRG calculation similarly as in real space. Unlike in real space, the interaction of particles, represented using these single-particle basis states, is generally nonlocal and off-diagonal. For example, the Hubbard interaction is local and diagonal in real space. In momentum space, on the other hand, each momentum point in the first Brillouin zone serves as a basis site, and the Hubbard interaction becomes off-diagonal, which involves the coupling from two to four momentum points, and long-ranged.

For two reasons, implementation of DMRG in momentum or other non-local basis space is technically more challenging than in real space. First, the basis states are not naturally ordered. One should determine the order of these basis states that optimizes the final results [56, 258, 259]. Local optimization of orbital ordering could be achieved by taking an adaptive scheme to update the active basis states with a unitary transformation for the fermion operators on the two middle sites in the superblock of DMRG [259]. This unitary transformation optimizes the basis states by minimizing the entanglement between the augmented system and environment blocks. It defines a new set of basis states and orders them optimally. One could further improve this optimization scheme by sorting the orbitals according to their mutual entanglement structures [258]. The purpose is to reduce long-range correlations by placing strongly entangled orbitals close to one another. Second, as the interaction becomes nonlocal and off-diagonal, there are more operators whose matrix elements need to be evaluated and stored, leading to a dramatic increase in computational time and memory space. This problem could be significantly ameliorated by factorizing the Hamiltonian and introducing the so-called *comple-mentary operators* that take partial sums over certain combinations of operators [56] within each subblock.

(i) Momentum space calculations

The application of momentum-space DMRG benefits from momentum con-servation. It allows the Hamiltonian to be block diagonalized according to the values of the total momentum so that more basis states can be retained at each DMRG iteration. Furthermore, in a given interacting fermion model, different scattering processes, such as the forward, backward, or Umklapp scattering, could be readily identified in momentum space. This provides a unique scheme, not feasible in other basis spaces, to understand the effect of each scattering process by screening other scattering terms in the model.

The momentum-space DMRG has been mainly applied to explore the ground-state properties of the Hubbard model [56, 260, 261]. In both one and two dimensions, the bipartite entanglement entropy is found to satisfy the volume law and scale quadratically with the Hubbard interaction in the weak coupling limit. Nevertheless, the momentum-space DMRG can still provide accurate results for the two-dimensional Hubbard model with moderate system sizes [261]. A one-dimensional Hubbard model with long-range hopping is also investigated with DMRG [260]. This model is difficult to investigate with the real-space DMRG because the hopping is long-ranged.

One possibility to combat the volume-law increase of the entanglement entropy is to carry out DMRG calculations using a hybrid real- and momentum-space representation. This approach aims to study a two-dimensional cylindrical system by taking a real-space representation in the direction along the axis of the cylinder and a momentum space representation in the direction around the circumference. In this way, the translational invar-iance and good momentum quantum numbers are preserved in the transverse direction. This hybrid approach was used to study the interacting fermionic Hofstadter model [262] and the Hubbard model [263] in two dimensions. It leads to a considerable reduction in computation time and memory space compared with the pure real-space approach [262, 263].

(ii) Fractional quantum Hall effect

The fractional quantum Hall effect occurs in a strong magnetic field where the kinetic energy of electrons is quenched into highly degenerate Landau levels and the electron–electron interaction is the only relevant term in the Hamil-tonian. This leads to the emergence of highly entangled and nonperturbative ground states with fractionalized particles.

Due to the nonperturbative nature of the fractional quantum Hall effect, numerical methods have played a crucial role in revealing its microscopic picture. In the Landau gauge, the noninteracting Landau orbitals are Gaussian localized [264], and the system could be mapped onto an effective "one-dimensional" chain with long-range interaction. It was shown that certain model wave functions of fractional quantum Hall effects, such as the Laughlin [265] and Moore–Read states [266], can be exactly represented by MPS [267]. This exact MPS representation has an infinite bond dimension (as the virtual space has support on conformal towers of states), but it can be truncated to a finite-dimensional MPS with high fidelity, which can be used for efficiently computing physical quantities [267, 268]. Furthermore, the topological order of fractional quantum Hall liquids can be identified from the entanglement spectrum obtained by DMRG through the conjecture made by Li and Haldane [269].

Like in the Hubbard model, the total momentum on a torus (or angular momentum on a sphere) of the fractional quantum Hall system is conserved. Therefore, the momentum-space DMRG can be extended to apply to the fractional quantum Hall system. It yields a powerful numerical tool for accessing a fractional quantum Hall system whose size is significantly larger than what could be handled by exact diagonalizations. Shibata and Yoshioka [59] made the first attempt along this line, emphasizing the investigation of the physical properties of electrons in higher Landau levels. More DMRG calculations were followed by a number of groups [60, 61, 270, 271]. As a result, the efficiency of DMRG has been significantly improved, which allows more than twenty thousand states to be retained [61]. In addition, DMRG has also been applied for studying fractional quantum Hall effects in a bosonic system [272], as well as in an extended Kagome Heisenberg model [273].

(iii) Quantum chemistry calculations

The DMRG implementation in quantum chemistry was first presented in the study of π-electrons of conjugated polyenes whose electronic properties are modeled by an extended Hubbard [62, 63] or the Parisier–Parr–Pople [64, 65] Hamiltonian. The fully *ab initio* DMRG determination of the electronic structure of molecules, on the other hand, started from the work of White and Martin in 1999 for the calculation of the ground-state energy of a water molecule with 25 active molecular orbitals [57]. Their work demonstrated the potential of DMRG for *ab initio* quantum chemistry calculations. The early applications focused on the multiconfigurational calculations for small molecules [66, 274, 275, 276, 277]. It was also applied to solve full configuration interaction problems for quasi-one-dimensional molecules, including linear hydrogen chains [278, 279], polyenes [278, 280] and other

π-conjugated organic systems [281], transition-metal complexes or clusters [282, 283]. DMRG allows more than 100 active orbitals to be included in the calculation. This size of active basis space is not reachable by full configuration interaction algorithms.

In the quantum chemical DMRG, molecular orbitals play the role of lattice sites. As the Coulomb interaction is inherently long-ranged and off-diagonal in the orbital space, the lattice formed by the active orbitals are often far from one-dimensional-like and relatively large bond dimensions are required to use DMRG for diagonalizing the Hamiltonian. As the number of complementary operators whose matrix elements need to be evaluated and stored scales quadratically with the dimension of the active space, not many active orbitals can be used in the DMRG iteration. Luo *et al.* introduced a useful scheme to tame this problem, which allows DMRG to optimize the orbitals by exchanging one or two least-active orbitals in the active space with an equal number of inactive orbitals from a larger pool of orbitals after each DMRG sweep [67]. An orbital is chemically inactive if it is almost empty or fully occupied. The occupation number of an active orbital is determined by the eigenvalue of the single-particle density matrix obtained from the ground-state wave function [67, 284, 285].

The rapid development of DMRG has turned it into a reference approach for large-scale multiconfigurational calculations [147, 286], which dramatically broadens its range of applications in quantum chemistry. To more efficiently encode the entanglement structure of active orbitals, a higher-dimensional tensor network extension, based on the tree tensor network states [287, 288], is used to represent the ground-state wave function. This kind of tensor network state encodes entanglement in a tree-like structure, allowing for a more feasible description of molecules. Furthermore, the relativistic effects have also been explored [289, 290, 291].

Besides the static properties, DMRG has also been extended to study dynamical correlation functions. However, dynamical correlations involve the contribution of excited states whose wave functions are described by some orbitals not included in the active space. For the inclusion of those omitted orbitals, DMRG has to be combined with an approach that can capture these contributions, including perturbation [292, 293], coupled-cluster [294], or other methods [295, 296].

(iv) Nuclear structures

The DMRG scheme also provides a good practical truncation strategy for large-scale nuclear structure calculations. It works in the framework of nuclear shell models with an effective Hamiltonian in which a nucleus is modeled by filled core shells and partially filled valence orbitals of protons and neutrons. The core is assumed to be inert. This significantly reduces the dimension of the active shell subspace.

The potential of DMRG in the nuclear structure calculation was first exemplified in a large angular momentum shell interacting through a pairing and a quadrupolar force in an oblate nucleus [58]. Several schemes were then proposed to implement DMRG in different symmetry channels, including the particle-hole [297], the z-component of the total angular momentum [298], and the total angular momentum symmetries [299, 300]. Their applications led to accurate treatment for quite a number of nuclei, including ^{28}Si [301], ^{56}Ni [298, 301], and ^{48}Cr [299, 301]. Further improvement to the accuracy could be achieved by optimally arranging the order of the proton and neutron orbitals according to the criterion that minimizes the sum of the distance between any two orbitals weighted by their quantum mutual information [302]. It reduces the DMRG error for the ground-state energy of ^{56}Ni by one order of magnitude [301, 302].

Classical Statistical Models

In the formalism of path integral, a d-dimensional quantum system is mapped onto a $(d + 1)$-dimensional classical statistical model. This correspondence between quantum and classical models implies that a method developed in a d-dimensional quantum system can be extended, with certain modifications, to a $(d + 1)$-dimensional classical system and vice versa.

As the partition function of a classical statistical model can be always represented as a network product of local tensors, various tensor network RG methods, including TMRG [68, 70, 303], coarse-graining TRG [128, 133, 134], and SRG [129, 131], have been developed in studying these systems in the past decades. Among them, TMRG [68] is a direct generalization of DMRG. In a classical system, the transfer matrix plays a similar role to the Hamiltonian in a quantum system. TMRG [68] is a method for diagonalizing transfer matrices based on the reduced thermal density matrix. As an exactly soluble system, the Ising model often serves as an ideal system for testing each method.

Tensor network methods have been applied to investigate nearly all two-dimensional classical statistical models, especially their critical behaviors around the phase transition points. This includes, for example, the Ising model [303, 304], Potts model [305, 306, 307, 308], clock model [309, 310, 311], vertex model [312, 313], self-avoiding-walk model [314], XY model [315, 316], Heisenberg model [317], and lattice gauge models with different symmetries [203, 204]. A number of three-dimensional classical statistical models have also been studied with tensor network methods [110, 131, 135, 318, 319, 320].

Quantum Information

The reduction of the Hilbert space carried out in DMRG or other tensor network renormalization group methods is a problem of quantum data compression. In the language of quantum information, it is to find an optimal scheme to minimize the quantum information loss or to carry out an optimal lossy quantum data compression. The von Neumann entropy, or the entanglement entropy, is the most fundamental measure in quantum information. It quantifies the nonlocal quantum correlations between a subsystem and its complement. In fact, the entanglement entropy was used in the DMRG calculations even before this terminology was formally introduced into the field of DMRG [34, 258].

By exploring the entanglement structures of quantum states, we understand why tensor network states approximate ground states of quantum lattice models with short-range interactions so well [47, 111]. At the heart of this insight is the area law of entanglement entropy [37], which confines the space of physical interest to a small corner of the whole Hilbert space. In other words, the area law places strong bounds on quantum entanglement that a many-body system can generate in its ground state. It translates directly to the amount of memory and time required to compute a quantum state in actual calculations.

In one dimension, if the entanglement of a bipartite system is bounded or grows logarithmically with its size, an efficient simulation with tensor network methods is possible. This explains why DMRG works so well in one dimension because the entanglement of one-dimensional ground states is bounded by the area law in a gapped system or grows logarithmically with the lattice size in a critical system [47]. However, the simulation of the time evolution may not be efficient even in one dimension since the bipartite entanglement grows linearly with time.

In two dimensions, the boundary grows with the system size. It is impossible to catch this fast-growing entanglement by using an MPS or other one-dimensional tensor network state to represent a ground-state wave function when the system size becomes sufficiently large. To resolve this problem, a two-dimensional representation of quantum states, like PEPS [111] or PESS [114], has to be introduced to spread the entanglement onto all bonds on the boundary.

Quantum information also plays a vital role in the DMRG calculations of quantum lattice models with nonlocal interactions, such as the Hubbard model in momentum space [56], and quantum chemical systems [57]. In treating nonlocal quantum models with DMRG, the ordering of the lattice sites or molecular orbitals seriously affects the accuracies of converged results. An optimal scheme is to order the single-particle states or orbitals by minimizing their overall distances determined by their mutual quantum information [258].

On the other hand, tensor networks provide a natural framework for a fundamental understanding of inherent entanglement structures of physical systems. Particularly, tensor networks, such as PEPS, could be used to explore the relationship between the edge and bulk states through the so-called entanglement Hamiltonians [321]. The entanglement Hamiltonian H_e is defined from the reduced density matrix. Its eigenvalues encode essential information about the boundary states. If the system is gapped and not topologically ordered, H_e is usually a boundary Hamiltonian with local interactions. In a critical system, however, H_e is generally a Hamiltonian with long-range interactions.

Tensor networks also provide a neat and unique representation for classifying topological phases in a gapped system [322]. As a short-range correlated state or a gapped state is well described by an MPS in one dimension, it was shown that all quantum states are equivalent to trivial product states in the absence of any symmetry [323]. This means no topological order without symmetry protection in one dimension. A topologically nontrivial phase may exist in a phase space restricted by symmetries. This symmetry-protected topological state is short-range entangled. However, by definition, an intrinsic topologically ordered state is long-ranged entangled.

Machine Learning

Tensor networks were initially introduced to provide an efficient tool for attacking quantum many-body problems. One emerging application direction of tensor networks in a seemingly unrelated field is machine learning. Deep neural networks can characterize complex learning tasks such as image classification or speech recognition. The reason why the neural networks work so well could be understood using the language of quantum entanglement [324]. On the other hand, quantum wave functions could be modeled by making use of fully connected neural networks and restricted Boltzmann machines [325]. It was shown that a Boltzmann machine is essentially equivalent to a tensor network [326].

Tensor networks, including MPS [327], PEPS [328] and the tree tensor network [329, 330], provide a natural way to parameterizing machine learning models. It was used for a variety of tasks, such as supervised learning for images with MPS [327, 331], MPO [332], or PEPS [328], mixed unsupervised and supervised learning with multiscale tensor networks [329]. Compared with other machine learning approaches, tensor networks offer clear theoretical insight and interpretation, more sophisticated training algorithms and strategies originally developed for solving quantum many-body problems, a dramatic reduction in memory needed, and a few other advantages.

However, caution should be taken in applying tensor networks to machine learning since neural networks are characterized by nonlinear functions, whereas tensor networks are generally linear. It is still unclear how the entanglement is embedded in the data set. In a study on the supervised image classification using the MNIST data set of handwritten digits, it was found that the entanglement grows very fast with the number of images in the training data set, and the bond dimension of MPS used for training should be as high as the image number in order to capture the actual image structures [333]. Nevertheless, other kinds of tensor networks can be exploited to resolve this problem by block isolating or squeezing entanglement in these kinds of data sets [333].

2

Basic Algebra of Tensors

2.1 Diagrammatic Representation of Tensors

A tensor is a multidimensional array of numerical values. It is used to describe linear relations between vectors, scalars, and other tensors. Elementary examples of such links include the dot product and linear maps. The *order* of a tensor is the number of indices needed to label the array. For example, a vector is a one-dimensional array and a first-order tensor. A matrix is a two-dimensional array and therefore is a second-order tensor. Scalars are pure numbers and are zeroth-order tensors.

A tensor network is a network of tensors whose indices are connected by bonds according to some pattern. A bond that connects two tensors is an internal one. Summing over an index is to contract the corresponding bond in the graphic representation of the tensor network. A single tensor is a tensor network without any internal bonds.

The mathematical formula of a tensor or a tensor network can be readily represented using diagrammatic notations in terms of tensor diagrams. In these diagrams, a tensor is represented by a vertex, and the lines connected to this vertex are the indices of this tensor. For example, a scale, a vector, a matrix, and a fourth-order tensor can be represented as

(a) scale x : $\quad \bullet^x = (x)$ $\qquad$ (b) vector V_α : $\qquad (V)\!\!-\!\alpha$

(c) matrix $M_{\alpha\beta}$: $\quad \alpha\!-\!(M)\!-\!\beta$ $\qquad$ (d) tensor $T_{\alpha\beta\gamma\delta}$: $\qquad \alpha\!-\!\boxed{T}\!-\!\gamma$ with δ above and β below

A diagrammatic representation of a tensor network is simply a collection of a set of vertices interconnected by lines. A line connects at most two vertices or tensors.

A line connected by two vertices is the line or index to be contracted. Conversely, an open line connects just one vertex and corresponds to an index without contraction.

The contractions or multiplications of two or more tensors can be also represented using tensor diagrams. For instance, the scalar product of a matrix, product of two matrices, trace of the product of four matrices, and contraction of two order-three tensors with four open indices can be represented by the diagrams

(a) TrA

(b) $C = AB$

(c) Tr$(ABCD)$

(d) $C_{\alpha\sigma_1\sigma_2\beta} = \sum_\gamma A_{\alpha,\sigma_1,\gamma} B_{\gamma,\sigma_2,\beta}$

A tensor network contraction is to identify and sum over a bond that connects two tensors. After contraction, the unconnected legs remain as indices of the resulting new tensor. If there are no open bonds, the network corresponds to a scalar.

To differentiate from other matrices, we use a rounded square to represent a unitary matrix

$$U = -\boxed{U}- \;. \tag{2.1}$$

A truncated unitary matrix or an isometry is commonly used in the basis transformation to compress a higher dimensional basis space to a lower one. Graphically, an isometry is more conveniently represented by a deformed square with an orientation indicating the direction of transformation

$$-\!\!\langle\!A\rangle\!\!- \qquad \text{or} \qquad -\!\!\langle\!A\rangle\!\!- \;. \tag{2.2}$$

Here the thick bond has a higher dimension than the thinner one. In DMRG or TRG, these kind of matrices correspond to a third-order tensor with two indices in and one index out, which can be also represented as

$$=\!\!\langle\!A\rangle\!\!- \qquad \text{or} \qquad -\!\!\langle\!A\rangle\!\!= \;. \tag{2.3}$$

2.2 QR and LQ Decompositions

A QR decomposition (also called QR factorization) is an efficient algorithm to factorize a matrix A into a product of a unitary matrix Q and an upper triangular matrix R:

$$A = QR = \quad -\!\!\boxed{Q}\!-\!\boxed{R}\!- \quad .\tag{2.4}$$

This kind of decomposition is often used to solve linear equations as well as eigenvalue problems. If A has n linearly independent columns, then the first n columns of Q form an orthonormal basis for the column space of A. This decomposition, in particular the upper triangular form of R, indicates that the kth column of A only depends on the first k columns of Q. If A is a square and invertible matrix, then the QR decomposition is unique if we gauge R so that its diagonal elements are all positive.

If A is an $m \times n$ rectangular matrix with $m > n$, the QR decomposition can be represented as

$$A = QR = \begin{pmatrix} Q_1 & Q_2 \end{pmatrix} \begin{pmatrix} R_1 \\ 0 \end{pmatrix} = Q_1 R_1 = \quad -\!\boxed{Q_1}\!\!-\!\boxed{R_1}\!- \tag{2.5}$$

where R_1 is an $n \times n$ upper triangular matrix, Q_1 is an $m \times n$ matrix, and Q_2 is an $m \times (m - n)$ matrix. In this case, Q_1 is an isometric matrix:

$$Q_1^\dagger Q_1 = \quad -\!\langle Q_1^\dagger |\!-\!| Q_1 \rangle\!- \quad = I_{n \times n}, \tag{2.6}$$

$$Q_1 Q_1^\dagger = \quad -\!| Q_1 \rangle\!-\!\langle Q_1^\dagger |\!- \quad \neq I_{m \times m}. \tag{2.7}$$

Here $Q_1 Q_1^\dagger$ is not an identity matrix. Instead it is a projection operator:

$$\left(Q_1 Q_1^\dagger \right)^2 = Q_1 Q_1^\dagger. \tag{2.8}$$

Similarly, one can also take a LQ decomposition to factorize matrix A into a product of a lower triangular matrix L and a unitary matrix Q:

$$A = LQ = \quad -\!\boxed{L}\!-\!\boxed{Q}\!- \quad . \tag{2.9}$$

In fact, from the transpose of A, we have

$$A^T = Q^T L^T. \tag{2.10}$$

Thus, we can also find L and Q by taking a QR decomposition for A^T.

2.3 LU Decomposition with Partial Pivoting

An LU decomposition or factorization factors a square matrix A as the product of a lower triangular matrix, L, and an upper triangular matrix, U:

$$A = LU. \tag{2.11}$$

It is easy to verify that

$$A_{1,1} = L_{1,1} U_{1,1}. \tag{2.12}$$

If $A_{1,1} = 0$, then at least one of $L_{1,1}$ and $U_{1,1}$ has to be zero, which implies that either L or U is singular. This is impossible if A is full-ranked or invertible. If A is invertible, this singularity can be removed by simply reordering the rows (or columns) of A so that its diagonal matrix elements are all nonzero.

Assume P to be a row permutation matrix, which, when left-multiplied to A, reorders the rows of A so that the diagonal matrix element is the largest matrix element in the absolute value in that row. In this case, PA is safe to be LU decomposed:

$$PA = LU. \tag{2.13}$$

This LU factorization with partial pivoting (abbreviated as LUP) is numerically stable. Clearly, L and U obtained from this equation are not uniquely determined. We can remove this ambiguity by requiring that all the diagonal matrix elements of L are equal to 1.

If A is a Hermitian positive-definite matrix, one can always set U to be the Hermitian conjugate of L and write A as

$$A = LL^\dagger. \tag{2.14}$$

This is also called the Cholesky decomposition.

The LUP decomposition is usually employed to solve a system of linear equations or to compute the determinant of a matrix. It offers a simple decomposition scheme to fix the gauge in the bi-orthogonalization of MPS discussed in §17.1.

2.4 Singular Value Decomposition

Singular value decomposition (SVD) is used to diagonalize a matrix in terms of two unitary matrices, hence to factorize a matrix A into a product of two unitary matrices, U and V, and a singular matrix Λ:

$$A_{n \times m} = U_{n \times n} \Lambda_{n \times m} V_{m \times m}^\dagger = \ -\!\!\boxed{U}\!\!-\!\!\diamondsuit\!\!-\!\!\boxed{V^\dagger}\!\!-\ . \tag{2.15}$$

Both U and V are square matrices. The columns of U and V,

$$U = (u_1, u_2, \ldots u_n), \qquad V = (v_1, v_2, \ldots v_m) \tag{2.16}$$

are called left- and right-singular vectors of A, respectively. $\Lambda_{n \times m}$ is a rectangular diagonal matrix with non-negative diagonal matrix elements:

$$\Lambda_{i,j} = \Lambda_i \delta_{i,j}, \qquad \Lambda_i \geq 0. \tag{2.17}$$

The diagonal entries Λ_i of Λ are the singular values of A. The number of non-zero singular values is the rank of A, denoted as $r(A)$. It is simple to show that $r(A) \leq \min(n, m)$. A common convention is to list the singular values in descending order (i.e. $\Lambda_1 \geq \Lambda_2 \geq \cdots \geq \Lambda_{r(A)}$). In this case, the diagonal matrix Λ is uniquely determined. A singular value for which we can find two or more left (or right) singular vectors that are linearly independent is called degenerate. Nondegenerate singular values have unique left- and right-singular vectors up to arbitrary phase factors. A SVD is unique if all singular values are nondegenerate.

Degenerate singular values, by definition, have nonunique singular vectors. If u_i and u_j are two left-singular vectors corresponding to the same singular value, then any normalized linear combination of these two vectors is also a left-singular vector corresponding to this singular value. A similar statement is true for right-singular vectors.

SVD is closely related to the canonical (or eigenvalue) decomposition. The left-singular vectors of A are the eigenvectors of $AA^\dagger$ with the corresponding eigenvalues equal to $\Lambda\Lambda^\dagger$:

$$AA^\dagger U = U\Lambda V^\dagger V\Lambda^\dagger U^\dagger U = U\left(\Lambda\Lambda^\dagger\right). \tag{2.18}$$

Similarly, the right-singular vectors of A are the eigenvectors of $A^\dagger A$ with the eigenvalues equal to $\Lambda^\dagger\Lambda$:

$$A^\dagger A V = V\Lambda^\dagger U^\dagger U\Lambda V^\dagger V = V\left(\Lambda^\dagger\Lambda\right). \tag{2.19}$$

Therefore, the nonzero singular values of A are the square roots of the nonzero eigenvalues of both $AA^\dagger$ and $A^\dagger A$.

If $A_{i,j}$ is a wave function of a two-body quantum system whose basis states are labeled by i and j, its SVD is also known as the Schmidt decomposition of the wave function.

From the property of SVD, it is simple to show that the rank of a product of two matrices, A and B, equals the minimal rank of these two matrices, that is:

$$r(AB) = \min[r(A), r(B)]. \tag{2.20}$$

In the implementation of DMRG or TRG, we often need to find an optimal low-rank approximation for a matrix, say $A_{n\times m}$. This approximation is determined by SVD. To find a matrix B of rank $k < r(A)$ that gives the best approximation to A, one needs to discard the singular values Λ_i with $i > k$ (this is also called basis truncation). B is then given by

$$B = U_{n\times k}\Lambda'_{k\times k}V^\dagger_{k\times m}, \tag{2.21}$$

where Λ' is a diagonal matrix that contains only the k largest singular values of Λ. The reason that this approximation is optimal is because the difference, measured in terms of Frobenius norm, between A and B is simply given by the square sum of discarded singular values:

$$\parallel A - B \parallel^2 = \mathrm{Tr}\left(AA^\dagger - AB^\dagger - BA^\dagger + BB^\dagger\right) = \sum_{i>k} \Lambda_i^2. \tag{2.22}$$

2.5 Polar Decomposition

Given a matrix $M_{m \times n}$ with $m \geq n$, one can find its left polar decomposition

$$M = AP = \quad \boxed{A}\!-\!\!\bigcirc\!\!\!P\!\!- \quad, \tag{2.23}$$

where $A_{m \times n}$ is an isometric matrix, satisfying

$$A^\dagger A = I_{n \times n}. \tag{2.24}$$

$P_{n \times n}$ is a positive symmetric matrix.

This polar decomposition can be obtained by first taking SVD

$$M = U \Lambda V^\dagger. \tag{2.25}$$

A and P are then given by

$$A = UV^\dagger, \tag{2.26}$$

$$P = V \Lambda V^\dagger. \tag{2.27}$$

If $m \leq n$, a right polar decomposition of M is similarly defined.

2.6 Higher-Order Singular Value Decomposition

For a higher-order tensor, there does not exist a general multilinear decomposition method with all the properties of the matrix singular value decomposition, where the singular values with the left and right singular vectors can be simultaneously determined. Nevertheless, a number of decomposition schemes for tensors exist. This includes, for example, the Tucker decomposition and the canonical decomposition. A particularly useful scheme for the study of tensor renormalization is HOSVD, which is a generalization of SVD and a special kind of Tucker decomposition. It bears the closest resemblance in decomposition form to SVD.

HOSVD of an order-n tensor, $T_{i_1, i_2, \ldots, i_n}$, is defined by the orthogonal decomposition of the tensor [334]:

$$T_{i_1, i_2, \ldots, i_n} = \sum_{j_1 j_2 \cdots j_n} S_{j_1, j_2, \ldots, j_n} (U_1)_{i_1 j_1} (U_2)_{i_2 j_2} \cdots (U_n)_{i_n j_n}, \tag{2.28}$$

where U's are the unitary matrices (also called singular matrices). Graphically, this decomposition can be represented, for example, for an $n = 4$ tensor, as

$$(2.29)$$

In this equation, S is the core tensor of T, which plays a similar role as the singular matrix in SVD. But S is not a diagonal tensor. Instead, it possesses the following two properties for any index, say the second index j_2:

(i) all orthogonality,

$$\langle S_{:j_2:\cdots:} \mid S_{:j_2':\cdots:} \rangle = 0, \qquad \text{if } j_2 \neq j_2', \tag{2.30}$$

where $\langle S_{:j_2:\cdots:} \mid S_{:j':\cdots:} \rangle$ is the inner-product of these two subtensors defined by

$$\langle S_{:j_2:\cdots:} \mid S_{:j_2':\cdots:} \rangle = \sum_{j_1 j_3 \cdots j_n} S^*_{j_1,j_2,j_3,\ldots,j_n} S_{j_1,j_2',j_3,\ldots,j_n}, \tag{2.31}$$

(ii) pseudo-diagonal: the subtensor $S_{:j_2:\cdots:}$ is ordered according to its Frobenius norm:

$$|S_{:j_2:\cdots:}| \geq |S_{:j_2':\cdots:}|, \qquad \text{if } j_2 < j_2', \tag{2.32}$$

where $|S_{:j_2:\cdots:}|$ is defined by

$$|S_{:j_2:\cdots:}| = \sqrt{\langle S_{:j_2:\cdots:} \mid S_{:j_2:\cdots:} \rangle}. \tag{2.33}$$

This norm plays a similar role as the singular value of a matrix.

The unitary matrices on the right-hand side of Eq. (2.28) can be obtained by taking a sequence of SVD. To determine U_1, for example, we first convert $T_{i_1,i_2,\ldots,i_n}$ into a matrix $A_{i_1, i_2 \cdots i_n}$:

$$A_{i_1, i_2 i_3 \cdots i_n} = T_{i_1,i_2,i_3,\ldots,i_n}, \tag{2.34}$$

with the first index i_1 as the row index and the rest $(i_2, \ldots, i_n)$ as the column index. From the theory of HOSVD, U_1 is nothing but the left singular matrix of A under SVD. Hence, U_1 is determined by the canonical transformation of $AA^\dagger$:

$$AA^\dagger = U_1 \Lambda U_1^\dagger, \tag{2.35}$$

where Λ is the eigenvalue of $AA^\dagger$. Furthermore, it can be shown that

$$|S_{i_1:\cdots:}|^2 = \Lambda_{i_1}. \tag{2.36}$$

HOSVD provides a natural and effective scheme to truncate tensors. In many cases, it provides an optimal or at least nearly optimal low-rank approximation for tensors [335, 336]. But unlike the matrix SVD, it may not always be the best low-rank approximation for an arbitrary tensor. A Tucker decomposition may generate a better approximation in case the result of HOSVD is not optimal, but its computational cost is higher.

2.7 Low-Rank Approximation of Tensors

HOSVD provides a simple scheme to truncate tensors. However, the truncated tensor obtained may not be optimal. To find an optimal low-rank approximation to a tensor, we need to find a set of isometric matrices $U = \left(U^{(1)} \cdots U^{(n)}\right)$ such that the tensor difference before and after the truncation

$$\Delta(U) = \|T - U \circ S\| \tag{2.37}$$

is minimized. In this expression, $U \circ S$ and $\|A\|$ are defined by

$$(U \circ S)_{i_1,i_2,\ldots,i_n} = \sum_{j_1 j_2 \cdots j_n} S_{j_1,j_2,\ldots,j_n} U^{(1)*}_{i_1 j_1} U^{(2)*}_{i_2 j_2} \cdots U^{(n)*}_{i_n j_n}, \tag{2.38}$$

$$\|A\| = \mathrm{Tr}\left(A^\dagger A\right) = \sum_{i_1 \ldots i_n} A^*_{i_1,\ldots,i_n} A_{i_1,\ldots,i_n}. \tag{2.39}$$

T is the original tensor and S is a low-rank approximation to T defined by

$$S_{j_1,j_2,\ldots,j_n} = \sum_{i_1,i_2,\ldots,i_n} T_{i_1,i_2,\ldots,i_n} U^{(1)}_{i_1 j_1} U^{(2)}_{i_2 j_2} \cdots U^{(n)}_{i_n j_n}. \tag{2.40}$$

Please note that $U^{(i)}$ is an isometric matrix, not a unitary matrix. It satisfies the conditions

$$\left(U^{(i)}\right)^\dagger U^{(i)} = I_{D_i}, \qquad U^{(i)} \left(U^{(i)}\right)^\dagger \neq I_{N_i}, \tag{2.41}$$

where N_i and $D_i \leq N_i$ are the dimensions of the ith component before and after truncation.

Using these definitions, it is straightforward to show that

$$\Delta(U) = \|T\| - \|S\|. \tag{2.42}$$

As $\|T\|$ does not depend on U, minimizing $\Delta(U)$ is equivalent to maximizing $\|S\|$. $\|S\|$ is a quadratic function of each $U^{(i)}$.

We determine the isometric matrices by taking an orthogonal iteration [335]. To determine $U^{(1)}$, for example, we first fix all U-matrices except $U^{(1)}$, $\|S\|$ then becomes

$$\|S\| = \text{Tr}\left(M^{(1)}U^{(1)}\right), \tag{2.43}$$

where

$$M^{(1)}_{j_1 i_1} = \sum_{j_2 \ldots j_n} S^*_{j_1 j_2 \ldots j_n} X_{j_2 \ldots j_n}[i_1], \tag{2.44}$$

$$X_{j_2 \ldots j_n}[i_1] = \sum_{i_2 \ldots i_n} T_{i_1, i_2, \ldots, i_n} U^{(2)}_{i_2 j_2} \cdots U^{(n)}_{i_n j_n}. \tag{2.45}$$

$M^{(1)}$ is nothing but the derivative of $\|S\|$ with respect to $U^{(1)}$. From its SVD

$$M^{(1)} = U \lambda V^\dagger, \tag{2.46}$$

we find

$$U^{(1)} = V U^\dagger. \tag{2.47}$$

This is a solution that is obtained by fixing $(U^{(1)})^\dagger$. In order to find a self-consistent solution of $U^{(1)}$, we need to repeat this calculation by substituting $(U^{(1)})^\dagger$ in $M^{(1)}$ with the solution (2.47) until $U^{(1)}$ is converged.

Similarly, one can find other isometric matrices. At each iteration, all the isometric matrices except the one to find, say $U^{(i)}$, are fixed. A matrix similar to $M^{(1)}$ is defined by swapping index 1 in the expressions of Eq. (2.44) with i. Repeating the above iterations until all $U^{(i)}$ are fully converged, we obtain optimized isometric matrices and subsequently the truncated tensor S.

2.8 Automatic Differentiation

In the variational determination of local tensors, either by minimizing the ground state energy [111] or by full update [117], we often need to evaluate the derivatives of the Hamiltonian matrices or wave function overlaps in the framework of tensor network states. This is technically a rather challenging task. First, it often involves a summation of an infinite number of products of tensor network states with different kinds of configurations, which is difficult to simplify even if we assume the system is translation invariant. Second, the optimization of tensor states is a highly nonlinear problem, and it often causes severe instability in numerical simulations. Moreover, it is cumbersome and error prone to derive the gradients of tensor network states analytically. Automatic differentiation, on the other hand, provides an elegant and efficient solution to these problems by composing the whole tensor network program in a fully differentiable manner [120].

Automatic differentiation refers to a general way of taking a program that computes the value of a function and automatically constructing a procedure for computing derivatives of that function at a given point. It works based on the fact that any computer program, no matter how complicated, is to execute a sequence of elementary arithmetic operations and functions.

Automatic differentiation is also called algorithmic differentiation or computational differentiation. It differs from symbolic differentiation and conventional numerical differentiation. Symbolic differentiation intends to obtain rigorous expressions for derivatives. Unfortunately, these expressions are generally complex and redundant, hence difficult to program and compute. Numerical differentiation is based on the method of finite differences, which may induce large round-off errors in the discretization. Moreover, it is difficult for symbolic and numerical differentiation to calculate higher derivatives, where complexity and errors increase dramatically and are slow at computing partial derivatives of a function with many input parameters. Automatic differentiation solves all of these problems at the expense of introducing more software engineering. Unlike symbolic differentiation, it does not generate a symbolic expression of the gradient function. It computes the value of derivatives to the machine precision without introducing the round-off errors as in numerical differentiation. Automatic differentiation is, in principle, numerically stable and efficient. The cost in computing a derivative is of the same order as in computing the function itself [337, 338], which implies that the computational resource (including both computational time and memory space) needed scales linearly with that in the calculation of the function value itself.

Automatic differentiation computes the gradient of a composite function,

$$y = f_n(f_{n-1} \cdots f_2(f_1(x)) \cdots), \tag{2.48}$$

with respect to the input parameter, x, through the *chain rule of derivatives*:

$$\frac{\partial y}{\partial x} = \frac{\partial f_n}{\partial f_{n-1}} \frac{\partial f_{n-1}}{\partial f_{n-2}} \cdots \frac{\partial f_2}{\partial f_1} \frac{\partial f_1}{\partial x}. \tag{2.49}$$

If we set $y_0 = x$ and y_i the value of the function obtained up to the ith level,

$$y_i = f_i(f_{i-1}(\cdots f_2(f_1(x)) \cdots)) = f_i(y_{i-1}) \qquad (i = 1, \ldots, n), \tag{2.50}$$

this composite function can be graphically represented as

where $y = y_n$, nodes represent variables (scalars, vectors, matrices, tensors, etc.), and edges represent some mathematical operations. Each derivative in the chain rule is a Jacobian matrix $\partial y_i/\partial y_{i-1}$. The product of these Jacobians gives the gradient of the composite function.

The arrowed graph above sets up a particular operation scheme of automatic differentiation, called *forward accumulation* or *forward mode*, which traverses the chain rule following the same order as the function evaluation itself, from inputs to outputs. In this scheme, one first fixes the independent variable to which the differentiation is performed and recursively computes the derivative of each subfunction according to the chain rule. The computational complexity of one forward accumulation is proportional to the complexity of computing the composite function. It is computationally efficient if the dimension of the input parameters, $\mathrm{Dim}(x)$, is significantly smaller than that of the output parameters, $\mathrm{Dim}(y)$.

If, however, $\mathrm{Dim}(x) \gg \mathrm{Dim}(y)$, the forward accumulation may become less efficient because $\mathrm{Dim}(x)$ sweeps are needed in order to evaluate all partial derivatives. This often happens in deep learning or tensor network optimizations. In this case, a *reverse accumulation* or *reverse mode* is more efficient to traverse the chain rule since only $\mathrm{Dim}(y)$ sweeps are necessary. The reverse mode automatic differentiation was first proposed by Seppo Linnainmaa in 1976 [339].

Reverse accumulation is a backpropagation algorithm. It is to evaluate the gradient by multiplying terms from left to right using a series of Jacobian products, namely from the output back to the input. To carry out this backpropagation, it is instructive to introduce the *adjoint variable*

$$\bar{y}_i = \frac{\partial y}{\partial y_i} \tag{2.51}$$

to denote the gradient of the final output y with respect to variable y_i. The reverse mode is simply to propagate the adjoint from

$$\bar{y}_n = \bar{y}\frac{\partial y}{\partial y_n}, \qquad \bar{y} = 1 \tag{2.52}$$

all the way back to

$$\bar{y}_i = \bar{y}_{i+1}\frac{\partial y_{i+1}}{\partial y_i} \tag{2.53}$$

with $i = n - 1, \ldots, 1$, and finally to

$$\frac{\partial y}{\partial x} = \bar{x} = \bar{y}_1\frac{\partial y_1}{\partial x}. \tag{2.54}$$

At each step, the adjoint is updated by multiplying a local Jacobian matrix. Graphically, the reverse accumulation can be represented as

Both the forward and the adjoint backpropagation modes can be generalized to more complex functions or computational graphs to allow a node to branch into two or more paths. For example, in the following graph,

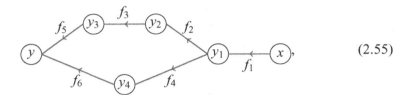

$$(2.55)$$

node y_1 is split into two paths through the actions of f_2 and f_4. All the nodes generated from y_1, y_2, and y_4, are called the children of y_1. In the forward mode, the chain rule can be implemented just following the arrows in the graph.

In the reverse mode, the computational graph has exactly the same topology as in the original one, except that the nodes are replaced by their adjoints and the arrows are reversed. For example, the reverse mode graph corresponding to graph (2.55) is

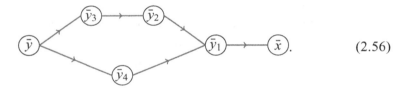

$$(2.56)$$

In general, the adjoint variable is defined by accumulating all the contributions from its child nodes:

$$\bar{y}_i = \sum_{j \in \text{children of } i} \bar{y}_j \frac{\partial y_j}{\partial y_i}. \tag{2.57}$$

The building blocks of a computational graph are called *primitives*. A primitive is an elementary mathematical operation, such as addition, multiplication, or an arbitrary function. Each has an associated backward function, which generates the associate adjoint with the Jacobian matrices from its child nodes. It is not necessary to store all Jacobian matrices. One can group many elementary computation steps as a primitive. In this way, the forward pass of these clustered primitives can be operated as a black box. This can save computational time as well as memory space.

2.9 Trotter–Suzuki Decomposition

The Trotter–Suzuki decomposition was developed by Suzuki [115], who applied the Trotter product formula [73]

$$e^{A+B} = \lim_{n \to \infty} \left(e^{\frac{A}{n}} e^{\frac{B}{n}} \right)^n, \qquad [A, B] \neq 0 \qquad (2.58)$$

to map a d-dimensional quantum system into an equivalent $(d + 1)$-dimensional classical one. This approach is widely used in the path integral representation of quantum systems as well as in quantum Monte Carlo simulations. It is also a useful tool for constructing the tensor network representation of the partition function or density matrix for a quantum lattice model.

The simplest Trotter–Suzuki decomposition reads

$$e^{\varepsilon(A+B)} = e^{\varepsilon A} e^{\varepsilon B} + O(\varepsilon^2), \qquad (2.59)$$

where ε is a small parameter, and A and B are two arbitrary operators that do not commute with each other. This is the leading approximation to the Trotter product formula with an error in the second order of ε.

More generally, the Trotter–Suzuki decomposition takes the form [115, 340, 341]

$$e^{\varepsilon(A+B)} = e^{\varepsilon p_1 A} e^{\varepsilon p_2 B} e^{\varepsilon p_3 A} e^{\varepsilon p_4 B} \cdots e^{\varepsilon p_M B} + O(\varepsilon^{M+1}), \qquad (2.60)$$

where $(p_1, p_2, \cdots p_M)$ is a set of parameters that can be determined such that the correction is less than or of the order of ε^{M+1}. The second-order Trotter–Suzuki decomposition, for example, reads

$$e^{\varepsilon(A+B)} = e^{\varepsilon A/2} e^{\varepsilon B} e^{\varepsilon A/2} + O(\varepsilon^3). \qquad (2.61)$$

A frequently used fourth-order Trotter–Suzuki decomposition reads [115, 340]

$$e^{\varepsilon(A+B)} = \prod_{i=1}^{5} e^{p_i \varepsilon A/2} e^{p_i \varepsilon B} e^{p_i \varepsilon A/2} + O(\varepsilon^5), \qquad (2.62)$$

where all $p_i = 1/(4 - 4^{1/3})$, except $p_3 = 1 - 4p_1 < 0$.

For the nth-order Trotter–Suzuki decomposition, the error is of order ε^{n+1}. To reach a given imaginary time β, one has to perform β/ε imaginary time steps, such that the error grows (at worst) linearly in β and the resulting error is bounded by $\varepsilon^n \beta$.

3

Tensor Network Representation of Classical Statistical Models

3.1 Tensor Network Models

Classical statistical models can always be represented as tensor network models [128, 130]. This is a ubiquitous and important property of classical statistical models, which allows us to probe these models from the perspective of tensor network algorithms.

There are many ways to represent a given classical lattice model as a tensor network model. A tensor network model can be defined on the original lattice. It can also be defined on the dual or other geometrically related lattices. Three approaches are commonly used to construct a tensor network representation for a classical statistical model.

The first approach works for a classical statistical model with only nearest-neighbor interactions. In this case, the partition function is written as a product of matrices defined by the local Boltzmann factors on all bonds connecting two neighboring sites. This leads to a simple tensor representation of classical statistical models called the *matrix-network model*.

The second approach defines the local tensors on the original lattice sites by factorizing the local bond matrices with SVD. This approach works in all kinds of lattices. The order of the local tensor is equal to the coordination number of the lattice if there are only nearest-neighbor interactions. This representation is more appropriate for a lattice whose coordination number is small.

The third approach is to take a duality transformation for the physical variables and define the local tensors in the dual lattice. It allows a classical statistical model to be represented as a tensor network model in the corresponding dual space. The order of the local tensor is the coordination number of the dual lattice. This kind of transformation is particularly useful if the coordination number of the dual lattice is smaller than the coordination number of the original lattice. For example, a classical model defined on a triangular lattice whose coordinate number is 6 can be represented as a tensor network model on a honeycomb lattice whose coordinate number is 3. The dual lattice of the dice lattice is the Kagome lattice. The square lattice is self-dual.

If the original statistical model includes not only the nearest neighboring two-site interactions but also many-site interactions within each building unit cell, a tensor network model can still be defined in the dual lattice without enlarging the bond dimension. This is an advantage of representing the partition function in the dual space.

For a vertex-sharing lattice, for example, a Kagome lattice, there is a simple way to define a tensor network model. In this case, one can set the physical variables as the bond variables and define the local tensor by collecting all the Boltzmann factors at each minimal simplex. For example, for the Kagome lattice, the minimal simplex is a triangle and the tensor network model is defined on the honeycomb lattice formed by connecting all the centers of triangles. This approach is simple to implement. But it works only on vertex-sharing lattices.

3.2 Matrix-Network Models

A classical statistical model with two-body interactions can always be represented as a matrix-network model. Let us take the Ising model,

$$H = \sum_{\langle ij \rangle} K(S_i, S_j), \qquad K(S_i, S_j) = -JS_iS_j, \tag{3.1}$$

defined on a square lattice as an example to show how to construct such a matrix-network model. In (3.1), $\langle \rangle$ denotes summation over nearest neighbors only. $K(S_i, S_j)$ is the interaction between two local spins, S_i and S_j. S_i takes two values ± 1. This model is rigorously soluble on the square as well as triangular lattices [342, 343, 344].

Thermodynamics of this model is determined by the partition function:

$$Z = \mathrm{Tr}e^{-\beta H} = \mathrm{Tr} \prod_{\langle ij \rangle} e^{-\beta K(S_i,S_j)} = \mathrm{Tr} \prod_{\langle ij \rangle} M_{S_i,S_j}, \tag{3.2}$$

where Tr is to sum over all spin configurations and

$$M_{S_i,S_j} = e^{-\beta K(S_i,S_j)} \tag{3.3}$$

is a matrix defined on the bond linking sites i and j. Equation (3.2) is just the matrix-network representation of the Ising model.

3.3 Tensor Network Representation in the Original Lattice

A classical statistical model with local interactions can be expressed as a tensor network model through SVD. This kind of tensor network model is defined on the original lattice. As an example, let us consider the Ising model defined by Eq. (3.1).

From the matrix representation of the Boltzmann weight, we first take SVD to decouple it into a product of two unitary matrices, U and V, and a semi-positive diagonal matrix λ:

$$-\!\!\boxed{M}\!\!- \quad = \quad -\!\!\boxed{U}\!\!-\!\!\langle\lambda\rangle\!\!-\!\!\boxed{V}\!\!- \ . \tag{3.4}$$

We then express it as a product of two matrices

$$-\!\!\boxed{M}\!\!- \quad = \quad -\!\!\boxed{Q^a}\!\!-\!\!\boxed{Q^b}\!\!- \ , \tag{3.5}$$

where

$$Q^a = U\lambda^{1/2}, \qquad Q^b = \lambda^{1/2}V. \tag{3.6}$$

Now let us group all Q's that connect to site i. A local tensor at site i is defined by tracing out S_i from the product of these Q's

$$T^i_{x,y,z,\dots} = \sum_{S_i} Q^{\alpha_1}_{S_i,x} Q^{\alpha_2}_{S_i,y} Q^{\alpha_3}_{S_i,z}\cdots . \tag{3.7}$$

The order of this tensor is equal to the number of sites interacting with site i. The bond dimension equals the dimension of the local basis states $d = 2$. This defines a tensor network representation of the partition function

$$Z = \mathrm{Tr}\prod_i T^i_{x_i,y_i,z_i\dots}. \tag{3.8}$$

Every tensor index is shared by two sites connected by a bond. The trace is to sum over all bond indices.

On a bipartite lattice – for example, a honeycomb lattice – the partition function can be simply expressed as

$$Z = \mathrm{Tr}\prod_{i\in a, j\in b} T^a_{x_i,y_i,z_i} T^b_{x_j,y_j,z_j}, \tag{3.9}$$

where the superscripts a and b stand for the two sublattices of the honeycomb lattice, and

$$\boxed{T^a}\!\!- \quad = \quad \begin{matrix} \boxed{Q^a} \\ \diagup \ \ \ \boxed{Q^a}\!\!- \\ \boxed{Q^a} \end{matrix} \ , \qquad \boxed{T^b}\!\!- \quad = \quad \begin{matrix} \boxed{Q^b} \\ \diagup \ \ \ \boxed{Q^b}\!\!- \\ \boxed{Q^b} \end{matrix} \ . \tag{3.10}$$

Moreover, if M is positive definite and symmetric, then $V = U^\dagger$ and $(Q^b)^\dagger$. In this case, $T \equiv T^a = T^b$ and the partition function can be simply expressed as

$$Z = \mathrm{Tr} \prod_i T_{x_i, y_i, z_i}. \tag{3.11}$$

3.4 Tensor Network Representation in the Dual Space

Let us take the Ising model defined on a triangular lattice as an example to illustrate how to represent its partition function as a tensor network in its dual lattice. The partition function of the model is

$$
\begin{aligned}
Z &= \mathrm{Tr} e^{-\beta H} \\
&= \mathrm{Tr}_S \prod_{\triangle_{ijk}} e^{\beta J (S_i S_j + S_j S_k + S_k S_i)/2},
\end{aligned} \tag{3.12}
$$

where Tr_S is to sum over all spin configurations and the product is taken over all small triangles. The factor $1/2$ is introduced in Eq. (3.12) because each bond is shared by two triangles. This partition function can be expressed as a tensor network in terms of bond variables through a *duality transformation*. The triangular lattice is dual to the honeycomb lattice, and vice versa

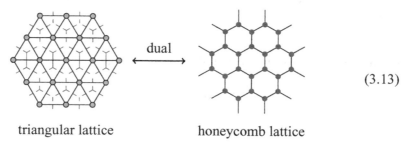

$$\tag{3.13}$$

triangular lattice   honeycomb lattice

Given a bond on the triangular lattice, let us define the bond spin by the product of two end spins:

$$\sigma_{ij} = S_i S_j. \tag{3.14}$$

This bond spin also takes two values $\sigma_{ij} = 1$ or -1, corresponding to a state of two parallel or antiparallel spins. Graphically, this dual transformation can be represented as

$$\tag{3.15}$$

In the dual space, the partition function becomes

$$Z = \text{Tr}_\sigma \text{Tr}_S \prod_{\langle ij \rangle} \delta(\sigma_{ij} - S_i S_j) \prod_{\triangle_{ijk}} e^{\beta J(\sigma_{ij} + \sigma_{jk} + \sigma_{ki})/2}, \tag{3.16}$$

where Tr_σ is to sum over all σ variables. The number of the bonds in the triangular lattice, N_{bond}, equals the sum of the total number of sites of the dual lattice, N_{site}, and the number of triangles, N_{triangle}. This indicates that the N_{bond}-delta functions in the equation can be equivalently written as a product of N_{site}-delta functions defined on all lattice sites and N_{triangle}-delta functions defined on all triangles. On each triangle, it is simple to show that the product of the three bond spins is equal to one:

$$\sigma_{ij} \sigma_{jk} \sigma_{ki} = S_i S_j S_j S_k S_k S_i = 1. \tag{3.17}$$

This is the constraint defined within each triangle, independent of the original spin variables. It allows the summation for the N_{site}-delta functions to be carried out. The partition function then becomes

$$Z = \text{Tr}_\sigma \prod_{\triangle_{ijk}} \frac{1 + \sigma_{ij} \sigma_{jk} \sigma_{ki}}{2} e^{\beta J(\sigma_{ij} + \sigma_{jk} + \sigma_{ki})/2}. \tag{3.18}$$

Using i to label the site position of the dual lattice, we can express the partition function Eq. (3.18) as

$$Z = \text{Tr} \prod_i T^i_{\sigma_x \sigma_y \sigma_z}, \tag{3.19}$$

where the trace is to sum over all bond indices, and

$$T^i_{\sigma_x \sigma_y \sigma_z} = \quad \overset{\sigma_x \qquad \sigma_z}{\underset{\sigma_y}{\boxed{T^i}}} \quad = \frac{1 + \sigma_x \sigma_y \sigma_z}{2} e^{\beta J(\sigma_x + \sigma_y + \sigma_z)/2} \tag{3.20}$$

is a third-order tensor defined on the hexagonal lattice. Here σ_x, σ_y, and σ_z are the three integer bond indices of dimension $d = 2$, defined on the three bonds emitted from site i along the x, y, and z links, respectively.

This derivation can be readily extended to other lattices. A square lattice is self-dual. Again, a dual variable is defined by the product of the two spins linked by a bond, as illustrated by Eq. (3.21):

$$\begin{array}{c} \sigma_{li} \\ S_i \quad | \quad S_l \\ \sigma_{ij} \, \text{——}\!\!\bullet\!\!\text{——} \, \sigma_{kl} \\ S_j \quad | \quad S_k \\ \sigma_{jk} \end{array} \quad . \tag{3.21}$$

For the Ising model, it is straightforward to show that the tensor network representation of the partition function on the square lattice is given by

$$Z = \text{Tr}_\sigma \prod_i T^i_{\sigma_x \sigma_y \sigma_{x'} \sigma_{y'}}, \tag{3.22}$$

where

$$T^i_{\sigma_x \sigma_y \sigma_{x'} \sigma_{y'}} = \sigma_x - \boxed{T^i} - \sigma_{x'} \;\; \underset{\sigma_{y'}}{\overset{\sigma_y}{}} = \frac{1 + \sigma_x \sigma_y \sigma_{x'} \sigma_{y'}}{2} e^{\beta J(\sigma_x + \sigma_y + \sigma_{x'} + \sigma_{y'})/2} \tag{3.23}$$

is a fourth-order tensor defined on the dual square lattice. Here σ_x, σ_y, $\sigma_{x'}$, and $\sigma_{y'}$ are the four bond variables connecting site i.

The dual lattice of the honeycomb lattice is a triangular lattice. In the dual space, each site has six neighbors. Hence, the order of each local tensor is six. In this case, there is no advantage to representing a honeycomb lattice model in its dual space.

It is simple to extend the above duality transformation to other statistical models. As an example, let us consider the *q-state Potts model* defined on the triangular lattice:

$$H = J \sum_{\langle i,j \rangle} \delta(\theta_i - \theta_j), \qquad \theta_i = \frac{2\pi m}{q} \quad (m = 0, \ldots, q-1), \tag{3.24}$$

where $\delta(\theta)$ is the *Kronecker delta function* that equals 1 if $\theta = 0$ or 0 otherwise. The bond variable is now defined by

$$\sigma_{ij} = \text{mod}[m_i - m_j, q], \qquad m_i = \frac{q\theta_i}{2\pi}. \tag{3.25}$$

Following the same step of derivation given for the Ising model, it is simple to show that the partition function of this model is a network product of the local tensor:

$$T_{xyz} = \delta(\text{mod}[x + y + z, q]) e^{-\beta J[\delta(x) + \delta(y) + \delta(z)]/2}, \tag{3.26}$$

where the bond variable x (similarly y or z) takes integer values from 0 to $q - 1$. When $q = 2$, the Potts model reduces to the spin-1/2 Ising model.

However, it should be pointed out that Eq. (3.14) is not a one-to-one mapping. The derivation holds, for example, for the Ising model in the absence of an applied magnetic field. If a Zeeman field term or a multisite interaction term is added to the Hamiltonian, the dual transformation (3.14) is no longer valid. Nevertheless, a tensor network model can still be defined in the dual lattice. But the dimension of local tensors has to be extended from d ($d = 2$ for the Ising model) to d^2 to distinguish all spin configurations spanned by S_i and S_j.

To understand this more clearly, let us consider an extended Ising model in an applied magnetic field defined on a square lattice:

$$H = -J \sum_{\langle ij \rangle} S_i S_j - h \sum_i S_i + J_\square \sum_{ijkl \in \square} S_i S_j S_k S_l, \tag{3.27}$$

where $ijkl \in \square$ means that $ijkl$ are the four vertex indices on each minimal square. The partition function of this model can still be expressed as a tensor network defined by Eq. (3.22). But the local tensor now changes to

$$T_{\sigma_{il}\sigma_{ij}\sigma_{jk}\sigma_{lk}} = \exp(-\beta H_\square), \tag{3.28}$$

where

$$H_\square = -\frac{J}{2} \left(S_i S_l + S_i S_j + S_j S_k + S_l S_k \right)$$
$$- \frac{h}{4} \left(S_i + S_j + S_l + S_k \right) + J_\square S_i S_j S_k S_l \tag{3.29}$$

is a function of bond variables $(\sigma_{il}, \sigma_{ij}, \sigma_{jk}, \sigma_{lk})$ that are implicitly defined from the physical variables (S_i, S_j, S_k, S_l) through the expression

$$\sigma_{ij} = (S_i + 1) + \frac{1}{2}(S_j + 1). \tag{3.30}$$

Here σ_{ij} runs from 0 to 3. Hence, the dimension of σ_{ij} is 4.

3.5 Vertex-Sharing Lattice Models

Bond variables are introduced in the discussion for the tensor network representation of a classical lattice model, either in the original lattice or in the dual lattice. Since the number of bonds (N_{bond}) is always larger than the physical variables defined on the lattice sites (N_{site}), these kinds of tensor network representations have redundant degrees of freedom. This leads to additional gauge symmetry in the definition of local tensors.

However, on a *vertex-sharing lattice*, such as the Kagome lattice (Fig. 3.1(a)), one can always define a tensor network model without introducing extra redundant degrees of freedom if there are only nearest-neighbor interactions. This kind of tensor network model is defined by taking the physical variables defined on the lattice sites as the bond variables. To understand this, let us consider the partition function of the Ising model:

$$Z = \text{Tr} e^{-\beta H} = \text{Tr}_S \prod_{\triangle_{ijk}} e^{-\beta [K(S_i, S_j) + K(S_j, S_k) + K(S_k, S_i)]}, \tag{3.31}$$

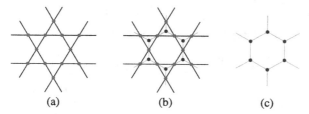

Figure 3.1 (a) Kagome lattice as an example of vertex-sharing lattice in two dimensions. The local tensor is defined at the center of each triangle, as shown in (b). (c) Honeycomb lattice on which the tensor network model is defined.

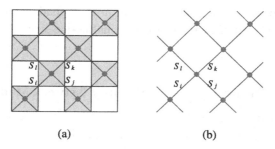

Figure 3.2 (a) Square lattice as a vertex-sharing lattice. The local tensor is defined on each shaded plaquette. (b) tensor network model defined on the 45-degree rotated square lattice formed by connecting the centers of all shaded squares.

where $\prod_{\triangle_{ijk}}$ is a product of all building triangles in the Kagome lattice. If we take the physical variables S_i defined at each vertex of the triangle as the bond variables (Fig. 3.1(b)), Eq. (3.31) defines a tensor network model on the honeycomb lattice (Fig. 3.1(c)):

$$Z = \mathrm{Tr}_S \prod_{\triangle_{ijk}} T_{S_i, S_j, S_k}, \qquad (3.32)$$

where

$$T_{S_i, S_j, S_k} = \quad \underset{S_j}{\overset{S_i \diagdown \underset{T}{\bigcirc} \diagup S_k}{}} \quad = e^{-\beta[K(S_i, S_j) + K(S_j, S_k) + K(S_k, S_i)]} \qquad (3.33)$$

is the local tensor. A three-spin interaction term among S_i, S_j, and S_k, H_{ijk} can also be added to the exponent.

The square lattice is edge-sharing. It can also be regarded as a vertex-sharing lattice if we consider only the shaded squares shown in Fig. 3.2(a). The local tensor is defined at the center of each shaded square. The physical variables defined at the vertices are the bond variables of local tensors. Similar to the Kagome lattice, the partition function can be expressed as a tensor network model:

$$Z = \mathrm{Tr}e^{-\beta H} = \mathrm{Tr}_S \prod_{\square_{ijkl}} T_{S_i,S_j,S_k,S_l}, \tag{3.34}$$

where

$$T_{S_i,S_j,S_k,S_l} = e^{-\beta[K(S_i,S_j)+K(S_j,S_k)+K(S_k,S_l)+K(S_l,S_i)]} \tag{3.35}$$

is the local tensor defined on a 45-degree rotated square lattice (Fig. 3.2(b)). One can also add any other interaction terms within each shaded square to the model. The number of bond variables equals the original lattice size.

3.6 Duality Properties of Tensor Network Models

A classical statistical model can be represented as a tensor network model either in the original lattice or in the dual lattice. If the lattice is self-dual, these two kinds of representations have the same lattice as well as tensor structures. Here we discuss the duality properties for the q-state clock model and the q-state Potts model based on their tensor network representations on the square lattice, which is self-dual.

3.6.1 *q*-State Clock Model

The ferromagnetic q-state clock model is defined by the Hamiltonian:

$$H = -\sum_{\langle ij \rangle} \cos\left(\theta_i - \theta_j\right), \qquad \theta_i = \frac{2\pi m}{q}, \ (m = 0,\ldots,q-1). \tag{3.36}$$

This model is also called the *vector Potts model*. It possesses a Z_q-symmetry because the Hamiltonian is invariant if all spin variables θ_i are changed by $2\pi/q$. If $q = 2$, it reduces to the Ising model. The $q = 3$ clock model is equivalent to the three-state Potts model. It becomes the classical XY model in the $q \to \infty$ limit.

Despite its simplicity, the clock model serves as a prototypical system in the study of phase transitions. For example, the clock model of $2 \leq q \leq 4$ exhibits a second-order phase transition at finite temperature on the square lattice. When $q \geq 5$, this model possesses three phases: a paramagnetic phase at high temperatures, a magnetically long-range-ordered phase at low temperatures, and a Kosterlitz–Thouless phase in the intermediate temperature regime. It undergoes two phase transitions, one from the paramagnetic phase to the KT phase and the other from the KT phase to the magnetically ordered phase at a lower temperature.

In the original lattice, the partition function can be expressed as a tensor network with each local tensor defined at each original lattice site:

$$Z = \text{Tr} e^{-\beta H} = \text{Tr} \prod_i T_{l_i, r_i, d_i, u_i}, \tag{3.37}$$

where (l_i, r_i, d_i, u_i) denote the bond indices linking site i from (left, right, down, up) directions, respectively. As discussed in §3.3, local tensor T is determined by factorizing the Boltzmann factor

$$e^{-\beta H_{ij}} = e^{\beta \cos(\theta_i - \theta_j)} \tag{3.38}$$

at each bond. In order to impose the Z_q symmetry, we do this factorization by inserting an identity

$$\sum_\theta \delta \left(\theta_i - \theta_j - \theta\right) = \frac{1}{q} \sum_\theta \sum_{m=1}^q e^{im(\theta_i - \theta_j - \theta)} = 1 \tag{3.39}$$

into the above expression. This allows the Boltzmann factor to be factorized as

$$e^{-\beta H_{ij}} = \sum_m V_{\theta_i, m} I_m V^*_{\theta_j, m}, \tag{3.40}$$

where V is a unitary matrix

$$V_{\theta, m} = \frac{1}{\sqrt{q}} e^{im\theta}, \tag{3.41}$$

and I_m is a bond vector

$$I_m = I_{-m} = \sum_\theta e^{-im\theta} e^{\beta \cos \theta}. \tag{3.42}$$

The local tensor is then found to be [345]:

$$T_{l_i, r_i, d_i, u_i} = \sum_{\theta_i} \sqrt{I_{l_i} I_{r_i} I_{d_i} I_{u_i}} V^*_{\theta_i, l_i} V_{\theta_i, r_i} V^*_{\theta_i, d_i} V_{\theta_i, u_i}$$

$$= \frac{\sqrt{I_{l_i} I_{r_i} I_{d_i} I_{u_i}}}{q} \delta \left(\text{mod}[l_i - r_i + d_i - u_i, q]\right). \tag{3.43}$$

It is zero if $\text{mod}[l_i - r_i + d_i - u_i, q] \neq 0$. This is a consequence of the Z_q-symmetry.

The local tensor can also be defined in the dual space to satisfy the Z_q-symmetry. A square lattice is an ensemble of small squares and the partition function can be written as a product of the Boltzmann factor on all these squares:

$$Z = \text{Tr} \prod_\square e^{-\beta H_\square}, \tag{3.44}$$

where $H_\square$ is the Hamiltonian defined on each small square

$$H_\square = \frac{1}{2} \sum_{\langle ij \rangle \in \square} H_{ij}. \tag{3.45}$$

For a small square, we label the values of spins on its four vertices by $\theta_1, \theta_2, \theta_3, \theta_4$ clockwise. On its neighboring squares, we label their vertex variables anticlockwise.

Now we introduce the dual variables:

$$\sigma_1 = \text{mod}[\theta_2 - \theta_1, 2\pi] \tag{3.46}$$
$$\sigma_2 = \text{mod}[\theta_3 - \theta_2, 2\pi] \tag{3.47}$$
$$\sigma_3 = \text{mod}[\theta_4 - \theta_3, 2\pi] \tag{3.48}$$
$$\sigma_4 = \text{mod}[\theta_1 - \theta_4, 2\pi]. \tag{3.49}$$

Clearly these dual variables satisfy the constraint

$$\text{mod}[\sigma_1 + \sigma_2 + \sigma_3 + \sigma_4, 2\pi] = 0. \tag{3.50}$$

It can also be expressed as

$$\frac{1}{q} \sum_{m=1}^{q} e^{i(\sigma_1 + \sigma_2 + \sigma_3 + \sigma_4)m} = 1. \tag{3.51}$$

The local tensor is then defined by

$$\begin{aligned} T'_{\sigma_1,\sigma_2,\sigma_3,\sigma_4} &= \frac{1}{q} \sum_m e^{i(\sigma_1 + \sigma_2 + \sigma_3 + \sigma_4)m} e^{\frac{1}{2}\beta(\cos\sigma_1 + \cos\sigma_2 + \cos\sigma_3 + \cos\sigma_4)} \\ &= \frac{1}{q} \sum_m W_{m,\sigma_1} W_{m,\sigma_2} W_{m,\sigma_3} W_{m,\sigma_4}, \end{aligned} \tag{3.52}$$

where

$$W_{m,\sigma} = e^{i\sigma m} e^{\frac{1}{2}\beta \cos\sigma}. \tag{3.53}$$

The local tensor is a product of four W's. Each bond is described by $WW^\dagger$. Taking SVD for this matrix, we find that

$$WW^\dagger = qUJU^\dagger, \tag{3.54}$$

where U is a unitary matrix

$$U_{m,\sigma} = \frac{1}{\sqrt{q}} e^{im\sigma}, \tag{3.55}$$

and

$$J_\sigma = e^{\beta \cos\sigma} \tag{3.56}$$

is the bond spectrum in the dual space. The local tensor in the dual lattice is

$$T^{\text{dual}}_{\sigma_1,\sigma_2,\sigma_3,\sigma_4} = q^2 \sum_m \sqrt{J_{\sigma_1} J_{\sigma_2} J_{\sigma_3} J_{\sigma_4}} U_{m,\sigma_1} U_{m,\sigma_2} U_{m,\sigma_3} U_{m,\sigma_4}$$

$$= q\sqrt{J_{\sigma_1} J_{\sigma_2} J_{\sigma_3} J_{\sigma_4}} \delta\left(\text{mod}[\sigma_1 + \sigma_2 + \sigma_3 + \sigma_4, 2\pi]\right). \quad (3.57)$$

It is also Z_q-symmetric.

This derivation indicates that the partition function of the clock model is determined purely by the bond vector in both the original and dual lattices. Thus, the models are dual to each other if the normalized bond vector at an inverse temperature β_1 in the original lattice is equal to that at an inverse temperature β_2 in the dual lattice [345]:

$$\frac{I_m(\beta_1)}{I_0(\beta_1)} = \frac{J'_m(\beta_2)}{J'_0(\beta_2)}, \quad (m = 1, \ldots, q - 1). \quad (3.58)$$

Here J'_m is relabeled J_σ by an integer m, which is the argument of σ in unit $2\pi/q$:

$$J'_m = J_\sigma, \quad m = \frac{q\sigma}{2\pi}. \quad (3.59)$$

If $\beta_1 = \beta_2 = \beta_c$, the model is self-dual:

$$\frac{I_m(\beta_c)}{I_0(\beta_c)} = \frac{J'_m(\beta_c)}{J'_0(\beta_c)}, \quad (m = 1, \ldots, q - 1). \quad (3.60)$$

I_m is the Fourier transformation of J'_m. For a system with only one critical point, the self-dual solution β_c is just equal to the inverse of the transition temperature. On the other hand, if the system undergoes more than one transition, β_c does not correspond to any of the inverse critical temperatures.

For the q-state clock model, it can be shown that Eq. (3.60) does have a solution at an inverse temperature β_c when $2 \leq q \leq 5$. For $2 \leq q \leq 4$, the solution is given by

$$\beta_c = \begin{cases} \dfrac{q-1}{q} \ln(\sqrt{q} + 1), & q = 2, 3 \\[3mm] \ln\left(\sqrt{2} + 1\right), & q = 4 \end{cases}. \quad (3.61)$$

As the clock model with $2 \leq q \leq 4$ has only one phase transition, these solutions of β_c are just the inverse critical transition temperatures.

For $q = 5$, Eq. (3.60) has a self-dual solution. The value of the self-dual β_c is determined by the equation [345]:

$$\frac{e^{5\beta_c/4}}{\cosh\left(\frac{\sqrt{5}}{4}\beta_c\right)} = \sqrt{5} + 1. \quad (3.62)$$

There is no analytic solution for this equation. By solving this equation numerically, we find that $\beta_c \approx 1.07631807160464779$. However, the five-state clock model has two transition points. This self-dual temperature $1/\beta_c$ does not correspond to any of these two transition temperatures. If T_{c_1} and T_{c_2} are the two critical transition temperatures for the five-state clock model, we conjecture that the equation

$$T_{c_1} T_{c_2} \beta_c^2 = 1 \tag{3.63}$$

holds.

For the clock model with $q \geq 6$, there is no solution for Eq. (3.60), which implies no exact self-dual point.

3.6.2 q-State Potts Model

The derivation for the q-state clock model can be readily extended to the q-state Potts model (3.24) on the square lattice. In two dimensions, this model exhibits a first-order phase transition if $q > 4$ and a second-order phase transition if $q \leq 4$.

In the original lattice, it can be shown that the local tensor still has the form defined by Eq. (3.43), but the bond vector changes to

$$I_m = \sum_\theta e^{-im\theta} e^{\beta \delta_{\theta,0}} = \begin{cases} e^\beta + q - 1 & m = 0 \\ e^\beta - 1 & m \neq 0 \end{cases}. \tag{3.64}$$

In the dual lattice, the local tensor is defined by Eq. (3.57), and the bond vector is replaced by

$$J'_m = e^{\beta \delta_{m,0}} = \begin{cases} e^\beta & m = 0 \\ 1 & m \neq 0 \end{cases}. \tag{3.65}$$

The q-state Potts model is self-dual for all q. There is a phase transition from the high-temperature paramagnetic phase to the low-temperature ferromagnetic phase. The inverse transition temperature is determined by the self-dual equation (3.60). Substituting these expressions of the bond spectra in the original and dual lattices into Eq. (3.60) yields

$$\frac{e^{\beta_c} - 1}{e^{\beta_c} + q - 1} = e^{-\beta_c}. \tag{3.66}$$

The solution is [346]:

$$\beta_c = \ln\left(1 + \sqrt{q}\right). \tag{3.67}$$

4

Tensor Network Representation of Operators

4.1 Matrix Product Operators (MPO)

In one dimension, a quantum Hamiltonian, generally a sum of many local terms, can be represented as a matrix product operator [78, 347]. Below we take the one-dimensional Heisenberg model as an example to show how to construct its MPO representation.

The Heisenberg model reads

$$H = J \sum_i \mathbf{S}_i \cdot \mathbf{S}_{i+1} - B \sum_i S_{iz}, \tag{4.1}$$

where $\mathbf{S}_i = (S_{ix}, S_{iy}, S_{iz})$ is the spin operators at site i, J is the exchange constant, and B is a magnetic field applied along the z-axis. If we count sites from left to right and use H_i to represent all the terms in the Hamiltonian from the left end to site i, we can rewrite (4.1) into a recursion formula

$$H_i = H_{i-1} + J\mathbf{S}_{i-1} \cdot \mathbf{S}_i - BS_{iz}, \tag{4.2}$$

which can also be written as

$$\begin{pmatrix} H_i & \mathbf{S}_i & 1 \end{pmatrix} = \begin{pmatrix} H_{i-1} & \mathbf{S}_{i-1} & 1 \end{pmatrix} \begin{pmatrix} 1 & 0 & 0 \\ J\mathbf{S}_i^T & 0 & 0 \\ -BS_{iz} & \mathbf{S}_i & 1 \end{pmatrix}. \tag{4.3}$$

Here the three components of $\mathbf{S}_i$ are ordered as a row vector. This equation defines an MPO whose local tensor is simply defined by the above recursion matrix:

$$h_i = \begin{pmatrix} 1 & 0 & 0 \\ J\mathbf{S}_i^T & 0 & 0 \\ -BS_{iz} & \mathbf{S}_i & 1 \end{pmatrix}. \tag{4.4}$$

The full Hamiltonian can then be represented as

$$H = \left(\prod_{i=1}^{N} h_i \right) h_{N+1}^T, \tag{4.5}$$

$$= \boxed{h_1} - \boxed{h_2} - \cdots - \boxed{h_N} - \boxed{h_{N+1}^T}$$

where h_1 is a vector

$$h_1 = \left(\begin{array}{ccc} -BS_{1z} & \mathbf{S}_1 & 1 \end{array} \right). \tag{4.6}$$

Here h_{N+1} is also a vector, which depends on the boundary conditions:

$$h_{N+1} = \left(\begin{array}{ccc} 1 & \mathbf{0} & 0 \end{array} \right) \tag{4.7}$$

if an open boundary condition is used or

$$h_{N+1} = \left(\begin{array}{ccc} 1 & J\mathbf{S}_1 & 0 \end{array} \right) \tag{4.8}$$

if the periodic boundary condition, $\mathbf{S}_{N+1} = \mathbf{S}_1$, is used. $\mathbf{0} = (0,0,0)$ is the zero vector.

This scheme can be generalized to a Hamiltonian with long-range interactions [348]. As an example, let us consider the following extended Heisenberg model with a special kind of long-range interaction:

$$H = J \sum_{i,j} a^{|i-j|} \mathbf{S}_i \cdot \mathbf{S}_j - B \sum_i S_{iz}. \tag{4.9}$$

This Hamiltonian has the following recursion form:

$$\begin{aligned} H_i &= H_{i-1} + J\mathbf{S}_i \cdot \sum_{m=1} a^m \mathbf{S}_{i-m} - BS_{iz} \\ &= H_{i-1} + Ja\mathbf{S}_i \cdot \mathcal{S}_{i-1} - BS_{iz}, \end{aligned} \tag{4.10}$$

where

$$\mathcal{S}_i = \sum_{m=0} a^m \mathbf{S}_{i-m} = \mathbf{S}_i + a\mathcal{S}_{i-1}. \tag{4.11}$$

It can also be written as

$$\left(\begin{array}{ccc} H_i & \mathcal{S}_i & 1 \end{array} \right) = \left(\begin{array}{ccc} H_{i-1} & \mathcal{S}_{i-1} & 1 \end{array} \right) \left(\begin{array}{ccc} 1 & 0 & 0 \\ Ja\mathbf{S}_i^T & aI_3 & 0 \\ -BS_{iz} & \mathbf{S}_i & 1 \end{array} \right), \tag{4.12}$$

where I_3 is a 3×3 identity matrix. Now the local tensor of the Hamiltonian is

$$h_i = \begin{pmatrix} 1 & 0 & 0 \\ Ja\mathbf{S}_i^T & aI_3 & 0 \\ -h\mathbf{S}_{iz} & \mathbf{S}_i & 1 \end{pmatrix}. \tag{4.13}$$

For these two examples, h_i ($i = 2, \dots, N$) are 5×5 matrices with operator entries (referred to as "operator matrices" hereafter). More generally, the dimensions of these operator matrices are not equal to 5 for other interacting spin models. For example, for the so-called $J_1 - J_2$ model, whose Hamiltonian is defined by

$$H = \sum_i (J_1\mathbf{S}_{i-1} \cdot \mathbf{S}_i + J_2\mathbf{S}_{i-2} \cdot \mathbf{S}_i), \tag{4.14}$$

the recursive relation between H_i and H_{i-1} can be represented as

$$\begin{aligned} & \begin{pmatrix} H_i & \mathbf{S}_i & \mathbf{S}_{i-1} & 1 \end{pmatrix} \\ & = \begin{pmatrix} H_{i-1} & \mathbf{S}_{i-1} & \mathbf{S}_{i-2} & 1 \end{pmatrix} \begin{pmatrix} 1 & 0 & 0 & 0 \\ J_1\mathbf{S}_i^T & 0 & I_3 & 0 \\ J_2\mathbf{S}_i^T & 0 & 0 & 0 \\ 0 & \mathbf{S}_i & 0 & 1 \end{pmatrix} \end{aligned} \tag{4.15}$$

if $i \geq 2$, and

$$\begin{pmatrix} H_2 & \mathbf{S}_2 & \mathbf{S}_1 & 1 \end{pmatrix} = \begin{pmatrix} 0 & \mathbf{S}_1 & 1 \end{pmatrix} \begin{pmatrix} 1 & 0 & 0 & 0 \\ J_1\mathbf{S}_2^T & 0 & I_3 & 0 \\ 0 & \mathbf{S}_2 & 0 & 1 \end{pmatrix} \tag{4.16}$$

if $i = 2$. Here open boundary condition is assumed. The local operator matrices are then defined by

$$h_i = \begin{pmatrix} 1 & 0 & 0 & 0 \\ J_1\mathbf{S}_i^T & 0 & I_3 & 0 \\ J_2\mathbf{S}_i^T & 0 & 0 & 0 \\ 0 & \mathbf{S}_i & 0 & 1 \end{pmatrix} \quad (i > 2), \tag{4.17}$$

and

$$h_2 = \begin{pmatrix} 1 & 0 & 0 & 0 \\ J_1\mathbf{S}_2^T & 0 & I_3 & 0 \\ 0 & \mathbf{S}_2 & 0 & 1 \end{pmatrix}, \tag{4.18}$$

$$h_1 = \begin{pmatrix} 0 & \mathbf{S}_1 & 1 \end{pmatrix}. \tag{4.19}$$

Now h_i ($i > 2$) is an 8×8 matrix, h_2 is a 5×8 matrix, and h_1 is a five-dimensional vector.

These examples show that a Hamiltonian can be represented as a tensor network operator. We can always construct an MPO representation for a given Hamiltonian [349], whether it contains long-range interactions or not, although a simple analytic expression of MPO may not be available. Similarly, we can construct a higher-dimensional representation of tensor operators, no matter if the Hamiltonian is defined in one or higher dimensions.

4.2 Imaginary Time Evolution Operator

To find the ground state, one can apply the thermal density matrix $\exp(-\beta H)$, to an arbitrary initial state $|\Psi_0\rangle$ that is not orthogonal to the ground state

$$|\Psi\rangle = \lim_{\beta \to \infty} e^{-\beta H}|\Psi_0\rangle. \tag{4.20}$$

This is to perform an evolution in the imaginary time direction, which cools down the temperature. In a quantum system, as the eigenstates of H are unknown, it is impossible to do this imaginary time evolution in one step. In the framework of path integral, however, one can carry out the evolution by dividing β into many small steps and express the thermal density matrix ρ as

$$\rho = e^{-\beta H} = \left(e^{-\varepsilon H}\right)^M, \tag{4.21}$$

where $\varepsilon = \beta/M$ and M is a large integer number, also called the Trotter number. For small ε, we can evaluate $\exp(-\varepsilon H)$ by utilizing the Trotter–Suzuki decomposition formula introduced in §2.9.

Let us consider a one-dimensional Hamiltonian with only nearest-neighboring interactions:

$$H = \sum_i H_{i,i+1}. \tag{4.22}$$

This Hamiltonian can be decomposed into two parts, $H = H_1 + H_2$, with each a sum of commuting terms:

$$H_1 = \sum_{i=\text{odd}} H_{i,i+1}, \tag{4.23}$$

$$H_2 = \sum_{i=\text{even}} H_{i,i+1}. \tag{4.24}$$

Using the Trotter–Suzuki formula, we can express the small-step evolution operator approximately as

$$e^{-\varepsilon H} = e^{-\varepsilon H_2}e^{-\varepsilon H_1} + o(\varepsilon^2) = \prod_{i=\text{even}} v_{i,i+1} \prod_{i=\text{odd}} v_{i,i+1} + o(\varepsilon^2), \tag{4.25}$$

where $v_{i,i+1}$ is the local evolution operator

$$v_{i,i+1} = e^{-\varepsilon H_{i,i+1}}. \tag{4.26}$$

Its matrix elements can be graphically represented as

$$\langle s'_i s'_{i+1} | v_{i,i+1} | s_i s_{i+1} \rangle = \quad \text{(4.27)}$$

The evolution operator $\exp(-\varepsilon H_1)$ is a direct product of these local evolution operators. For example, for a system of $N = 8$,

$$e^{-\varepsilon H_1} = \quad \text{(4.28)}$$

If periodic boundary condition is assumed, the one-step evolution operator for this $N = 8$ system is

$$e^{-\varepsilon H} = \quad \begin{matrix} e^{-\varepsilon H_2} \\ e^{-\varepsilon H_1} \end{matrix} \quad . \quad \text{(4.29)}$$

The density matrix is an evolution of the product of $\exp(-\varepsilon H)$ along the imaginary time or Trotter direction. The partition function is the trace of the density matrix:

$$Z = \text{Tr} e^{-\beta H} = \lim_{\varepsilon \to 0} \text{Tr} \left(e^{-\varepsilon H_2} e^{-\varepsilon H_1} \right)^M . \quad \text{(4.30)}$$

The trace implies that the periodic boundary condition must be taken along the Trotter direction. If $M = 4$ and $N = 8$, the partition function can be pictorially represented as

$$Z = \quad \text{(4.31)}$$

4.3 Quantum Transfer Matrix

In Eq. (4.31), the partition function is formulated as a flow of the evolution operator along the Trotter direction. It can also be formulated as a flow of a quantum transfer matrix along the spatial direction. In this case, we group all the v-operators in each column as one operator and define it as a transfer matrix. However, v is a local evolution operator along the imaginary time. It does not directly represent the evolution of basis states along the spatial direction. In order to reflect more precisely this evolution procedure along the spatial direction, we need to redefine the in-coming and out-going states, and set

$$\tau(\sigma_i^m \sigma_i^{m+1} | \sigma_{i+1}^m \sigma_{i+1}^{m+1}) \equiv \langle s_i^m, s_{i+1}^m | v_{i,i+1} | s_i^{m+1}, s_{i+1}^{m+1} \rangle, \qquad (4.32)$$

where the superscripts represent the coordinates in the Trotter space and

$$\sigma_i^m \equiv (-1)^{i+m} s_i^m \qquad (4.33)$$

is the quantum number of spin operators in the Trotter representation. Graphically, Eq. (4.32) can be represented as

$$(4.34)$$

The arrows indicate the flow directions of basis states.

There is one-to-one correspondence between matrix elements v and τ. σ_i^m differs from s_i^m by a sign factor, which is introduced to more conveniently describe the spin conservation in the Trotter space. τ is referred to as the local transfer matrix, defined from the matrix elements of v. However, τ and v have different physical meanings.

The transfer matrix is defined by grouping all the local transfer matrices in one column between two neighboring lattice sites. For example, the matrix elements of the transfer matrix from site $2i - 1$ to site $2i$, T_{2i-1}, are defined by

$$\langle \sigma_{2i-1}^1 \cdots \sigma_{2i-1}^{2M} | T_{2i-1} | \sigma_{2i}^1 \cdots \sigma_{2i}^{2M} \rangle \equiv \prod_{m=1}^{M} \tau_{2i-1,2i}^{2m-1,2m}. \qquad (4.35)$$

Similarly, the matrix elements of the transfer matrix between sites $2i$ and $2i + 1$ are

$$\langle \sigma_{2i}^1 \cdots \sigma_{2i}^{2M} | T_{2i} | \sigma_{2i+1}^1 \cdots \sigma_{2i+1}^{2M} \rangle \equiv \prod_{m=1}^{M} \tau_{2i,2i+1}^{2m,2m+1}. \qquad (4.36)$$

The partition function is simply a product of all the transfer matrices along the spatial direction

$$Z = \lim_{\varepsilon \to 0} \text{Tr}(T_1 T_2 \cdots T_{N-1} T_N). \tag{4.37}$$

It can be intuitively understood as an evolution of a quantum state in the Trotter space at the first lattice site by the subsequent action of the transfer matrix T_i to the final states at the last lattice site. The trace is a result of periodic boundary conditions in the spatial direction.

4.3.1 Translation-Invariant System

It isn't easy to evaluate the partition function directly from the chain product of the transfer matrices one after another. However, if the Hamiltonian is invariant under the translation of two lattice sites; that is,

$$H_{i+2,i+3} = H_{i,i+1}, \tag{4.38}$$

then the transfer matrix T_i is also invariant under the translation of two lattice sites, satisfying the equation

$$T_{i+2} = T_i. \tag{4.39}$$

In this case, the partition function can be expressed as

$$Z = \lim_{\varepsilon \to 0} \text{Tr}(T_1 T_2)^{N/2} = \lim_{\varepsilon \to 0} \text{Tr}\,(\mathcal{T}_M)^{N/2}, \tag{4.40}$$

with $\mathcal{T}_M$ the translation-invariant transfer matrix

$$\mathcal{T}_M \equiv T_1 T_2. \tag{4.41}$$

A graphical representation of the partition function is

$$(4.42)$$

In Eq. (4.40), the lattice size, N, is assumed to be even. If N is odd, the partition function reads

$$Z = \lim_{\varepsilon \to 0} \text{Tr}\,(\mathcal{T}_M)^{(N-1)/2}\,T_1, \tag{4.43}$$

The transfer matrix $\mathcal{T}_M$ above is not symmetric. By diagonalizing this matrix, we can find its left and right eigenvectors:

$$\mathcal{T}_M|\psi_l^R\rangle = \lambda_l|\psi_l^R\rangle, \tag{4.44}$$

$$\langle\psi_l^L|\mathcal{T}_M = \lambda_l\langle\psi_l^L|, \tag{4.45}$$

where λ_l is the eigenvalue, and $|\psi_l^R\rangle$ and $\langle\psi_l^L|$ are the corresponding right and left eivenvectors, respectively. We assume λ_l to be in descending order such that $|\lambda_1| \geq |\lambda_2| \geq |\lambda_3| \geq \ldots$. The left and right eigenvectors satisfy the biorthonormal condition:

$$\langle\psi_l^L|\psi_{l'}^R\rangle = \delta_{l,l'}. \tag{4.46}$$

$\mathcal{T}_M$ can be represented using these eigenvalues and eigenvectors as

$$\mathcal{T}_M = \sum_l \lambda_l|\psi_l^R\rangle\langle\psi_l^L|. \tag{4.47}$$

In the thermodynamic limit $N \to \infty$, the partition function is determined purely by the largest eigenvalue of $\mathcal{T}_M$, λ_1, and the corresponding eigenvectors. If λ_1 is nondegenerate, then

$$Z_{N\to\infty} = \lambda_1^{N/2} \tag{4.48}$$

if N is even, and

$$Z_{N\to\infty} = \lambda_1^{(N-1)/2}\langle\psi_1^L|T_1|\psi_1^R\rangle \tag{4.49}$$

if N is odd. If the largest eigenvalue is degenerate, then the partition function should be a sum of all these degenerate terms.

This discussion indicates that to calculate the partition function, only the largest eigenvalue and the corresponding eigenvectors of the transfer matrix $\mathcal{T}_M$ need to be determined. This is an important property of the quantum transfer matrix. It allows us to adopt the idea of DMRG to find the leading eigenstates of the quantum transfer matrix approximately, in analogy to the diagonalization of the ground state from the Hamiltonian [72, 75]. The application of the quantum transfer matrix allows us to evaluate thermodynamic quantities directly in the thermodynamic limit, which eliminates the finite-size effect generally encountered in the exact diagonalization as well as quantum Monte Carlo simulations.

Under the periodic boundary condition, if the Hamiltonian is invariant under the translation of a single lattice site, we have

$$P_T^{-1}T_1P_T = T_2, \tag{4.50}$$

$$P_T^{-1} T_2 P_T = T_1, \tag{4.51}$$

where P_T is the translation operator in the Trotter space. Using these equations, we can show that

$$\left[T_1, P_T^2\right] = \left[T_2, P_T^2\right] = 0. \tag{4.52}$$

Moreover, according to the definition, $P_T^{2M} = I$. If the lattice size $N = 2ML$, with L an integer number, we can also write the partition function as

$$Z = \lim_{\varepsilon \to 0} \text{Tr} \, (T_1 T_2)^{ML} = \lim_{\varepsilon \to 0} \text{Tr} \, (P_T T_1)^N. \tag{4.53}$$

This expression may not be of much practical use since it is numerically difficult to implement the action of the translation operator P_T.

In case both T_1 and T_2 are symmetric and semi-positive definite – that is, all eigenvalues of T_1 and T_2 are larger than or equal to zero – the transfer matrix can be symmetrized. To see this, let us rewrite Eq. (4.40) as

$$Z = \lim_{\varepsilon \to 0} \text{Tr} \left(T_1^{1/2} T_2 T_1^{1/2} \right)^{N/2} = \lim_{\varepsilon \to 0} \text{Tr} \left(\mathcal{T}_M^S \right)^{N/2}. \tag{4.54}$$

This defines a new transfer matrix:

$$\mathcal{T}_M^S = T_1^{1/2} T_2 T_1^{1/2}. \tag{4.55}$$

As $T_1^{1/2}$ is a symmetric matrix when T_1 is symmetric and semi-positive definite, $\mathcal{T}_M^S$ is a real symmetric transfer matrix. This symmetrized transfer matrix is a product of three matrices. For a spin-S Heisenberg model with either ferromagnetic or anti-ferromagnetic coupling, the transfer matrix can be symmetrized in the absence of an external field. But, in general, T_1 and T_2 are neither symmetric nor semi-positive definite. In that case, the transfer matrix cannot be symmetrized.

4.3.2 Local Transfer Matrix

We take the spin-1/2 XY model in a transverse field in one dimension as an example to show how to construct the local transfer matrix. The local Hamiltonian reads:

$$H_{i,i+1} = -J \left(S_i^x S_{i+1}^x + S_i^y S_{i+1}^y \right) - h(S_i^z + S_{i+1}^z). \tag{4.56}$$

This model has an exact solution. A simple way to solve this model is to take the Jordan–Wigner transformation to represent spin operators by fermion operators. For the spin-1/2 operators, the Jordan–Wigner transformation reads:

$$S_i^+ = c_i^\dagger \exp \left(i\pi \sum_{k<i} c_k^\dagger c_k \right), \tag{4.57}$$

$$S_i^- = \exp\left(-i\pi \sum_{k<i} c_k^\dagger c_k\right) c_i, \tag{4.58}$$

$$S_i^z = c_i^\dagger c_i - \frac{1}{2}, \tag{4.59}$$

where c_i is the fermion annihilation operator at site i. The XY-model can be expressed as a noninteracting fermion model using these fermion operators. This fermion model is diagonal in momentum space:

$$H = -\frac{J}{2} \sum_i \left(c_i^\dagger c_{i+1} + h.c.\right) - 2h \sum_i \left(c_i^\dagger c_i - \frac{1}{2}\right)$$

$$= -J \sum_k \cos k c_k^\dagger c_k - 2h \sum_k \left(c_k^\dagger c_k - \frac{1}{2}\right), \tag{4.60}$$

from which the partition function is found to be

$$Z = \prod_k e^{-\beta h} \left[1 + e^{\beta(J\cos k + 2h)}\right]. \tag{4.61}$$

The corresponding free energy density is

$$f = -\frac{1}{2\pi\beta} \int_0^{2\pi} dk \ln\left[2\cosh\frac{\beta(J\cos k + 2h)}{2}\right]. \tag{4.62}$$

Next, we derive these formulas using the method of quantum transfer matrices. This derivation is more complicated but allows us to calculate the correlation length readily.

To find the local transfer matrix, we first diagonalize the local Hamiltonian $H_{i,i+1}$:

$$H_{i,i+1} = U^\dagger E U, \tag{4.63}$$

where U is a unitary matrix and E is the diagonal eigenvalue matrix. In the representation of $|s_i, s_{i+1}\rangle = (|\uparrow\uparrow\rangle, |\uparrow\downarrow\rangle, |\downarrow\uparrow\rangle, |\downarrow\downarrow\rangle)$, the local evolution matrix is then found to be

$$v_{i,i+1} = U^\dagger e^{-\varepsilon E} U = \begin{pmatrix} e^{\varepsilon h} & 0 & 0 & 0 \\ 0 & \cosh\frac{\varepsilon J}{2} & \sinh\frac{\varepsilon J}{2} & 0 \\ 0 & \sinh\frac{\varepsilon J}{2} & \cosh\frac{\varepsilon J}{2} & 0 \\ 0 & 0 & 0 & e^{-\varepsilon h} \end{pmatrix}. \tag{4.64}$$

Here $v_{i,i+1}$ is block-diagonal and symmetric because $H_{i,i+1}$ is Hermitian and the z-component of the total spin at sites i and $i+1$ is conserved:

$$s_i^m + s_{i+1}^m = s_i^{m+1} + s_{i+1}^{m+1}. \tag{4.65}$$

Inserting (4.64) into (4.32) and using the definition (4.33), we find the expression of the local transfer matrix in the basis space $|\sigma^m, \sigma^{m+1}\rangle = (|\uparrow\uparrow\rangle, |\uparrow\downarrow\rangle, |\downarrow\uparrow\rangle, |\downarrow\downarrow\rangle)$ to be

$$\tau^{m,m+1} = \begin{pmatrix} \sinh\frac{\varepsilon J}{2} & 0 & 0 & 0 \\ 0 & \cosh\frac{\varepsilon J}{2} & e^{-\varepsilon h} & 0 \\ 0 & e^{\varepsilon h} & \cosh\frac{\varepsilon J}{2} & 0 \\ 0 & 0 & 0 & \sinh\frac{\varepsilon J}{2} \end{pmatrix}. \tag{4.66}$$

Here τ is block-diagonal but not symmetric at a finite magnetic field. From Eq. (4.65), it is easy to show that

$$\sigma_i^m + \sigma_i^{m+1} = \sigma_{i+1}^m + \sigma_{i+1}^{m+1}. \tag{4.67}$$

Thus the σ-spin in the Trotter space is conserved. In general, there is a one-to-one correspondence between the quantum transfer matrix and the Hamiltonian, including their symmetries.

From this local conservation law, it is simple to show that the total σ-spin,

$$Q = \sum_m \sigma_i^m, \tag{4.68}$$

is a good quantum number in the Trotter space, independent of i. Thus, $\mathcal{T}_M$ is block-diagonal according to the value of Q. For the spin-1/2 XY model, it can be shown that the leading eigenvector of $\mathcal{T}_M$ is nondegenerate and lies in the $Q = 0$ subspace, irrespective of coupling constants J and h [157].

The quantum transfer matrix from site 1 to site 2 is defined by

$$T_1 = \tau^{12}\tau^{34}\cdots\tau^{2M-1,2M}. \tag{4.69}$$

The corresponding partition function is

$$Z = \mathrm{Tr}\,(T_1 T_2)^{N/2} = \mathrm{Tr}\,(P_T T_1)^N,$$

where P_T is the translation operator along the Trotter direction. Setting $N = 2Mn$ with n an integer, $P_T T_1$ corresponds to the vertex–vertex transfer matrix of the six-vertex model whose matrix elements are given by

$$\langle \underline{\sigma}^1 \cdots \underline{\sigma}^{2M} | P_T T_1 | \sigma^1 \cdots \sigma^{2M} \rangle = \prod_{m=1}^M \langle \underline{\sigma}^{2m-2}\underline{\sigma}^{2m-1} | \tau | \sigma^{2m-1}\sigma^{2m} \rangle, \tag{4.70}$$

where $\underline{\sigma}^0 = \underline{\sigma}^{2M}$.

From the analytic solution of the six-vertex model, we find that the maximal eigenvalue of $P_T T_1$ in the limit $M \to \infty$ is

$$\lambda_1 = 2\exp\left(\sum_{l=1}^\infty \frac{1}{2\pi}\int_0^{2\pi} dk\,\ln\left[1 + \left(\frac{\beta\,(J\cos k + 2h)}{2\pi\,(l - 1/2)}\right)^2\right]\right)$$

$$= \exp\left[\frac{1}{2\pi}\int_0^{2\pi} dk\,\ln\left(2\cosh\frac{\beta\,(J\cos k + 2h)}{2}\right)\right]. \tag{4.71}$$

The free energy obtained with this largest eigenvalue is the result shown in Eq. (4.62).

The largest eigenvalue in the subspace of $Q = 1$ can also be found from the exact solution. It is given by

$$\lambda_2 = \frac{\beta J}{2} \exp\left(\frac{1}{2\pi} \int_0^{2\pi} dx \ln\left(\frac{\sinh\left[\beta\left(J\cos x + 2h\right)/2\right]}{\beta\left(J\cos x + 2h\right)/2}\right)\right). \tag{4.72}$$

The transverse correlation length ξ_{xx} is then found to be

$$\xi_{xx}^{-1} = \ln\frac{\lambda_1}{\lambda_2} = \frac{1}{2\pi} \int_0^{2\pi} dx \ln\left(\coth|\beta(J\cos x + 2h)/2|\right), \tag{4.73}$$

if $2|h| \leq J$, and

$$\xi_{xx}^{-1} = \ln\left(\left|\frac{2h}{J}\right| + \sqrt{\left(\frac{2h}{J}\right)^2 - 1}\right)$$
$$+ \frac{1}{2\pi} \int_0^{2\pi} dx \ln\left(\coth\left|\frac{\beta\left(J\cos x + 2h\right)}{2}\right|\right), \tag{4.74}$$

if $2|h| > J$.

The discussion on the conservation law can be readily extended to an interacting electron model, for example, the one-dimensional Hubbard model:

$$H = -t\sum_{i\sigma}\left(c_{i+1,\sigma}^\dagger c_{i,\sigma} + h.c.\right) + U\sum_i n_{i\uparrow}n_{i\downarrow}, \tag{4.75}$$

where $c_{i\sigma}$ is the annihilation operator of electrons at site i with spin σ, and $n_{i\sigma} = c_{i,\sigma}^\dagger c_{i,\sigma}$. In this case, both charge and spin are conserved. If we use $n_{i\sigma}^m$ to denote the number of electrons with spin σ at site i and time slice m, then the spin and charge conservation laws for this system are described by the equation

$$n_{i,\sigma}^m + n_{i+1,\sigma}^m = n_{i,\sigma}^{m+1} + n_{i+1,\sigma}^{m+1} \tag{4.76}$$

in real space. In the Trotter space, the corresponding conservation law becomes

$$\tilde{n}_{i,\sigma}^m + \tilde{n}_{i,\sigma}^{m+1} = \tilde{n}_{i+1,\sigma}^m + \tilde{n}_{i+1,\sigma}^{m+1}, \tag{4.77}$$

where

$$\tilde{n}_{i,\sigma}^m = (-1)^{i+k}(n_{i,\sigma}^m - 1/2). \tag{4.78}$$

When $i+k$ is odd, this transformation between n_σ^i and $\tilde{n}_\sigma^i$ is equivalent to a particle-hole transformation: $c \leftrightarrow c^\dagger$.

4.3.3 Open Boundary Systems

It is straightforward to extend the derivation for the quantum transfer matrix representation of the partition function to an open boundary system in which there is no interaction between the two ends. Along the Trotter direction, the transfer matrix still satisfies the periodic boundary condition. Following the steps leading to Eq. (4.37), we find the partition function in an open boundary system to be

$$Z = \lim_{\varepsilon \to 0} \langle \Psi_1 | T_1 T_2 \cdots T_{N-1} | \Psi_N \rangle, \tag{4.79}$$

where $\langle \Psi_1 |$ is a bra defined at the left end of the chain

$$\langle \Psi_1 | = \sum_{\sigma_1^m} \langle \sigma_1^1, \sigma_1^2, \cdots \sigma_1^{2M} | \phi(\sigma_1^1, \sigma_1^2, \cdots \sigma_1^{2M}), \tag{4.80}$$

and

$$\phi(\sigma_1^1, \sigma_1^2, \cdots \sigma_1^{2M}) = \prod_{m=1}^{M} \delta_{\sigma_1^{2m}, \sigma_1^{2m+1}}. \tag{4.81}$$

Similarly, $|\Psi_N\rangle$ is a ket defined at the right end of the chain:

$$|\Psi_N\rangle = \sum_{\sigma_N^m} \phi(\sigma_N^1, \sigma_N^2, \cdots \sigma_N^{2M}) | \sigma_N^1, \sigma_N^2, \cdots \sigma_N^{2M} \rangle. \tag{4.82}$$

If the local Hamiltonian is invariant under the translation of two lattice sites $H_{i+2,i+3} = H_{i,i+1}$, the transfer matrix T_i should also be two-site translation invariant, $T_{i+2} = T_i$. In this case, the partition function can be represented simply using T_1 and T_2, as in a periodic system. If the number of sites N is even,

$$Z = \lim_{\varepsilon \to 0} \langle \Psi_1 | \mathcal{T}_M^{N/2-1} T_1 | \Psi_N \rangle. \tag{4.83}$$

In the thermodynamic limit, $N \to \infty$, it becomes

$$Z = \lambda_1^{N/2-1} \langle \Psi_1 | \psi_1^R \rangle \langle \psi_1^L | T_1 | \Psi_N \rangle. \tag{4.84}$$

Here $|\psi_1^R\rangle$ and $\langle \psi_1^L |$ are the right- and left leading eigenvectors of $\mathcal{T}_M$, respectively.

4.3.4 Single Impurity System

If the system contains an impurity located at site $i = 0$, and the local Hamiltonians on the sites not connected to the impurity are site independent, then the partition function for an even N system can be represented as

$$Z = \mathrm{Tr} \left(\mathcal{T}_M^{N/2-1} \mathcal{T}^{\mathrm{imp}} \right), \tag{4.85}$$

where $\mathcal{T}^{\text{imp}}$ is the impurity transfer matrix. The impurity couples with its left and right sites, so $\mathcal{T}^{\text{imp}}$ is a product of two impurity transfer matricies, one from site -1 to 0, that is, T_{-1}^{imp}, and another from site 0 to 1, that is, T_0^{imp}:

$$\mathcal{T}^{\text{imp}} = T_{-1}^{\text{imp}} T_0^{\text{imp}}. \tag{4.86}$$

T_{-1}^{imp} and T_0^{imp} can be similarly defined from T by replacing the local Hamiltonian with the impurity Hamiltonian.

In the thermodynamic limit, the above expression becomes

$$Z = \lambda_1^{N/2-1} \text{Tr}\langle \psi_1^L | \mathcal{T}^{\text{imp}} | \psi_1^R \rangle. \tag{4.87}$$

Thus, the thermodynamics of this single-impurity system can be found from the expectation value of the impurity transfer matrix on the leading eigenvectors of $\mathcal{T}_M$.

4.4 MPO Representation of Quantum Transfer Matrix

The imaginary-time evolution operator V introduced in §4.2 is defined using two-site local transfer matrix v. One can convert it into an MPO by decomposing each local tensor v using SVD:

$$\tag{4.88}$$

Note that the horizontal and vertical bonds have different dimensions. The horizontal bond dimension is d^2 if the vertical one is d.

By absorbing $\sqrt{\lambda}$ into u and w, v is reexpressed as

$$v = \bar{u}\,\bar{w}, \tag{4.89}$$

where

$$\tag{4.90}$$

In the imaginary time evolution operator $\exp(-\varepsilon H)$, there are two local operators v connecting a lattice site. Using the tensor operators, the evolution operator becomes a zigzag MPO:

$$e^{-\varepsilon H} = \quad \text{(4.91)}$$

The partition function is simply a stacking of this zigzag MPO:

$$e^{-\beta H} = \quad \text{(4.92)}$$

Here we have taken $M = 3$. From the periodicity of the stacking zigzag MPO along the imaginary-time direction, we can choose a path to reexpress the partition function as a uniform tensor network state of width M (here $M = 3$) in the thermodynamic limit:

$$e^{-\varepsilon H} = \quad , \quad \text{(4.93)}$$

where τ is a tensor obtained by contracting two neighboring tensors, $\bar{u}$ and $\bar{w}$, along the vertical direction:

$$\quad \text{(4.94)}$$

The zigzag paths shown in Eq. (4.92) are along the diagonal direction. For example, the path linked by the dashed line corresponds to the top column MPO in Eq. (4.93).

The tensor network representation (4.93) suggests that we can express the partition function as a product of an effective evolution operator V along the imaginary time direction:

$$Z = \mathrm{Tr}\, e^{-\beta H} = \mathrm{Tr}\, V^M, \quad \text{(4.95)}$$

where

$$V = \quad . \quad \text{(4.96)}$$

The length of V is just the lattice size N.

Similarly, the partition function can be written as a product of a transfer matrix T along the spatial direction:

$$Z = \lim_{N \to \infty} \mathrm{Tr} T^N, \tag{4.97}$$

where

$$T = \begin{array}{c} -\boxed{\tau}- \\ -\boxed{\tau}- \\ -\boxed{\tau}- \end{array} . \tag{4.98}$$

The width of the transfer matrix equals the Trotter number M.

One can also write the evolution operator as an MPO, like the Hamiltonian itself, without invoking the Trotter–Suzuki formalism [348]. For example, for the Heisenberg model (4.1), the evolution operator can be expressed as

$$e^{-\varepsilon H} = I - \varepsilon H + o(\varepsilon^2)$$
$$= [\cdots \tau(\mathbf{S}_1)\tau(\mathbf{S}_2) \cdots \tau(\mathbf{S}_N) \cdots]_{11} + o(\varepsilon^2), \tag{4.99}$$

where subscript "11" indicates that the evolution operator is determined by the most upper-left element of the tensor product. The local tensor is defined by

$$\tau(\mathbf{S}_i) = \begin{pmatrix} I + \varepsilon h S_{iz} & -\sqrt{\varepsilon J}\mathbf{S}_i \\ \sqrt{\varepsilon J}\mathbf{S}_i^T & \mathbf{0} \end{pmatrix}. \tag{4.100}$$

Graphically, Eq. (4.99) can be represented as

$$e^{-\varepsilon H} = \left(\cdots \boxed{\tau}-\boxed{\tau}-\boxed{\tau}-\boxed{\tau}-\boxed{\tau}-\boxed{\tau}\cdots \right)_{11} + O(\varepsilon^2). \tag{4.101}$$

In obtaining the above expressions, all the terms of order ε^2 or higher are neglected. τ is not a symmetric operator matrix; its horizontal dimension is 4, and the vertical dimension, which acts on the physical basis states, is 2.

The local tensor τ can be similarly constructed for other models. The vertical dimension of τ is the total number of basis states at each site. However, the horizontal dimension is model-dependent. For the one-dimensional Hubbard model (4.75), for example, the vertical dimension of τ is 4, and the horizontal dimension is 5.

5

Tensor Network Ansatz of Wave Functions

5.1 Area Law of Entanglement Entropy

In a quantum many-body system, the dimension of the Hilbert space grows exponentially with the number of constituents, giving rise to an exponential growth of the resources needed for the simulation with the system size. However, if one is interested in the ground-state properties, it is believed that the number of required parameters is bounded for noncritical systems. It implies that the ground state only occupies a small subspace of the whole Hilbert space, and it is possible to represent ground states more efficiently.

The renormalization group is a systematic method for selecting the most relevant basis space of the ground state through local basis transformations. A common assumption adopted by DMRG or other RG methods is that the ground state or any other target state can be accurately expanded using a small fraction of the whole basis states. This assumption is justified by the discovery of the area law (or boundary law) of entanglement entropy.

Entanglement entropy is an entanglement measure of correlations of a system about another. More specifically, it measures how correlations between a subsystem, called the system block, and the rest of the system, called the environment block, scale with the size of the whole system. The system and environment blocks are complementary. We denote them by "sys" and "env," respectively. For a system described by the density matrix ρ, the entanglement entropy S of the system block is defined as the *von Neumann entropy* of the reduced density matrix ρ^{sys}:

$$S = -\mathrm{Tr}\rho^{sys} \log \rho^{sys}. \tag{5.1}$$

Here ρ^{sys} is defined from the density matrix by integrating out all degrees of freedom of the environment block,

$$\rho^{sys} = \mathrm{Tr}_{env}\, \rho. \tag{5.2}$$

At zero temperature,

$$\rho = |\Psi\rangle\langle\Psi|, \tag{5.3}$$

where $|\Psi\rangle$ is the ground-state wave function.

The area law of entanglement entropy is an inherent property of quantum many-body systems. It states that the leading term of the bipartite entanglement between the system and environment blocks grows to reflect the area of the boundary that separates these two blocks instead of the whole volume in a noncritical state (see Ref. [37] for a review). That is, in a n-dimensional hypercube of $N \times N \cdots \times N$, the entanglement entropy scales as

$$S \approx aN^{n-1} \tag{5.4}$$

in the limit $N \to \infty$, where a is a model- and geometry-dependent constant, determined by the correlation length. The interest in this area law originates from the insight that the entropy of black holes scales with the size of the event horizon. A similar scaling law of entanglement entropy is observed in non-critical quantum lattice systems. Both are in sharp contrast to the behavior of most states in Hilbert space, which exhibit a volume scaling rather than an area law.

The entanglement entropy is a coarse-grained measure quantifying the number of modes in a subsystem that is entangled with the other subsystem. The area law indicates that the entanglement entropy is not an extensive property, and most of the entanglement must concentrate around the boundary. It implies that the ground state is compressible, and the minimal number of basis states D that is needed for accurately describing its wave function is of the order

$$D \sim \exp\left(aN^{n-1}\right), \tag{5.5}$$

which is dramatically smaller than the total number of degrees of freedom.

In one dimension ($n = 1$) where the boundary is just a point, the entanglement entropy is a constant, independent of the lattice size. Thus, the entanglement entropy is upper bounded when the lattice size becomes sufficiently large, and the minimal number of basis states needed for accurately describing a ground state does not depend on the lattice size. This saturation of the entanglement entropy accounts for the success of DMRG in one dimension. In a periodic chain, the bipartite entanglement entropy is doubled since one needs two cuts to divide the whole system into two parts. Hence, the number of states retained in DMRG should be the square of the number of states used in an open chain in order to reach the same accuracy. In two dimensions, the entanglement entropy scales as N. In this case, the DMRG algorithm should keep a number of states that grows exponentially with N, and the simulation becomes inefficient for large N, but is still feasible for small N.

Intuitively, it is believed that the area law results from the existence of a finite characteristic length scale, namely the correlation length, on which two-point correlations decay. In other words, localized correlations would imply an entanglement area law. The entanglement of ground states is small because a large entanglement state does not gain energy for a physical Hamiltonian with short-range correlations. This heuristic picture or argument, which was initially just an expectation, has now been proven in some cases; for example, in the ground states of one-dimensional gapped systems with local interactions [36], where an area law was proven for all systems with an energy gap above the ground state. It has also been proven that all connected correlation functions between two blocks in a gapped system have to decay exponentially as a function of the distance [350]. In two dimensions, proof of the area law requires a hypothesis on the eigenvalue distribution of the Hamiltonian [351, 352]. At finite temperature, an area law is generally implied by a finite correlation length measured in terms of the mutual information [112], and similar links to topological order persist [353]. However, a firm and general proof for the connection between the decay of correlations and the area law is still absent.

Two kinds of corrections to the entanglement area law exist. One is the logarithmic correction at a critical point. The other is an additive correction if the system possesses a topological order. For example, the area law is mildly violated in one-dimensional critical systems described by conformal field theory where the leading term of the entanglement entropy scales as [46, 47, 48]

$$S \simeq \frac{c + \bar{c}}{6} \ln N, \qquad (5.6)$$

under periodic boundary conditions. Here c and $\bar{c}$ are the holomorphic and antiholomorphic central charges of the corresponding conformal field theory. This correction is still small in comparison with the total volume of the system. The entanglement entropy of two-dimensional free fermions with a finite Fermi surface is known to scale as $N \ln N$ [354, 355], but there are no such logarithmic corrections in certain gapless two-dimensional spin systems [356]. It is generally not clear to what extent such a logarithmic correction will occur in higher-dimensional systems.

This violation of an area law seems to be a pure quantum phenomenon as it occurs solely at zero temperature. In Ref. [112], it was shown that the mutual entropy $S(C)$, a measure of correlations for mixed states, is bounded by the correlation length for all local classical and quantum Hamiltonians. This area law of mutual entropy is valid at arbitrary temperatures, including zero temperature. The violation of this area law implies immediately that the correlation length is divergent. However, the converse may not be true since there are critical lattice systems that obey the area law [356, 357]. The area law may hold even in systems with algebraically decaying two-point correlations.

A topological order can lead to an additive correction to the entanglement entropy. In two dimensions, it was shown that the entanglement entropy of a disc-like region with a smooth boundary of length N scales as [358, 359]

$$S = aN - \gamma + \cdots , \tag{5.7}$$

where the ellipsis represents terms that vanish in the limit $N \to \infty$. The term γ is a universal additive correction to the entanglement entropy, characterizing the long-range entanglement in the ground state that can be quantified as

$$\gamma = \ln D_Q \tag{5.8}$$

with D_Q the total quantum dimension of anyons.

In case the von Neumann entropy is difficult to calculate, one can also determine the value of γ by evaluating the *Renyi entropy*. Given a reduced density matrix ρ^{sys}, the nth order Renyi entropy is defined by

$$S_n = \frac{1}{1-n} \ln \text{Tr} \left(\rho^{\text{sys}} \right)^n , \tag{5.9}$$

where $\text{Tr} \rho^{\text{sys}} = 1$ is assumed. The Renyi entropy becomes the von Neumann entropy in the limit $n \to 1$.

In terms of the Renyi entropy [360, 361, 362], the area law can be expressed as

$$S_n \simeq a_n N - \gamma + \cdots \tag{5.10}$$

in two dimensions for any given n in the limit $N \to \infty$. Again the terms represented by the ellipsis vanish in the thermodynamic limit; a_n is an n-dependent constant.

The area law sets a guideline for designing classes of wave functions to approximately but faithfully represent ground states. It has led to a deep understanding of DMRG. The wave function generated by DMRG is an MPS, which satisfies the one-dimensional area law. On the other hand, MPS can be generally regarded as a variational wave function [20], independent of DMRG. It can be determined either by DMRG or by other variational approaches.

The area law has also motivated several other tensor network ansatz in an attempt to replicate the success of DMRG. Some examples include two- or higher-dimensional PEPS or PESS, which can be thought of as a higher-dimensional generalization of MPS, or one-dimensional MERA [51], which is a wave function more suitable for describing a one-dimensional critical system whose entanglement entropy contains a logarithmic correction term. A proof that PEPS satisfies the area law is given in Ref. [112]. All these wave functions are summarized under the name of *tensor network states*. They constitute variational ansatz for characterizing the role of entanglement in quantum many-body states.

5.2 Matrix Product States (MPS)

As already mentioned, in a quantum many-body system, the Hilbert space grows exponentially with the system size. However, for a gapped system with local inter-actions, the ground state obeys the entanglement area law, which implies that only a tiny portion of the Hilbert space is relevant to the ground state. An important fact that has been found is that this relevant corner of Hilbert space can be parameterized efficiently. In one dimension, this parametrization is achieved by MPS.

MPS factorizes a wave function tensor into a product of many auxiliary tensors (or matrices) connected by the so-called virtual bond indices

$$\Psi_{\sigma_1\cdots\sigma_N} = A_1[\sigma_1]A_2[\sigma_2]\cdots A_N[\sigma_N], \tag{5.11}$$

$$= \begin{array}{ccccc} \sigma_1 & \sigma_2 & \sigma_3 & & \sigma_N \\ \mid & \mid & \mid & & \mid \\ \!\!-\!\!\boxed{A_1}\!\!-\!\!\!\!\boxed{A_2}\!\!-\!\!\!\!\boxed{A_3}\!\!-\!\!\cdots\!\!-\!\!\boxed{A_N}\!\!-\!\! \end{array},$$

where $A_i[\sigma_i]$ is a local tensor or matrix defined at site i. The corresponding physical state is

$$|\Psi\rangle = \sum_{\sigma_1\cdots\sigma_N} A_1[\sigma_1]A_2[\sigma_2]\cdots A_N[\sigma_N]|\sigma_1\cdots\sigma_N\rangle. \tag{5.12}$$

We can also write it as

$$|\Psi\rangle = \hat{A}_1\hat{A}_2\cdots\hat{A}_N, \tag{5.13}$$

where

$$\hat{A}_i = \sum_{\sigma} A_i[\sigma]|\sigma\rangle \tag{5.14}$$

is a matrix vector that generates the local tensor at site i. The local tensor $A_i[\sigma]$ is simply the vector element of $\hat{A}_i$:

$$A_i[\sigma] = \langle\sigma|\hat{A}_i. \tag{5.15}$$

5.2.1 MPS Factorization

If there is no restriction to the tensor virtual bond dimensions, it is always possible to factorize $\Psi_{\sigma_1\cdots\sigma_N}$ into an MPS without taking any approximation. This can be achieved by reshaping Ψ into a matrix of dimension $d \times d^{N-1}$ (d is the dimension of basis states at each lattice site), $\tilde{\Psi}_{\sigma_1,\sigma_2\cdots\sigma_N} = \Psi_{\sigma_1\cdots\sigma_N}$, and taking a QR decomposition

$$\Psi_{\sigma_1\cdots\sigma_N} = \sum_{\alpha_1}^{D_1} Q^{(1)}_{\sigma_1,\alpha_1} R^{(1)}_{\alpha_1,\sigma_2\cdots\sigma_N}, \tag{5.16}$$

where D_1 is the rank of $Q^{(1)}$. After reshaping $R^{(1)}_{\alpha_1, s_2 \cdots s_N}$ into a new matrix $\tilde{R}^{(1)}_{\alpha_1 \sigma_2, \sigma_3 \cdots \sigma_N} = R^{(1)}_{\alpha_1, \sigma_2 \cdots \sigma_N}$ and taking the QR decomposition $\tilde{R}^{(1)} = Q^{(2)} R^{(2)}$, the wave function can then be written as

$$\Psi_{\sigma_1 \cdots \sigma_N} = \sum_{\alpha_1}^{D_1} \sum_{\alpha_2}^{D_2} Q^{(1)}_{\sigma_1, \alpha_1} Q^{(2)}_{\alpha_1 \sigma_2, \alpha_2} R^{(2)}_{\alpha_2, \sigma_3 \cdots \sigma_N}, \tag{5.17}$$

where D_2 is the rank of $Q^{(2)}$. To continue the above reshaping and QR decomposition steps, we will eventually obtain

$$\Psi_{\sigma_1 \cdots \sigma_N} = \sum_{\alpha_1 \cdots \alpha_2} Q^{(1)}_{\sigma_1, \alpha_1} Q^{(2)}_{\alpha_1 \sigma_2, \alpha_2} \cdots Q^{(N-1)}_{\alpha_{N-2} \sigma_{N-1}, \alpha_{N-1}} R^{(N-1)}_{\alpha_{N-1}, \sigma_N}. \tag{5.18}$$

To define

$$(A_1)_{\alpha_1} [\sigma_1] = Q^{(1)}_{\sigma_1, \alpha_1}, \tag{5.19}$$

$$(A_i)_{\alpha_{i-1}, \alpha_i} [\sigma_i] = Q^{(i)}_{\alpha_{i-1} \sigma_i, \alpha_i}, \quad 2 \leq i \leq N - 1, \tag{5.20}$$

$$(A_N)_{\alpha_{N-1}} [\sigma_N] = R^{(N-1)}_{\alpha_{N-1}, \sigma_N}, \tag{5.21}$$

we then obtain the MPS representation of Ψ. $A_i[\sigma]$ is a matrix function of physical variable σ. It is a $D_i \times D_{i+1}$ matrix, with D_{i+1} the dimension of the virtual basis states on the bond linking sites i and $i + 1$.

This discussion indicates that a quantum state can be rigorously represented as an MPS. The virtual index of MPS is the quantum number of basis states for a subsystem, and the local tensor $(A_i)_{\alpha_i, \alpha_{i+1}} [\sigma]$ is just a basis transformation matrix between $|\alpha_i, \sigma\rangle$ and $|\alpha_{i+1}\rangle$. This rigorous MPS representation of a quantum state without truncation is of no practical use because its total number of free parameters still grows exponentially with the system size. However, the basis dimensions of MPS can be truncated so that all D_i's are smaller than a maximal bond dimension D if the entanglement entropy of the corresponding quantum state, such as the ground state, is governed by the area law in one dimension. The MPS provides an approximate but efficient representation of the wave function in this case. The number of free parameters in this MPS grows linearly with the system size. Nevertheless, as discussed in Chapter 10, the result can converge exponentially fast towards the exact one with increasing D if the one-dimensional entanglement area law is obeyed.

5.2.2 MPS from Projection

One can also generate the MPS representation by imaginary time evolution for the ground-state wave function. As revealed by Eq. (4.20), one can project out the ground-state wave function by successively applying the evolution operator

$\exp(-\varepsilon H)$, represented by Eq. (4.91), to an arbitrary initial wave function $|\Psi_0\rangle$ that is not orthogonal to the ground state.

In a translation-invariant system, since the projection operators $\exp(-\varepsilon H)$ under the Trotter–Suzuki approximation is translation invariant by shifting two lattice sites, so is the wave function. Therefore, the wave function should have the form

$$\Psi_{\sigma_1...\sigma_N} = \qquad\qquad\qquad\qquad\qquad\qquad\qquad . \qquad (5.22)$$

There are many ways to set up the initial MPS. One way is to set $|\Psi_0\rangle$ as the state at infinite temperature. As all states are equally partitioned at infinite high temperature, Ψ_0 is simply a product state with

$$A[\sigma] = B[\sigma] = 1. \qquad (5.23)$$

For a quantum spin system, there is no polarization at sufficiently high temperature, and the average spin is zero at all lattice sites; that is,

$$\langle S_i \rangle = \langle \Psi | S_i | \Psi \rangle = 0. \qquad (5.24)$$

If the Hamiltonian is spin rotation invariant, then the total spin of the system will remain zero with the imaginary time evolution. However, the rounding error may break the symmetry between the up and down spins in a practical calculation. The ground state obtained from this evolution may exhibit a finite magnetic order as a manifestation of spontaneous symmetry breaking.

The first step of projection is to apply the imaginary time evolution operator represented by (5.22) to the initial state,

$$|\Psi'\rangle = \qquad\qquad\qquad\qquad\qquad\qquad\qquad$$

$$= \qquad\qquad\qquad\qquad\qquad\qquad\qquad . \qquad (5.25)$$

This yields a new MPS whose local tensors are defined by

$$\qquad\qquad\qquad\qquad\qquad\qquad . \qquad (5.26)$$

After the projection, the bond dimensions of A' and B' become $D = d^2 D_0$ if the initial bond dimension is D_0. Repeating this projection procedure, we can finally obtain an exact ground-state wave function in the MPS representation.

However, as the bond dimension increases by d^2 after each projection, we have to truncate the bond dimension once it becomes larger than what we can handle. It implies that the state thus obtained is still an approximation to the true ground state. It becomes exact only at the limit where the bond dimension approaches infinity ($D \to \infty$). The bipartite entanglement of the system determines the speed of convergence to the ground state. If the area law of entanglement entropy holds, an MPS with a sufficiently large D will accurately approximate the ground state.

An MPS is an optimal representation of a ground state in one dimension. It yields a local description of quantum many-body states based on their entanglement structures. As discussed in §10.1, it is the wave function that DMRG generates. It can be viewed as a trial wave function in which all matrix elements of local tensors are the variational parameters. The ground state can also be represented as an MPS in higher dimensions, but the bond dimension has to grow exponentially with the system size to satisfy the entanglement area law.

5.3 One-Dimensional AKLT States

In 1983, Haldane made a conjecture about the excitation spectra of quantum anti-ferromagnetic Heisenberg spin chains. It states that for a spin-S antiferromagnetic Heisenberg model defined by Eq. (4.1) in the absence of an external magnetic field, the excitation spectrum is gapless when S is a half-odd-integer and gapped when S is an integer [28, 29, 30].

To understand this conjecture, Affleck, Lieb, Kennedy, and Tasaki (AKLT) [26, 27] introduced a set of extended Heisenberg spin models, now called the AKLT models, which provide a prototypical framework for understanding the physics of the Haldane gap in integer quantum spin chains. The ground states of these models, known as the AKLT states, can be rigorously solved and represented as MPS in one dimension. They have also served as valuable examples for such concepts as valence bond solid order and symmetry protected topological order.

The AKLT model is an extension of the one-dimensional quantum Heisenberg spin model. The spin-1 AKLT model [26] reads

$$H_{AKLT} = \sum_i P_2(\mathbf{S}_i + \mathbf{S}_{i+1}), \tag{5.27}$$

$$P_2(\mathbf{S}_1 + \mathbf{S}_2) = \frac{1}{2}\mathbf{S}_1 \cdot \mathbf{S}_2 + \frac{1}{6}(\mathbf{S}_1 \cdot \mathbf{S}_2)^2 + \frac{1}{3}. \tag{5.28}$$

P_2 is a projection operator acting on two neighboring sites. It contains both bilinear and biquadratic terms. P_2 takes an eigenvalue 1 if the total spin $S_{12} = \mathbf{S}_1 + \mathbf{S}_2$ equals 2 or 0 if S_{12} equals 0 or 1. Hence, P_2 is to project out the state with $S_{12} = 2$ on the two neighboring sites. The ground state of this local Hamiltonian $H_{1,2}$ is four-fold degenerate and has a total spin $S_{12} = 0$ or 1. Since the full Hamiltonian is a sum of projection operators, the ground state energy has to be nonnegative. If we find a state with an energy eigenvalue 0, that is just a ground state.

The spin-1 AKLT model can be solved by regarding a spin-1 operator as a symmetrized product of two spin-1/2 operators. This decoupling is equivalent to fractionalizing each $S = 1$ spin into two virtual $S = 1/2$ spins and representing it as

$$|\sigma\rangle = \sum_{\alpha\beta} C_{\alpha\beta}[\sigma] |\alpha\beta\rangle_s, \qquad (5.29)$$

where $\sigma = (1, 0, -1)$ are the quantum numbers of $S = 1$ spins, $\alpha = (\uparrow, \downarrow)$ (similarly β) are the virtual spin-1/2 quantum numbers. The subscript s in $|\ \rangle_s$ stands for symmetrization and $|\alpha\beta\rangle_s$ is a symmetrized virtual spin basis state. The two virtual $S = 1/2$ spins have four basis states in total. After symmetrization, the two virtual spins are used to represent the three $S = 1$ triplet states. $C_{\alpha\beta}[\sigma]$ is a short notation for the Clebsch–Gordan coefficient

$$C_{\alpha\beta}[\sigma] = \overset{\sigma}{\underset{\alpha}{} \textcircled{C} \overset{}{\underset{}{\beta}}} = \langle 1\sigma | 1/2, \alpha; 1/2, \beta \rangle. \qquad (5.30)$$

Given σ, $C[\sigma]$ is a 2×2 matrix whose value is given by

$$C[1] = \begin{pmatrix} 1 & 0 \\ 0 & 0 \end{pmatrix}, \quad C[0] = \frac{1}{\sqrt{2}} \begin{pmatrix} 0 & 1 \\ 1 & 0 \end{pmatrix}, \quad C[-1] = \begin{pmatrix} 0 & 0 \\ 0 & 1 \end{pmatrix}. \qquad (5.31)$$

To solve the AKLT model, let us first consider a two-site system whose Hamiltonian equals P_2. For this system, the ground states are simply to pair two virtual spins on different sites into a spin singlet. This leaves two unpaired $S = 1/2$ virtual spins, one at each lattice site. The unpaired spins can form either a spin-0 or a spin-1 state. They are the ground states of P_2, and can be represented as

$$|\Psi\rangle = \sum_{\beta_1\alpha_2} \varepsilon_{\beta_1\alpha_2} |\alpha_1\beta_1\rangle_s |\alpha_2\beta_2\rangle_s, \qquad (5.32)$$

where ε is the antisymmetric matrix that is to project two $S = 1/2$ variables it connects to a spin singlet state:

$$\overset{}{\underset{}{-\textcircled{\varepsilon}-}} = \begin{pmatrix} 0 & 1 \\ -1 & 0 \end{pmatrix}, \quad \varepsilon^2 = -I. \qquad (5.33)$$

$|\Psi\rangle$ does not depend on the unpaired virtual quantum numbers α_1 and β_2. Thus, the total number of degeneracy is 4, resulting from the two free $S = 1/2$ spins, one on each site.

In the physical basis representation, the ground-state wave function can be written as

$$|\Psi\rangle = \sum_{\sigma_1\sigma_2} \Psi_{\sigma_1\sigma_2} |\sigma_1\sigma_2\rangle, \qquad (5.34)$$

where $\Psi_{\sigma_1\sigma_2}$ is a matrix defined by

$$\Psi_{\sigma_1\sigma_2} = C\,[\sigma_1]\,\varepsilon\,C\,[\sigma_2] = \;\raisebox{-0.5em}{\text{—}}\overset{\sigma_1}{\underset{}{\text{ⓒ}}}\text{—}\varepsilon\text{—}\overset{\sigma_2}{\underset{}{\text{ⓒ}}}\text{—}\;. \qquad (5.35)$$

It can also be expressed as

$$\Psi_{\sigma_1\sigma_2} = A\,[\sigma_1]\,A\,[\sigma_2]\,\varepsilon = \;\text{—}\overset{\sigma_1}{\text{Ⓐ}}\text{—}\overset{\sigma_2}{\text{Ⓐ}}\text{—}\varepsilon\text{—}\;. \qquad (5.36)$$

where

$$\text{—Ⓐ—} = \text{—ⓒ—}\varepsilon\text{—}\;. \qquad (5.37)$$

More explicitly,

$$A\,[1] = \begin{pmatrix} 0 & 1 \\ 0 & 0 \end{pmatrix}, \quad A\,[0] = \frac{1}{\sqrt{2}}\begin{pmatrix} -1 & 0 \\ 0 & 1 \end{pmatrix}, \quad A\,[-1] = \begin{pmatrix} 0 & 0 \\ -1 & 0 \end{pmatrix}. \qquad (5.38)$$

This solution can be extended to an arbitrary many-site system with open boundary conditions. The ground state is to pair all the neighboring virtual spins into singlets, and leave unpaired virtual spins only at the two ends of the chain. The wave function is again an MPS:

$$\Psi = \;\text{—ⓒ—}\varepsilon\text{—ⓒ—}\cdots\text{—}\varepsilon\text{—ⓒ—}\;. \qquad (5.39)$$

It is a VBS state formed by virtual spins. Each unpaired spin at the ends of the chain gives rise to a double degeneracy of the ground state. Thus, the total number of the degeneracy of the ground state is four. The unpaired spin-1/2 degrees of freedom in this spin-1 system actually reveals a generic mechanism for quantum number fractionalization.

On the other hand, if periodic boundary conditions are assumed and all spins are arranged in a ring, the two unpaired virtual spins will also form a spin singlet.

In this case, the ground state becomes nondegenerate and can be represented by a translation-invariant MPS:

$$|\Psi\rangle = \sum_{\sigma_1 \cdots \sigma_N} \text{Tr} \left(C[\sigma_1]\varepsilon \cdots C[\sigma_N]\varepsilon \right) |\sigma_1 \cdots \sigma_N\rangle$$

$$= \sum_{\sigma_1 \cdots \sigma_N} \text{Tr} \left(A[\sigma_1] \cdots A[\sigma_N] \right) |\sigma_1 \cdots \sigma_N\rangle. \tag{5.40}$$

Without unpaired edge states, the ground state becomes nondegenerate.

From this example, we see that an MPS could also be viewed as a trial wave function arising from virtual entangled pairs formed between two nearest-neighbor sites: a local tensor $A_{\alpha\beta}[\sigma]$ is to project two virtual bond variables, α and β, onto a physical one; two local tensors are connected by a virtual bond formed by a maximally entangled state of two virtual variables at the two ends of the bond.

For this AKLT state, the correlation function between two spins decays exponentially with their separation. The corresponding correlation length equals $1/\ln 3 < 1$ [26, 27], which is less than a lattice constant. Thus, the excitation of the AKLT model from the ground state is gapped. From this gapped ground state, one could gradually reduce the coupling constant of the biquadratic term in the AKLT model. It was found that there is no phase transition between the Heisenberg model and the AKLT model. Therefore, the ground states for these two models lie in the same phase, and the Haldane gap in the pure Heisenberg model also results from the spin fractionalization, the same as in the AKLT model.

The product of A-matrices satisfies the following properties:

$$A[-1]A[-1] = A[1]A[1] = 0, \tag{5.41}$$

$$A[0]A[-1] = -A[-1]A[0] = A[-1]/\sqrt{2}, \tag{5.42}$$

$$A[1]A[0] = -A[0]A[1] = A[1]/\sqrt{2}. \tag{5.43}$$

This means that in a squeezed space where all $S_{iz} = 0$ states are omitted, any two neighboring S_{iz} should be antiferromagnetically ordered, alternating between $+1$ and -1. For example, the spin-1 AKLT state in the original and squeezed spaces can take the following configurations, respectively:

Original space: ↑   0   0   ↓   0   ↑   ↓   0   0   ↑   0   ↓   0   ↑

Squeezed space: ↑   ↓   ↑   ↓   ↑   ↓   ↑

Thus, in the squeezed space, the spin-1 AKLT state contains a hidden antiferromagnetic order. In the original space, the state does not possess this order since one can arbitrarily insert many unpolarized states $|0\rangle$ between any two finite polarized states. This hidden antiferromagnetic order does not break any local spin rotation symmetry. One cannot define a conventional local order parameter to characterize this order. However, one can define a topological string order parameter to measure this hidden order [363]. A discussion on the string order is given in §11.4.3.

The AKLT model has four degenerate ground states, separated from the other states by a finite Haldane gap on an open chain in the thermodynamic limit. This degeneracy corresponds to a hidden $Z_2 \times Z_2$ topological symmetry [364, 365], manifesting as the degenerate edge states. The opening of the Haldane gap is due to the break of this discrete topological symmetry:

$$Z_2 \times Z_2 \to Z_1. \tag{5.44}$$

5.4 Multiscale Entanglement Renormalization Ansatz (MERA)

MPS provides an efficient and faithful representation of a ground state that satisfies the entanglement area law in a noncritical system in one dimension. However, in a critical system, there is a logarithmic correction to the entanglement entropy, $S = \alpha \ln N$, with α a coefficient proportional to the central charge and N the lattice size. If an MPS is used to represent this critical state, then the bond dimension of MPS, D, has to grow algebraically with the system size (i.e. $D \sim N^\alpha$). In other words, an MPS can accurately approximate short-range properties of a critical ground state but cannot accurately describe its asymptotic behavior at large distances.

A MERA is a hierarchical network on which quantum entanglements at different length scales are explicitly constructed. It is particularly useful for describing quantum critical systems in one dimension [51]. Local tensors in MERA are organized in layers, where each layer corresponds to a different length scale. A coarse-graining transformation connects two neighboring layers. Figure 5.1(a) shows, as an example, a graphical representation of a binary MERA where two sites are coarse-grained into one effective site. For this binary MERA, the total number of layers equals $\log_2 N$. To separate this MERA into two disconnected parts, one has to cut at least $\log_2 N$ bonds from top to bottom. It means that the maximal entanglement entropy that the MERA can carry is of the order of $\log_2 N$; that is,

$$S \sim \log_2 N. \tag{5.45}$$

Hence, MERA represents a critical ground state faithfully.

For the binary MERA, the coarse-graining transformation from the nth layer to the $(n + 1)$th one reduces the lattice size by 2. Each layer is comprised of two kinds of local tensors, called isometries (Fig. 5.1(b)) and disentanglers (Fig. 5.1(c)), respectively. At each layer, the lattice is partitioned into blocks of two sites. The disentanglers U are first applied across the boundaries of neighboring blocks to reduce the entanglement between these two blocks. The isometries V are then applied to map each block of two sites into a single effective site on the lattice one layer above.

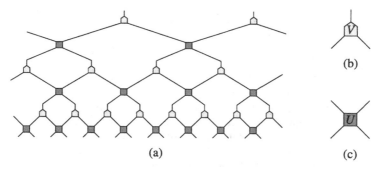

Figure 5.1 (a) A binary MERA based on a 2-to-1 coarse-graining scheme. It consists of two kinds of building blocks: (b) Isometry V, used to map two blocks into one block with a contraction to the number of basis states. V satisfies the isometric condition. (c) Disentangler U, used to eliminate the local entanglement between two blocks before coarse graining. U is a unitary matrix.

Let d_n denote the dimension of basis states at each site of the nth layer. The disentangler U is simply a unitary matrix of dimension $D = d_n^2$, satisfying the unitary condition $U^\dagger U = UU^\dagger = I_{D\times D}$. The isometry V is not a square matrix. Instead, it is a matrix of dimension $D' \times D$ ($D' = d_{n+1} < D$), satisfying the isometric constraint $VV^\dagger = I_{D'\times D'}$ and $V^\dagger V \neq I_{D\times D}$.

A critical system lacks a characteristic length scale and is invariant under changes of scale. In a scale-invariant system, correlation functions decay polynomially, contrasting sharply with a gapped system where correlation functions decay exponentially. This scale invariance could be implemented by assuming that all disentanglers and isometries are layer and site independent. This reduces the number of parameters, allowing an infinite system to be accessed.

A MERA provides a faithful representation of critical quantum states in one dimension. In practical calculation, however, the cost for optimizing a MERA is much higher than an MPS: the cost for contracting an MPS scales as D^3, while the cost for contracting a MERA scales as D^7. Thus, only a relatively small D can be handled in a MERA calculation.

5.5 Projected Entangled Pair State (PEPS)

An MPS describes the short-range entanglement of one-dimensional quantum states accurately. However, it is not well suited for describing ground states in two or higher dimensions for the following reasons. First, in order to use MPS to describe a two- or higher-dimensional quantum lattice system, one has to map this quantum system onto a one-dimensional lattice. This will unavoidably break the translational and rotational symmetries and effectively introduce long-range

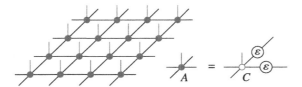

Figure 5.2 PEPS as a tensor network formed by fractionalizing physical states at each lattice site into four virtual states allowed by symmetries. C is a local tensor that maps four virtual states into a physical one. ε is a bond matrix that maximally entangles the two virtual states it connects. Generally, both C and ε, or equivalently A, can be used as variational parameters.

interactions even though there are only local interactions in the original lattice. To handle a long-range interaction and recover all the symmetries, one has to increase the virtual bond dimensions of MPS significantly. Second, MPS is not a natural representation that obeys the entanglement area law in two or higher dimensions because the entanglement entropy a virtual bond can carry is upper bounded by the bond dimension D. To accurately describe the entanglement property of the ground state, D has to grow exponentially with the system size, which is practically infeasible when the lattice size becomes large.

PEPS, as depicted in Fig. 5.2, constitutes a natural generalization of MPS to two or higher dimensions. This generalization, motivated by two-dimensional VBS states, again first introduced by Affleck et al. [26, 27], is obtained by distributing maximally entangled virtual states between any two nearest-neighbor sites. It keeps the locality of interactions and allows a good representation of local properties that are compatible with, for example, the translational symmetry. Furthermore, it captures the main feature of the entanglement area law since the minimal number of entangled bonds for separating the system into two blocks is proportional to the size of the interface. In other words, the minimal bond dimension needed for accurately describing a quantum state does not depend on the system size in a gapped system. Hence, PEPS yields an accurate parameterization of the ground state that satisfies the entanglement area law. It allows a many-body ground-state wave function, which contains exponentially many degrees of freedom, to be evaluated approximately but accurately on a polynomial time scale.

PEPS has been suggested in different variations by several groups. A tensor-product wave function was first proposed by Niggemann et al. [108] to study the ground-state properties of the Heisenberg model on the honeycomb lattice. A more general ansatz of tensor network states was suggested by Sierra and Martin-Delgado [109]. The use of the tensor network state as a variational wave function for the three-dimensional classical lattice model was suggested by Nishino et al. [110]. In 2004, the idea of tensor network states was discussed by Verstraete and

Cirac under the name of PEPS [111]. Their work has attracted broad interest because it reveals more clearly the physical picture embedded in the tensor network representation of quantum states.

From the discussion on the one-dimensional AKLT state in §5.3, we see that the local tensor in MPS, $A_{\alpha\beta}[\sigma]$, can be regarded as a transformation matrix that maps two virtual basis states, α and β, onto the physical state σ. This idea also holds in higher dimensions. On the square lattice, a local tensor in a PEPS can also be regarded as a mapping between four virtual basis states and one physical state. This is equivalent to fractionalizing a physical basis state into four virtual ones. The local correlation is enforced by maximally entangling virtual basis states between any two neighboring sites.

A PEPS wave function can be generally expressed as

$$|\Psi\rangle = \text{Tr} \prod_i A^{(i)}_{\alpha_i\beta_i\gamma_i\delta_i}[\sigma_i]\,|\sigma_i\rangle\,, \tag{5.46}$$

where Tr is to sum over all physical spin configurations and all virtual bond variables. Each bond variable runs from 1 to D, where D is the bond dimension that is the total degrees of freedom of each fractionalized bond variable. $A^{(i)}$ is a local tensor defined at site i.

To understand the physical meaning of local tensors, let us take an AKLT state defined on the square lattice [26, 27] as an example to show how to represent it as a PEPS. This AKLT state is the ground state of the Hamiltonian defined by

$$H = \sum_{\langle ij\rangle} P_4(\mathbf{S}_i + \mathbf{S}_j), \tag{5.47}$$

where $\mathbf{S}_i$ is the spin operator at site i. $P_4(\mathbf{S}_i + \mathbf{S}_j)$ is to project out $\mathbf{S}_i$ and $\mathbf{S}_j$ to the states of total spin $S = 4$.

The ground state of this AKLT model is a VBS whose wave function can be constructed in two steps:

(i) To fractionalize each $S = 2$ spin into four $S = 1/2$ spins. This is equivalent to writing the physical state $|\sigma\rangle$ as a superposition of four virtual spin states

$$|\sigma\rangle = \sum_{\alpha\beta\gamma\delta} C_{\alpha\beta\gamma\delta}[\sigma]\,|\alpha\beta\gamma\delta\rangle\,, \tag{5.48}$$

where $C_{\alpha\beta\gamma\delta}[\sigma] = \langle\alpha\beta\gamma\delta|\sigma\rangle$ is to project the four $S = 1/2$ spins onto a $S = 2$ spin state. $|2\rangle = |\uparrow\uparrow\uparrow\uparrow\rangle$ is just a fully polarized state, hence $C_{\alpha\beta\gamma\delta}[2] = \delta_{\alpha\beta\gamma\delta,1111}$. Here we denote the virtual up and down spin states by $\alpha = 1$ and -1, respectively. By applying the lowering spin operator $S^- = S^-_\alpha + S^-_\beta + S^-_\gamma + S^-_\delta$ recursively to $|2\rangle$, we find that the coefficients can be generally expressed as

$$C_{\alpha\beta\gamma\delta}[\sigma] = a_\sigma \delta_{\sigma,\alpha+\beta+\gamma+\delta}, \qquad (5.49)$$

where $a_\sigma = \left(1, 1/2, 1/\sqrt{6}\right)$ for $\sigma = (\pm 2, \pm 1, 0)$ are the normalization constants.

(ii) To pair a $S = 1/2$ spin with another $S = 1/2$ spin on one of the four neighboring sites to form a spin singlet. For example, an α spin at site i can pair with a γ spin at site $i + \hat{x}$. The resulting singlet wave function is

$$|\text{singlet}\rangle = \sum_{\alpha_i, \gamma_{i+\hat{x}}} \varepsilon_{\alpha_i, \gamma_{i+\hat{x}}} |\alpha_i\rangle |\gamma_{i+\hat{x}}\rangle. \qquad (5.50)$$

The $\varepsilon_{\alpha,\gamma}$ vanishes if $\alpha + \gamma \neq 0$. As two virtual spins are already frozen to a spin singlet, the total spin that can be formed by the other six $S = 1/2$ virtual spins on these two sites is always less than or equal to 3. Therefore, the states thus obtained are the ground states of P_4.

The VBS state is full coverage of these nearest-neighboring singlet pairs. In a periodic system, the corresponding wave function is just a translation-invariant PEPS defined by

$$|\Psi\rangle = \text{Tr} \prod_i A_{\alpha_i \beta_i \gamma_i \delta_i}[\sigma_i] |\sigma_i\rangle, \qquad (5.51)$$

where the local tensor $A_{\alpha\beta\gamma\delta}[\sigma]$ is determined by $C_{\alpha\beta\gamma\delta}[\sigma]$ and the antisymmetric matrix ε as shown in Fig. 5.2:

$$A_{\alpha\beta\gamma\delta}[\sigma] = \sum_{\alpha'\beta'} C_{\alpha'\beta'\gamma\delta}[\sigma] \varepsilon_{\alpha'\alpha} \varepsilon_{\beta'\beta}. \qquad (5.52)$$

$A_{\alpha\beta\gamma\delta}[\sigma]$ is finite only when $\sigma = \gamma + \delta - \alpha - \beta$.

5.6 Projected Entangled Simplex State (PESS)

PEPS emphasizes the entanglement between two nearest-neighbor sites of a lattice. In principle, it can be used to represent all quantum states satisfying the entanglement area law. In practical calculations, the bond dimension must be kept as small as possible in order to implement the calculation efficiently yet with sufficient accuracy. This is not a condition that can be readily fulfilled, especially in the following circumstances. First, it is difficult to use PEPS to represent a quantum state in which the local correlation or entanglement among all the basis states within a cluster (or simplex) containing more than two lattice sites, for example, the *simplex solid state* first introduced by Arovas [113], becomes important. Second, applying the PEPS algorithm on triangular or other lattices with high coordination numbers is technically difficult. A local tensor in PEPS on a triangular lattice contains seven indices, six for the virtual bond states and one for the physical basis states.

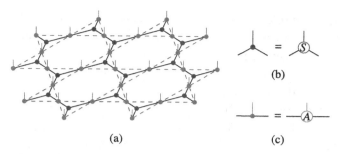

Figure 5.3 (a) A 3-PESS defined on a Kagome lattice (dashed lines). It contains two kinds of local tensors: (b) the simplex core tensor, S, defined at the center of each simplex (triangle), and (c) the projection tensor, $A[\sigma]$, defined at each node of the Kagome lattice. The local tensors form a decorated honeycomb lattice.

PESS extends pair correlations to simplex correlations. Here the word "simplex" refers to a cluster of lattice sites that constitute the basic unit, or "building block," of a two- or higher-dimensional lattice. For example, a triangle is a building block of a Kagome lattice and can be taken as a "simplex." One may also combine a number of simplices to form a larger simplex. Hence, the choice of a simplex is not unique. However, to represent a quantum state more efficiently, it should correctly reflect the system's symmetry.

Similar to PEPS, PESS is defined by introducing a number of virtual basis states at each node of the lattice. In addition to the local tensors, defined for projecting out the physical states from the virtual basis states at each node, the PESS contains a new type of local tensors called *entangled simplex tensors*. A simplex tensor describes the correlation, or entanglement, of virtual particles within a simplex. If a simplex contains n lattice sites, the corresponding PESS is referred to as an n-PESS. If we release the simplex definition and allow it to contain just two sites (i.e. $n = 2$), PESS is reduced to PEPS.

Figure 5.3 shows, as an example, a 3-PESS defined on a Kagome lattice. Like PEPS, PESS can be understood in terms of entangled simplex states of virtual systems that are locally projected onto the physical basis states. This class of states constitutes a natural generalization of PEPS. It yields a faithful representation of a quantum state obeying the entanglement area law and is an exact representation of the simplex solid state [113].

A simplex solid state is an extension of the AKLT state. It extends a bond singlet of the AKLT state to a simplex singlet. Similar to the AKLT states, the simplex solid states are extinguished by certain local projection operators. This feature allows one to construct a many-body Hamiltonian for which the simplex solid state is an exact ground state, usually with a gap to all low-energy excitations.

As an example, let us consider a $S = 2$ spin model defined on the Kagome lattice. The Kagome geometry is a two-dimensional network of corner-sharing triangles, each forming a three-site simplex. The centers of all simplices form a honeycomb lattice, on which the decorating sites form the Kagome lattice. A physical $S = 2$ state can be regarded as a symmetric superposition of two virtual $S = 1$ spins. On the Kagome lattice, two neighboring triangles (simplices) share a single site. We can assign each of the virtual $S = 1$ spins to one of the simplices associated with this site. There are then three $S = 1$ spins on each simplex triangle. The product of the three $S = 1$ virtual spins contains a unique spin singlet state. This allows us to define a virtual singlet on the simplex:

$$|0, 0\rangle = \frac{1}{\sqrt{6}} \sum_{\sigma_i \sigma_j \sigma_k} \varepsilon_{\sigma_i \sigma_j \sigma_k} |\sigma_i\rangle |\sigma_j\rangle |\sigma_k\rangle, \tag{5.53}$$

where $|\sigma_i\rangle$ ($\sigma_i = -1, 0, 1$) is a basis state of the $S = 1$ spin at site i and ε_{ijk} is the Levi–Civita antisymmetric tensor.

The product of all these simplex singlets forms a simplex solid state. Since the total physical spin at each simplex cannot exceed 3, this simplex solid state is the ground state of the Hamiltonian:

$$H = \sum_{\triangle} \left(J_4 P_{\triangle, 4} + J_5 P_{\triangle, 5} + J_6 P_{\triangle, 6} \right), \tag{5.54}$$

where subscript $\triangle$ represents a simplex triangle, J_4, J_5, and J_6 are nonnegative coupling constants, and $P_{\triangle, S}$ is the operator projecting a state at each simplex triangle onto a state with total spin S. The wave function of this state is a PESS, which can be expressed as

$$|\Psi\rangle = \text{Tr} \left(\cdots S_{abc} A_{aa'} \left[\sigma_i \right] A_{bb'} \left[\sigma_j \right] A_{cc'} \left[\sigma_k \right] \cdots \right) |\cdots \sigma_i \sigma_j \sigma_k \cdots\rangle, \tag{5.55}$$

where the trace is to sum over all spin configurations and all bond indices. S_{abc} is the entangled simplex tensor defined on the simplex honeycomb lattice. The physical basis states $\{\sigma_i, \sigma_j, \ldots\}$ are defined on the decorating sites of the honeycomb lattice $\{i, j, \ldots\}$ (i.e. on the Kagome lattice sites). The Roman letters $\{a, b, \ldots\}$ denote the virtual bond states. Because the virtual spins in each simplex triangle form a spin singlet, S_{abc} in this case is simply an antisymmetric Levi–Civita tensor,

$$S_{abc} = \quad\underset{\substack{\\ b \qquad c}}{\overset{\overset{a}{\mid}}{\triangle\!\!\!\!S}} \quad = \varepsilon_{abc}. \tag{5.56}$$

$A_{ab} [\sigma]$ is a 3×3 matrix that maps two virtual $S = 1$ spins onto an $S = 2$ physical spin. Its components are given by the Clebsch–Gordan coefficients of the SU(2) Lie algebra,

$$A_{ab}[\sigma] = \quad a - \!\!\!\!\!\!\! \underset{a \, -\!\!(\!A\!)\!-\, b}{\overset{\overset{\displaystyle\sigma}{|}}{}} \quad = \langle 2, \sigma | 1, a; 1, b \rangle. \tag{5.57}$$

PEPS and PESS are two types of trial wave functions. They arise naturally in constructing trial wave functions for quantum systems and form a comprehensive representation of tensor network states that satisfy the area law of entanglement entropy [37] on two- or higher-dimensional lattices. For a wide variety of systems, PESS provides an efficient and faithful representation of exponentially many coefficients by a small number of parameters describing the low-energy physics of many-body quantum states arising from local interactions.

6

Criterion of Truncation: Symmetric Systems

6.1 Density Matrix

The primary goal of real-space NRG is to find a limited set of basis states through a series of "local" basis transformations to accurately represent a target state, for example, the ground state of a Hamiltonian. In the calculation, we start from a small lattice system whose Hamiltonian can be fully diagonalized and then select a small set of basis states from this system according to specific criteria so that they can be used to accurately represent the ground-state wave function for a larger lattice system. Naturally, different RG schemes use different criteria.

In the Wilson NRG scheme [9], the basis states are retained or discarded according to their contribution to the thermal density matrix:

$$\rho = \frac{e^{-\beta H}}{\mathrm{Tr} e^{-\beta H}}. \tag{6.1}$$

It is the lowest energy eigenstates of the Hamiltonian H that are retained at each iteration. At zero temperature, the density matrix reduces to

$$\rho = |\Psi\rangle \langle \Psi|, \tag{6.2}$$

where Ψ is the ground-state wave function of the Hamiltonian.

This criterion of truncation works for the Kondo or Anderson impurity model, where the energy scales have an explicit hierarchy structure. However, it does not work for other quantum lattice models, for example, the Heisenberg model, because the lowest energy eigenstates in a smaller system may not be the most important states for representing the ground state in a larger system.

DMRG takes a different strategy to determine which states should be retained in the basis truncation. Instead of evaluating the thermal density matrix for the whole system, DMRG evaluates the *reduced density matrix* for a subsystem in a ground state or other particular target state. The reduced density matrix plays a central role in the DMRG calculation. Its eigenvalues measure the probabilities of the corresponding eigenstates in the target state. This avoids the blindness in the basis truncation encountered in the Wilson NRG. Next, we take the ground state as an example to discuss the criterion for optimally truncating basis states. The discussion, however, holds generally. It can be readily generalized to any other systems where the target states are obtained by diagonalizing Hamiltonians or other symmetric (or Hermitian) operators, such as a symmetric transfer matrix.

A DMRG iteration starts from a small system with, for example, N sites, then gradually increases its size to $N + 2$, $N + 4, \ldots$ until the desired length scale is reached. At each iteration, the collection of all the lattice sites is called a *superblock*. A superblock is divided into two parts: the *system* and the *environment*. The Hamiltonian of the superblock is diagonalized by utilizing a sparse matrix algorithm. An important step in DMRG is determining the basis states of the system with the largest weights in the ground state. This is achieved by taking the environment as a statistical bath, and the reduced density matrix of the system as a probe to obtain the desired information on the basis states. This is similar to performing an experimental measurement, where a physical quantity is determined by measuring the response function of the system to an external perturbation. Here the environment plays the role of external perturbation, and the reduced density matrix is the *response function* measured.

6.2 Reduced Density Matrix

In the study of a Hamiltonian or a symmetric transfer matrix, the reduced density matrix is symmetric and semi-positive definite. For example, in the conventional thermodynamic framework, the reduced density matrix of the system is defined from the thermal density matrix, Eq. (6.1), by integrating out all environment states:

$$\rho^{\text{sys}} = \frac{1}{\text{Tr} e^{-\beta H}} \text{Tr}_{\text{env}} e^{-\beta H}. \tag{6.3}$$

This is a Hermitian matrix. At zero temperature, $\beta \to \infty$, it becomes

$$\rho^{\text{sys}} = \text{Tr}_{\text{env}} |\Psi\rangle \langle\Psi|. \tag{6.4}$$

Here $|\Psi\rangle$ is the ground state.

The product of two basis states, one in the system $|s\rangle$ and the other in the environment $|e\rangle$, forms a basis state of the superblock. Suppose $\Psi_{s,e}$ to be the ground state wave function in this basis representation,

$$|\Psi\rangle = \sum_{s,e} \Psi_{s,e} |s\rangle |e\rangle, \tag{6.5}$$

then the matrix elements of the reduced density matrix are

$$\rho_{ss'}^{\text{sys}} = \sum_e \Psi_{s,e} \Psi_{s',e}^*. \tag{6.6}$$

A graphical representation of this equation is

$$\rho^{\text{sys}} = \Psi\Psi^\dagger = \quad \overbrace{\underbrace{\Psi^*}{\Psi}}^{}\Big] . \tag{6.7}$$

Correspondingly, the reduced density matrix of the environment is

$$\rho^{\text{env}} = \text{Tr}_{\text{sys}} |\Psi\rangle\langle\Psi| = \Psi^T \Psi^* = \Big[\overbrace{\underbrace{\Psi}{\Psi^*}} . \tag{6.8}$$

It is easy to show that the reduced density matrix is semi-positive definite. Hence, its eigenvalues are always larger than or equal to 0. An optimal truncation scheme is to find a subset of basis states in the system so that they have the largest overlap with the target state. As ρ^{sys} is symmetric (or Hermitian), we can diagonalize it using a unitary matrix U:

$$\rho^{\text{sys}} = U\Lambda U^\dagger = \quad -\boxed{U}-\diamondsuit-\boxed{U^\dagger}- , \tag{6.9}$$

where Λ is the diagonal eigenvalue matrix. The eigenvalues are arranged in descending order, $\Lambda_1 \geq \Lambda_2 \geq \cdots \geq \Lambda_{D_s}$, with D_s the dimension of the system. The trace of ρ^{sys} equals unity if the ground-state wave function is normalized:

$$\text{Tr}\rho^{\text{sys}} = \text{Tr}\Lambda = \langle\Psi|\Psi\rangle = 1. \tag{6.10}$$

The eigenvalue of ρ^{sys}, $\Lambda_m \geq 0$, measures the probability (or weight) of the corresponding eigenvector (or basis state),

$$|m\rangle = \sum_s U_{s,m}|s\rangle, \tag{6.11}$$

in the ground state. In these new basis states, the reduced density matrix becomes diagonal:

$$\rho^{\text{sys}} = \sum_m \Lambda_m |m\rangle\langle m|. \tag{6.12}$$

To truncate the basis space, we need only to keep the first D basis states. Clearly, these are the most probable states that should be retained in the basis truncation. The *truncation error* ε is the sum of the weights discarded, normalized by the total weight

$$\varepsilon = \sum_{m=D+1} \Lambda_m. \tag{6.13}$$

The truncation error is not the error for any physical observable. But it provides a measure to quantify the accuracy of the ground-state wave function after truncation. The error of the ground-state energy correlates with the truncation error. By extrapolating to the limit $\varepsilon \to 0$, one should get a better estimation for the ground-state energy and other physical variables.

6.3 Schmidt Decomposition

One can also understand the previous discussion from the entanglement structure of the wave function. To take a *Schmidt decomposition* is to take an SVD for the ground-state wave function $\Psi_{s,e}$:

$$\Psi = U\lambda V^\dagger = \ -\boxed{U}\!-\!\langle\lambda\rangle\!-\!\boxed{V^\dagger}\!- \ , \tag{6.14}$$

where U is the unitary matrix that diagonalizes the reduced density matrix in Eq. (6.9), and λ is the square root of the eigenvalue matrix of the reduced density matrix Λ: that is, $\lambda = \sqrt{\Lambda}$. If there is no degeneracy in λ, U and V are uniquely determined.

Here U and V define two sets of basis states whose dimensions are less than or equal to the minimum of the system and the environment dimensions. The probability or weight of $|m\rangle$ to appear in the ground state equals λ_m^2. To truncate the basis space by keeping the largest D basis states for both the system and the environment defines the best low-rank approximation for Ψ.

The singular value matrix λ determines the entanglement spectra between the system and environment in the ground state. The entanglement entropy is given by

$$S = -\mathrm{Tr}\rho^{\mathrm{sys}} \ln \rho^{\mathrm{sys}} = -\sum_m \Lambda_m \ln \Lambda_m. \tag{6.15}$$

The maximum of the function $f(x) = -x\ln x$ is located at $x = e^{-1}$. Here $f(x)$ increases monotonically with x for $0 < x < e^{-1}$, and decreases with x for $e^{-1} < x < 1$. No more than two Λ_m can be larger than e^{-1} since $\mathrm{Tr}\Lambda = 1$. Thus, if the contribution to the entropy from the largest eigenvalue of ρ^{sys} is larger than that from the largest discarded eigenvalue, the DMRG is just a maximum entanglement entropy method [34].

6.4 Variational Approach

Intuitively, one can take a variational approach to understand why the DMRG criterion is an optimal scheme for the basis truncation. To reduce the truncation error, one needs to find a small set of basis states:

$$|\alpha\rangle = \sum_s U_{s,\alpha}|s\rangle, \quad (\alpha = 1, \ldots, D \leq N_s), \tag{6.16}$$

to approximately represent the ground-state wave function $|\Psi\rangle$ by

$$|\Psi'\rangle = \sum_{\alpha=1}^{D} \sum_e \Phi_{\alpha,e}|\alpha\rangle|e\rangle, \tag{6.17}$$

so that the difference between Ψ and Ψ'

$$\big\||\Psi\rangle - |\Psi'\rangle\big\|^2 = 1 - \text{Tr}\left(\Psi^\dagger U\Phi + h.c\right) + \text{Tr}\Phi^\dagger\Phi, \tag{6.18}$$

is minimized. $|\Psi\rangle$ is assumed to be normalized. Taking a derivative for the above difference, we find that

$$\Phi = U^\dagger\Psi. \tag{6.19}$$

Substituting it into Eq. (6.18) gives

$$\big\||\Psi\rangle - |\Psi'\rangle\big\|^2 = 1 - \text{Tr}U^\dagger\rho^{sys}U. \tag{6.20}$$

According to the Rayleigh–Ritz principle, this equation is stationary if and only if the D column vectors of U are the first D largest eigenvectors of ρ^{sys}. In this case,

$$\big\||\Psi\rangle - |\Psi'\rangle\big\|^2 = \sum_{m>D} \Lambda_m \tag{6.21}$$

is nothing but the truncation error.

6.5 Edge and Bond Density Matrices

The ground-state wave function Ψ can be decomposed as a product of two matrices by taking, for example, a QR decomposition:

$$\Psi_{s,e} = \sum_\alpha S_{s,\alpha}E_{\alpha,e} = \overset{\displaystyle s}{\underset{\displaystyle (S)}{\big\downarrow}} \!\!-\!\!\alpha\!\!-\!\! \overset{\displaystyle e}{\underset{\displaystyle (E)}{\big\downarrow}} . \tag{6.22}$$

$S_{s,\alpha}$ and $E_{e,\alpha}$ can be regarded as the wave functions in these two blocks, linked by a virtual bond index α.

Using this representation, the reduced density matrix can be written as

$$\rho^{sys} = SEE^\dagger S^\dagger = SG_eS^\dagger = \begin{array}{c} \fbox{S^*}\!\!-\!\!\fbox{E^*} \\ | \qquad | \\ \fbox{S}\!\!-\!\!\fbox{E} \end{array} , \tag{6.23}$$

where

$$G_e = G_e^\dagger = EE^\dagger = \qquad (6.24)$$

is a symmetric and semi-positive matrix defined in the virtual basis space. The corresponding reduced density matrix of the environment block is

$$\rho^{\mathrm{env}} = E^T S^T S^* E^* = E^T G_s E^* = \qquad , \qquad (6.25)$$

and

$$G_s = G_s^\dagger = S^T S^* = \qquad . \qquad (6.26)$$

G_s is also a symmetric and semi-positive matrix defined in the virtual basis space.

From SVD, it is simple to show that S^* can be expressed as a product of the square root of G_s and an isometric matrix U:

$$S = U\sqrt{G_s^*}, \qquad (6.27)$$

where U satisfies the condition

$$U^\dagger U = G_s^{-1/2} S^\dagger S G_s^{-1/2} = I. \qquad (6.28)$$

Using this equation, we can write ρ^{sys} as [321, 366]:

$$\rho^{\mathrm{sys}} = SG_e S^\dagger = U\sqrt{G_s^*}G_e\sqrt{G_s^*}U^\dagger. \qquad (6.29)$$

Thus, ρ^{sys} has the same nonzero eigenspectrum as the operator:

$$\rho^E = \sqrt{G_s^*}G_e\sqrt{G_s^*}. \qquad (6.30)$$

Here ρ^E defines an edge matrix in the virtual basis space. It measures the entanglement spectrum between the system and environment of $|\Psi\rangle$ and can be regarded as the density matrix of the system block on the boundary (or interface) [321, 366].

As ρ^{sys} and ρ^E have the identical spectra, it is straightforward to show that

$$\mathrm{Tr}\left(\rho^{\mathrm{sys}}\right)^n = \mathrm{Tr}\left(\rho^E\right)^n = \mathrm{Tr}\left(G_e G_s^*\right)^n. \qquad (6.31)$$

In these expressions, the first "Tr" is to take trace over physical states, and the second and third "Tr"s are to take trace over virtual basis states.

For a translation-invariant MPS on an infinite lattice,

$$|\Psi\rangle = \sum_{\sigma_1 \cdots \sigma_N} \text{Tr} \left(\cdots A[\sigma_i] A[\sigma_{i+1}] \cdots \right) | \cdots \sigma_i \sigma_{i+1} \cdots \rangle, \tag{6.32}$$

G_s and G_e are simply given by

$$G_s \propto \lim_{N_s \to \infty} \langle 0| \mathcal{E}^{N_s} \propto X, \tag{6.33}$$

$$G_e \propto \lim_{N_e \to \infty} \mathcal{E}^{N_e} |0\rangle \propto Y, \tag{6.34}$$

where $|0\rangle$ is an arbitrary state that is not orthogonal to the leading eigenvector of $\mathcal{E}$. $\mathcal{E}$ is a transfer matrix defined by

$$\mathcal{E}_{\alpha\alpha',\beta\beta'} = \sum_{\sigma} A_{\alpha\beta}[\sigma] A^*_{\alpha'\beta'}[\sigma] = \tag{6.35}$$

X and Y are the dominant left and right eigenvectors of $\mathcal{E}$, respectively.

For the wave function defined by Eq. (6.22), if we cut the bond linking system and environment, this defines a wave function in which the system and environment blocks are completely disconnected:

$$\left| \Psi_{\alpha\beta} \right\rangle = \sum_{s,e} S_{s,\alpha} E_{\beta,e} | s, e \rangle . \tag{6.36}$$

From this, we can define a bond density matrix in the virtual space:

$$\rho^B_{\alpha\beta} \equiv \text{Tr} \left| \Psi_{\alpha\beta} \right\rangle \langle \Psi | = \left(G_s G^*_e \right)_{\alpha\beta} = \tag{6.37}$$

As G_s is a Hermitian matrix, this bond density matrix can be also written as

$$\rho^B = G_s G^*_e = G^{1/2}_s \left(\rho^E \right)^* G^{-1/2}_s. \tag{6.38}$$

Hence, ρ^B has the same eigen-spectrum as ρ^E.

7

Real-Space DMRG

7.1 Two Kinds of Algorithms

DMRG is an iterative method. It is implemented with two kinds of algorithms: infinite lattice algorithms and finite lattice algorithms.

The infinite lattice algorithm allows the size of the superblock to grow and can be used to study the thermodynamic limit directly. This approach is simple to implement. At each iteration, both system and environment are augmented. However, this may not be enough to yield accurate results up to the desired precision since the truncation error is accumulated with the increase of the superblock size. DMRG is canonical and generally works at a fixed particle number or spin value for a given system. In the application of this algorithm, be cautious in the following cases:

(i) a system with randomness or impurities since the Hamiltonian is not translation invariant;

(ii) an interacting electron system, for example, the Hubbard model, at a given electron or hole concentration not at half-filling, since it is not always possible to maintain the particle density constant during the growth of a superblock;

(iii) a system close to a first-order transition where two phases coexist and the RG flow generated with this algorithm might be trapped at a local minimum or metastable state favored only for small system sizes.

The finite lattice algorithm manages to eliminate these problems. It is more accurate for a system with a fixed lattice size. The idea is to stop the iteration of the infinite lattice algorithm at some preselected superblock length. At the subsequent iterations, one block is augmented while the other is shrunk. The reduced basis transformations are carried out only for the growing block. Thus, system and environment are zipped forward and backward. In the forward iteration, the system block grows at the expense of the environment. The backward iteration begins when the environment reaches some minimum size and becomes exact. The environment block now grows at the expense of the system. A full round of the forward and backward iterations; namely, a complete growth and shrinkage sequence for both blocks, is called a *sweep*. The basis states selected for both blocks can be improved by sweeping, which can significantly improve the accuracy of the final results.

We will take a one-dimensional quantum lattice model with open boundary conditions as an example to show how these two algorithms work.

7.1.1 Infinite Lattice Algorithm

The infinite lattice algorithm starts with a superblock that consists of two blocks, called left and right blocks, and two sites between these two blocks:

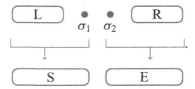

The four building blocks of this superblock are labeled, from left to right, as L, σ_1, σ_2, and R. Initially, L and R each contain just one site. The algorithm is to augment the left and right blocks by adding a lattice site to each, one at a time. The augmented left and right blocks are called system and environment, abbreviated as S and E, respectively. The *system* contains the left block plus the left added site; that is, $S = (L, \sigma_1)$, and the *environment* contains the right block plus the right added site; that is, $E = (\sigma_2, R)$. After the basis transformation, blocks S and E are set as the left and right blocks for the next iteration, respectively. The basis dimension for each block is increased by d, where d is the dimension of the physical basis states at each lattice site. It grows exponentially with the size of the block. If the dimension exceeds the maximum number of states D to retain, a truncation to the number of basis states is imposed.

For the model with only nearest-neighbor interactions, the Hamiltonian for the superblock can be expressed as

$$H = H_l + H_{l,\sigma_1} + H_{\sigma_1,\sigma_2} + H_{\sigma_2,r} + H_r, \qquad (7.1)$$

where H_l and H_r are the Hamiltonians for the left and right blocks, respectively, H_{σ_1,σ_2} is the Hamiltonian for the two middle sites, H_{l,σ_1} is the interaction between the left block and the left added site and $H_{\sigma_2,r}$ is the interaction between the right block and the right added site. The total Hilbert space of this superblock is the direct product of the individual spaces corresponding to each block and the added sites.

The DMRG calculation takes the following steps:

(i) Diagonalize the Hamiltonian to obtain the ground-state wave function $|\Psi\rangle$ of the superblock using a sparse matrix solver, for example, the Lanczos or the conjugate gradient method introduced in §A.2 or §A.3:

$$|\Psi\rangle = \sum_{l\sigma_1\sigma_2 r} \Psi_{l\sigma_1,\sigma_2 r} |l\sigma_1\sigma_2 r\rangle = \sum_{l\sigma_1\sigma_2 r} \begin{array}{c} l\ \sigma_1\ \sigma_2\ r \\ \boxed{\Psi} \end{array} |l\sigma_1\sigma_2 r\rangle, \qquad (7.2)$$

where $|l,\sigma_1\rangle$ and $|\sigma_2, r\rangle$ are the basis states of the system and environment blocks, respectively. $|\Psi\rangle$ is normalized to 1.

(ii) Factorize the wave function by the Schmidt decomposition:

$$\Psi_{l\sigma_1,\sigma_2 r} = \sum_k U_{l\sigma_1,k}\lambda_k V^*_{\sigma_2 r,k} \qquad (7.3)$$

$$\begin{array}{c} l\ \sigma_1\ \sigma_2\ r \\ \boxed{\Psi} \end{array} = \begin{array}{c} l\ \sigma_1 \qquad \sigma_2\ r \\ \boxed{U}\!-\!\langle\lambda\rangle\!-\!\boxed{V^\dagger} \end{array}, \qquad (7.4)$$

where the singular values $\lambda_k \geq 0$ are descending ordered. λ_k^2 measures the probability of the corresponding left and right singular vectors, u_k and v_k, in the ground-state wave function. Here u_k and v_k are the column vectors of U and V, respectively:

$$(u_k)_{l\sigma} = U_{l\sigma,k}, \qquad (v_k)_{\sigma r} = V_{\sigma r,k}. \qquad (7.5)$$

(iii) Truncate the basis space by keeping the first D singular values and singular vectors. The truncation error is

$$\varepsilon = \sum_{k>D} \lambda_k^2. \qquad (7.6)$$

(iv) Update the Hamiltonian and other operators involved in the interactions within or between blocks in the new basis space. Since the system (l,σ_1) is the left block for the next iteration, the updated Hamiltonian for the new left block is

$$H_{l,\text{new}} = U^\dagger \left(H_l + H_{l,\sigma_1} \right) U, \qquad (7.7)$$

or graphically

$$\boxed{H_{l,\text{new}}} = \begin{array}{c} \boxed{U^\dagger} \\ \boxed{H_l} \\ \boxed{U} \end{array} + \begin{array}{c} \boxed{U^\dagger} \\ \boxed{H_{l,\sigma_1}} \\ \boxed{U} \end{array}. \qquad (7.8)$$

U is the truncated left singular matrix. The Hamiltonian of the augmented right block, $H_{R,\text{new}}$, is renormalized by the truncated right singular matrix V. Other operators are similarly updated.

(v) Build a new superblock by adding a site to the system and another to the environment, and repeat steps (i)–(iv) until the lattice size becomes sufficiently large or the control error is below a predefined tolerance.

The static properties can be easily calculated by keeping the matrix elements of the relevant operators at each step and performing the corresponding basis reduction and transformation. For an N-site lattice with open boundary conditions, there are $N-1$ bonds. Thus, the ground-state energy per bond can be evaluated from the eigenenergy of the superblock $E_0(N)$ using the formula

$$e_0(N) = \frac{E_0(N)}{N-1}. \tag{7.9}$$

However, as the translation symmetry is broken, the energy density is more accurately evaluated at the center of the lattice. It is better to use the formula

$$e_0(N) = \frac{E_0(N) - E_0(N-2)}{2} \tag{7.10}$$

to determine the ground-state energy.

DMRG uses a finite set of basis states to approximately represent the full Hilbert space and does not involve any unphysical states. Thus, the ground-state energy evaluated with this method satisfies the *variational principle* and is an upper bound to the true ground-state energy $E_{0,\text{true}}(N)$:

$$E_0(N) \geq E_{0,\text{true}}(N). \tag{7.11}$$

From the ground-state wave function, we can evaluate the static n-point correlation functions $\langle O_{i_1} \cdots O_{i_n} \rangle$. The most relevant cases include $n=1$ for the local density or magnetization and $n=2$ for the density–density, spin–spin, and other two-point correlators. However, it should be pointed out that due to the basis truncation, the correlation between any two operators in the wave function generated by DMRG always decays exponentially with their separation in the long-distance limit. In order to evaluate the correlation functions in a power-law decay system, a careful finite-size scaling analysis should be done by extrapolation with respect to either the truncation error or the number of states retained.

To evaluate these correlation functions, let us rewrite the DMRG wave function, Eq. (7.2), as

$$|\Psi\rangle = \sum_{s,e} \Psi_{s,e}|s,e\rangle = \sum_{s,e} \overset{s}{\underset{\Psi}{\bigcirc}}\overset{e}{} \;|s,e\rangle. \tag{7.12}$$

For a single operator O_i defined in the system block, its expectation value is determined by

$$\langle \Psi | O_i | \Psi \rangle = \sum_{ss'e} \Psi^*_{s',e} \langle s' | O_i | s \rangle \Psi_{s,e}$$

$$= \quad\quad\quad\quad = \quad\quad\quad \rho^{\text{sys}} \; . \tag{7.13}$$

For a local observable in the environment, a similar formula exists.

For two operators, O_i and O_j, defined in the same block, say the system block, the formula that determines their correlation function is

$$\langle \Psi | O_i O_j | \Psi \rangle = \sum_{ss'e} \Psi^*_{s',e} \Psi_{s,e} \langle s' | O_i O_j | s \rangle = \sum_{ss'} \rho^{\text{sys}}_{ss'} \langle s' | O_i O_j | s \rangle. \tag{7.14}$$

On the other hand, if i and j belong to different blocks, their correlator is

$$\langle \Psi | O_i O_j | \Psi \rangle = \sum_{ss'ee'} \Psi^*_{s',e'} \Psi_{s,e} \langle s' | O_i | s \rangle \langle e' | O_j | e \rangle$$

$$= \quad\quad\quad\quad . \tag{7.15}$$

To measure these correlation functions, the matrix elements of each operator, together with the matrix elements of the product of $O_i O_j$ in the same block, have to be evaluated and stored. To store all the matrix elements of operator pairs $O_i O_j$ can be computationally expensive. It is more convenient if one only stores the matrix elements of each operator and calculates the correlations only when the operators lie in different blocks. Calculating the correlation of two operators inside the same block, for example, the correlator defined by Eq. (7.14), as the product of two individual operators in the truncated basis, that is,

$$\langle s' | O_i O_j | s \rangle \approx \sum_{s''} \langle s' | O_i | s'' \rangle \langle s'' | O_j | s \rangle \tag{7.16}$$

is a poor approximation and should be avoided since $|s''\rangle$ is an incomplete basis set.

7.1.2 Finite Lattice Algorithm

The finite lattice algorithm carries out the DMRG calculation by fixing the superblock size and augments just one block at each DMRG iteration.

Before applying the finite lattice algorithm, one should first run the infinite lattice algorithm to build up all left and right blocks. This process is referred to as a warmup. During this process, one needs to store the basis transformation matrices and all the matrix elements of operators for the left and right blocks. After that, a finite-lattice sweeping is carried out by zipping the system and environment, from left-to-right and then right-to-left, for a couple of times until the results become convergent, typically at a point when both left and right blocks have the same length. This sweeping process improves the solution and provides an optimized way to minimize the ground-state energy.

A step of the left-to-right sweeping can be graphically represented as

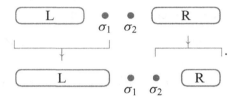

At each DMRG step, the left block is augmented by adding the left site, σ_1. However, the right block is shrunk one site at a time. The matrix elements of operators in the shrinking block are retrieved from the data stored in memory, obtained in the warmup or previous iterations.

A finite-lattice sweep takes about two to four times the CPU time of the initial infinite-system calculation, depending on whether the system has the reflection symmetry. To improve the performance, one can use the ground-state wave function obtained in the previous step as the initial wave function used in the Lanczos or other sparse matrix solver. In literature, this is referred to as the wave function prediction [228]. It can save the calculation time in the finite-lattice sweeping, often by more than an order of magnitude. The computational performance becomes much more pronounced if the predicted initial wave function is close to the final converged ground state.

Let us assume the ground state obtained in the previous iteration is given by

$$|\Psi^{\text{old}}\rangle = \sum_{l_a\sigma_1\sigma_2 r_b} \Psi^{\text{old}}_{l_a\sigma_1\sigma_2 r_b} |l_a\sigma_1\sigma_2 r_b\rangle, \qquad (7.17)$$

where a and b denote the lattice size of the left and right blocks, respectively. The full lattice size is equal to $a + b + 2 = N$, which remains fixed in the sweep. $|\Psi^{\text{old}}\rangle$ is a good approximation to the ground state in the current iteration. To construct the initial wave function, all we need to do is to project $|\Psi^{\text{old}}\rangle$ onto the new basis space:

$$\Psi_{l_{a+1}\sigma_2\sigma_3 r_{b-1}} = \langle l_{a+1}\sigma_2\sigma_3 r_{b-1}|\Psi^{\text{old}}\rangle, \qquad (7.18)$$

where $|l_{a+1}\rangle$ is related to the basis states $|l_a\sigma_1\rangle$ by

$$|l_{a+1}\rangle = \sum_{l_a,\sigma_1} U_{l_a\sigma_1,l_{a+1}} |l_a\sigma_1\rangle, \qquad (7.19)$$

and $U_{l_a\sigma_1,l_{a+1}}$ is the basis transformation matrix. Similarly, $|\sigma_3, r_{b-1}\rangle$ is related to $|r_b\rangle$ by

$$|r_b\rangle = \sum_{r_{b-1},\sigma_3} V_{r_{b-1}\sigma_3,r_b} |\sigma_3 r_{b-1}\rangle, \qquad (7.20)$$

and $V_{r_{b-1}\sigma_3,r_b}$ is the environment basis transformation matrix. Substituting the above two equations into Eq. (7.18) yields

$$\Psi_{l_{a+1}\sigma_2\sigma_3 r_{b-1}} = \sum_{l_a\sigma_1 r_b} U^*_{l_a\sigma_1,l_{a+1}} \Psi^{\text{old}}_{l_a\sigma_1\sigma_2 r_b} V^*_{r_{b-1}\sigma_3,r_b}. \qquad (7.21)$$

However, it should be pointed out that the wave function thus generated is not equal to $|\Psi^{\text{old}}\rangle$, since the spaces spanned by the basis $\{|l_{a+1}\rangle\}$ and $\{|r_b\rangle\}$ are just part of the spaces spanned by $\{|l_a,\sigma_1\rangle\}$ and $\{|\sigma_3, r_{b-1}\rangle\}$, respectively.

The total error accumulates with the lattice size in the infinite lattice algorithm and thus the precision is not just controlled by the truncation error. The finite-lattice sweep can reduce the accumulated error and significantly improve the accuracy of DMRG. Furthermore, as there is no other error accumulation in the finite-lattice sweep, the total error is strictly controlled by the truncation error. This is an advantage in the error analysis. In cases where computer power allows, one can generate well-converged results by fixing the truncation error within tolerance and allowing the number of states to be automatically adjusted. In this case, one can get an accurate estimation for any measurable quantity by extrapolating the result to the zero truncation error limit. Empirically, this is better than performing the extrapolation with respect to the number of states retained [34].

The accuracy of the DMRG calculation depends on the behavior of the eigenspectra of the reduced density matrix. Sweeping is essential to improve accuracy, especially in the study of electronic problems at an incommensurate filling or with a random potential. Typically, errors in the correlation functions are higher than in the energy. In some cases, it may happen that the DMRG wave function is trapped in a local minimum. This problem usually arises in the momentum space or quantum chemistry DMRG calculations.

7.2 DMRG in the MPO Language

As discussed in §4.1, a Hamiltonian can always be written as an MPO. We now reformulate the DMRG algorithm using the MPO language. Below we demonstrate how the method works by taking the Heisenberg model (4.1) as an example. The MPO representation of this model is given in Eq. (4.5). Corresponding to the four subblocks whose basis states are labeled by $(l, \sigma_i, \sigma_j, r)$, the Hamiltonian of the superblock is a product of four operator tensors,

$$H = \boxed{H_l} - \boxed{h_i} - \boxed{h_j} - \boxed{H_r} \quad , \tag{7.22}$$

where h_i is the local tensor, defined by Eq. (4.4). H_l and H_r are the effective Hamiltonian operator tensors of the left and right blocks, respectively. Initially, $i = 2$, $j = N - 1$, and the left block contains just the left-end site, and

$$H_l = h_1 = \begin{pmatrix} -hS_{1,z} & \mathbf{S}_1 & 1 \end{pmatrix}. \tag{7.23}$$

Similarly, the initial right block contains the right-end site, and

$$H_r = h_N h_{N+1}^\dagger = \begin{pmatrix} 1 & J\mathbf{S}_N & -hS_{N,z} \end{pmatrix}^T. \tag{7.24}$$

To carry out the DMRG calculation, we first diagonalize the Hamiltonian by solving the ground-state eigenequation:

$$\boxed{H_l} - \boxed{h_i} - \boxed{h_j} - \boxed{H_r} \underset{\Psi}{\bigcup} = E \underset{\Psi}{\bigcup} \ , \tag{7.25}$$

where E is the ground-state energy and Ψ is the corresponding wave function. To find the basis transformation matrices, U and V, we follow Eq. (7.4) to decouple the ground-state wave function Ψ by taking a Schmidt decomposition. The left block is then augmented to include site i. The corresponding Hamiltonian matrix is updated by the formula

$$\boxed{H_{l,\text{new}}} = \begin{array}{c} \boxed{U^\dagger} \\ \boxed{H_l} - \boxed{h_i} \\ \boxed{U} \end{array} \quad . \tag{7.26}$$

Similarly, the right block is augmented to include site j, and the corresponding Hamiltonian tensor, $H_{r,\text{new}}$, is updated by utilizing the transformation matrix V.

For the next iteration, we set $H_l = H_{l,\text{new}}$, $H_r = H_{r,\text{new}}$, and add two more sites into the superblock. Repeating this calculation until all the lattice sites are included in the superblock, we obtain the ground-state energy and the wave function for the whole system. One can then carry out the finite-lattice sweep to refine the ground-state energy and the corresponding wave function. At each step of the sweep, only one of the Hamiltonian matrices, H_l or H_r, is updated.

7.3 Error Analysis

To estimate the error in the energy before and after the basis truncation, let us take a Schmidt decomposition for the wave function before the truncation:

$$|\Psi_0\rangle = \sum_{\alpha=1} \lambda_\alpha |\alpha_s\rangle |\alpha_e\rangle, \tag{7.27}$$

where $|\Psi_0\rangle$ is the ground state of the Hamiltonian with the eigenenergy E_0

$$H|\Psi_0\rangle = E_0|\Psi_0\rangle. \tag{7.28}$$

λ_α is the singular value, and $|\alpha_s\rangle$ and $|\alpha_e\rangle$ are the corresponding singular vectors in the system and environment blocks, respectively. If $|\Psi_0\rangle$ is normalized, then

$$\sum_{\alpha=1} \lambda_\alpha^2 = 1. \tag{7.29}$$

After the truncation, the normalized wave function becomes

$$|\Psi\rangle = \frac{1}{\sqrt{1-\varepsilon}} \sum_{\alpha=1}^{D} \lambda_\alpha |\alpha_s\rangle |\alpha_e\rangle, \tag{7.30}$$

where D is the number of states retained and ε is the truncation error. Hence, the ground-state wave function before truncation can be expressed as

$$|\Psi_0\rangle = \sqrt{1-\varepsilon}|\Psi\rangle + \sqrt{\varepsilon}|\Phi\rangle, \tag{7.31}$$

where

$$|\Phi\rangle = \frac{1}{\sqrt{\varepsilon}} \sum_{\alpha>D} \lambda_\alpha |\alpha_s\rangle |\alpha_e\rangle. \tag{7.32}$$

The normalized wave function after the truncation can be also written as

$$|\Psi\rangle = \frac{1}{\sqrt{1-\varepsilon}}|\Psi_0\rangle - \frac{\sqrt{\varepsilon}}{\sqrt{1-\varepsilon}}|\Phi\rangle. \tag{7.33}$$

It should be noted that $|\Phi\rangle$ and $|\Psi_0\rangle$ are not orthogonal to each other:

$$\langle\Phi|\Psi_0\rangle = \sqrt{\varepsilon}. \tag{7.34}$$

From these expressions, it is simple to find the error of the wave function

$$||\Psi\rangle - |\Psi_0\rangle| = \sqrt{\langle\Psi - \Psi_0|\Psi - \Psi_0\rangle} \approx \sqrt{\varepsilon}, \tag{7.35}$$

and the error of the energy

$$\delta E = \langle\Psi|H|\Psi\rangle - E_0 = \frac{\varepsilon}{1-\varepsilon} \left(\langle\Phi|H|\Phi\rangle - E_0\right). \tag{7.36}$$

Now let us expand $|\Phi\rangle$ using the eigenstates of the Hamiltonian $|\Psi_l\rangle$:

$$|\Phi\rangle = \sqrt{\varepsilon}|\Psi_0\rangle + \sum_{l>0} a_l|\Psi_l\rangle, \qquad (7.37)$$

where $|\Psi_l\rangle$ is the lth eigenstate of H with the corresponding energy E_l. As $|\Phi\rangle$ is normalized, we have

$$\sum_{l>0} |a_l|^2 = 1 - \varepsilon. \qquad (7.38)$$

This gives

$$\delta E > \varepsilon (E_1 - E_0). \qquad (7.39)$$

Hence, the error in the energy is of the order of ε and lower bounded by its product with the energy gap, while the error in the wave function is of the order of $\sqrt{\varepsilon}$.

7.4 Heisenberg Spin Chains

We now take the one-dimensional Heisenberg model, defined by the Hamiltonian (4.1), to demonstrate the efficiency and accuracy of DMRG. It is also the first quantum system investigated with DMRG [18, 19].

In 1983, Haldane showed that in the limit of a large spin, the low-energy physics of the antiferromagnetic Heisenberg spin chain is described by the (1+1)-dimensional nonlinear σ model with a topological term proportional to $2\pi S$ with S the spin at each site [28, 29]. Based on the analysis of this effective field theory, he predicted that integer spin chains are gapped while half-integer spin chains are gapless. This prediction, now called Haldane conjecture, implies the failure of the Kubo–Anderson spin-wave theory in the description of antiferromagnetic spin chains. It also implies that proceeding from the quantum to the classical limit is highly nontrivial. As the integer and half-integer Heisenberg chains have very different low-energy physics, we discuss their physical properties separately.

7.4.1 Integer Spin Chains

An integer Heisenberg spin chain has a finite excitation gap with a finite correlation length ξ in the ground state. In this case, the entanglement area law holds, which implies that the entanglement entropy and other physical quantities, such as the ground-state energy, should converge exponentially with the system size N or the number of states retained D in the DMRG calculation, provided that N is significantly larger than ξ.

Figure 7.1 shows how the ground-state energy $e_0(N)$ varies with the lattice size for the spin-1 Heisenberg model obtained from DMRG by keeping $D = 100$

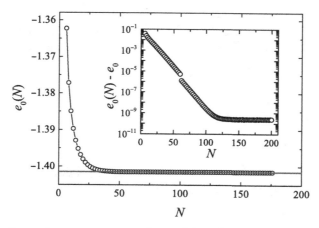

Figure 7.1 Ground-state energy per site, $e_0(N) = [E_0(N) - E_0(N-2)]/2$, as a function of the lattice length N for the spin-1 Heisenberg model with open boundary conditions. The ground-state energy e_0 obtained by White and Huse [19] with DMRG in the limit $D \to \infty$ and $N \to \infty$ (horizontal line) is also shown for comparison. The inset shows the size dependence of the energy difference between $e_0(N)$ and e_0. $D = 100$ basis states are retained in the DMRG calculation.

basis states. Open boundary conditions are assumed. As expected, $e_0(N)$ converges quickly with the system size. The converged value of $e_0(N)$ in the large-N limit is $e_0(D = 100) \approx -1.40148403872(1)$, consistent with the result reported in Ref. [19]. The true ground-state energy is determined by extrapolating $e_0(D)$ to the limit $D \to \infty$. This extrapolation was first done by White and Huse [19], who found the ground-state energy to be $e_0 \approx -1.401484038971(4)$. The difference between the converged ground-state energy and e_0 is reduced with increasing D.

The inset of Fig. 7.1 shows the size dependence of the difference between $e_0(N)$ and e_0, $e_0(N) - e_0$, for $D = 100$. This difference exhibits two distinguishing features in the small- and large-N regimes. It first drops exponentially with the system size N when $N \lesssim 100$, then tends to saturate with the further increase of N. This saturation results from the DMRG truncation of the basis states. It happens when the system size is significantly larger than the correlation length ξ, which is estimated to be $\xi \sim 6.03$ [19]. The difference $e_0(N) - e_0$ shows a pronounced drop at $N_t = 60$. This drop, as shown in Fig. 7.2, corresponds to the sudden drop in the entanglement entropy first found by Hu *et al.* [367]. This reveals an intriguing and universal feature of the Haldane gap state, or more generally, a quantum state that exhibits degenerate edge excitations in the thermodynamic limit.

Figure 7.2 shows the size dependence of the entanglement entropy $S(N)$ for the spin-1 Heisenberg model [367]. A similar drop was also observed in the quantum torus chain model [368]. At a given D, $S(N)$ increases quickly and tends to become saturated with increasing N. However, by further increasing N, $S(N)$

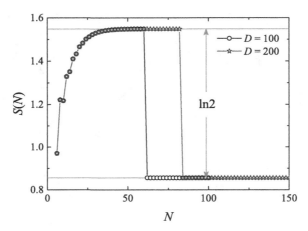

Figure 7.2 Lattice size dependence of the bipartite entanglement entropy $S(N)$ for the spin-1 Heisenberg model with open boundary conditions. $S(N)$ shows a sudden drop at $N_t = 62$ and 84 for $D = 100$ and 200, respectively. The size of the drop equals $\ln 2$.

suddenly drops and becomes almost N independent. This non-monotonic variance of the entanglement entropy N (except a weak $N = 4n$ to $4n + 2$ oscillation), which looks strange at first glance, reveals an important topological property of the integer spin chain. As discussed in §5.3, the Haldane gap in an integer spin chain could be understood from the picture of VBS, which is the exact ground states of the so-called AKLT Hamiltonian.

In a spin-1 system, each S=1 spin is fractionalized into two $S = 1/2$ spins, and each fractionalized spin forms a spin singlet with another fractionalized spin on one of the neighboring sites except at the two ends of the chain. In a short lattice system, the correlation between spins will mediate an effective interaction to couple the two $S = 1/2$ end spins to form a spin singlet, contributing a $\ln 2$ to the entanglement entropy. However, when the lattice size becomes larger than a threshold N_t ($N_t = 60$ for $D = 100$), which is significantly larger than the correlation length, the coupling between the two end spins is terminated by the basis truncation at a given D. In this case, the two end spins become independent and make no contribution to the entanglement entropy. As revealed by Fig. 7.3, N_t grows almost logarithmically with D, and diverges in the limit $D \to \infty$.

Thus, the drop in the entanglement entropy results from the basis truncation, which decouples the two fractionalized end spins, and the size of the drop equals $\ln 2$. This is just what we find from the DMRG calculation, which indicates that the entanglement entropy is a sensitive probe for detecting fractionalized excitations in DMRG, and the size of the drop measures the entanglement between the two end spins.

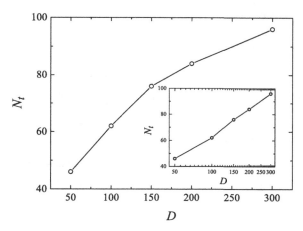

Figure 7.3 Bond dimension D dependence of the lattice size N_t, at which the bipartite entanglement entropy shows a sudden drop for the spin-1 Heisenberg model. The inset shows the semi-log plot of N_t versus D.

The excitation energy gap of the spin-1 Heisenberg model can be calculated from the difference between the lowest energy of the states with total spin 1 and the ground-state energy. This calculation was first done by White and Huse [19]. The gap value they obtained was 0.41050(2).

7.4.2 Half-Integer Spin Chains

The isotropic half-odd-integer Heisenberg spin chains are critical, and the spin-spin correlation length diverges in the ground state. Figure 7.4 shows the ground-state energy e_0 as a function of the lattice size N for the spin-1/2 Heisenberg model, obtained by keeping $D = 100$ states. Again, open boundary conditions are assumed. The spin-1/2 Heisenberg model is integrable by the Bethe ansatz. The exact ground-state energy obtained by solving the Bethe ansatz equation in the thermodynamic limit is [369]

$$e_{\mathrm{ex}} = -\ln 2 + \frac{1}{4}. \tag{7.40}$$

Compared to the exact result, we find that the ground-state energy obtained with DMRG converges quickly with the increase of the lattice size. As revealed by the inset of Fig. 7.4, the ground-state energy e_0 converges algebraically towards the exact result, and the error shows the following power-law behavior:

$$e_0 - e_{\mathrm{ex}} \sim N^{-\eta}, \tag{7.41}$$

where the exponent η is found to be approximately 1.96 for $D = 100$, close to the exact value 2, predicted by conformal field theory [370]:

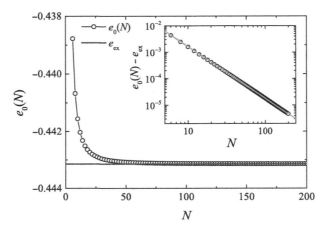

Figure 7.4 Ground-state energy per site, $e_0(N)$, as a function of the lattice length N for the $S = 1/2$ Heisenberg model with open boundary conditions, obtained with DMRG by keeping $D = 100$ basis states. The horizontal line is the exact ground-state energy, e_{ex}, derived from the Bethe ansatz [369]. The inset shows the log-log plot of the error, $e_0 - e_{\text{ex}}$, as a function of N. The line is the fitting curve to the large-N points using Eq. (7.41) with $\eta \approx 1.96$.

$$e_0(N) = e_{\text{ex}} + \frac{\alpha}{N^2} + \text{higher-order terms}, \tag{7.42}$$

where α is a constant that is proportional to the central charge c.

Figure 7.5 shows the size dependence of the entanglement entropy for the spin-1/2 Heisenberg model with open boundary conditions. As the ground state is critical, the entanglement entropy, as predicted by conformal field theory, diverges with the lattice size logarithmically [47, 48]:

$$S(N) = \frac{c}{6} \ln N + b, \tag{7.43}$$

where b is a non-universal constant. This logarithmic divergence is indeed observed in the DMRG calculation. By fitting the numerical data with Eq. (7.43), we find that $c \approx 1.0253(8)$, which is close to the value $c = 1$ predicted by conformal field theory [371]. Due to the truncation of the basis space in the DMRG calculation, the correlation length becomes finite for any given bond dimension D. Thus, the critical behavior exists only when the lattice size is not significantly larger than the correlation length. In other words, when the lattice size becomes significantly larger than the correlation length, the system becomes noncritical, and the area law of the entanglement entropy, which holds in a gapped system, is recovered. The correlation length should become larger in a larger lattice system thanks to the critical logarithmic correction to the entanglement entropy.

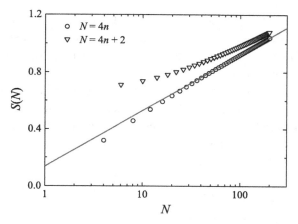

Figure 7.5 Semi-log plot of the entanglement entropy, $S(N)$, as a function of the lattice length N in the ground state of the $S = 1/2$ Heisenberg model with open boundary conditions, obtained with DMRG by keeping $D = 100$ basis states. The straight line is the fitting curve of the large-N data with $N = 4n$ using Eq. (7.43). From the fitting, the central charge is estimated to be $c = 1.0253(8)$.

7.5 Periodic System

In the study of bulk properties, it is better to use periodic (in some cases antiperiodic) boundary conditions to reduce the finite-size effect. This boundary condition can eliminate the edge states emerging at the two ends of an open chain and the effects induced by the edge state, such as the Friedel oscillation.

To implement DMRG with periodic boundary conditions, we should rearrange the superblock such that there is no direct interaction between the left and right blocks whenever possible. A typical configuration of superblock used in the study of the Heisenberg spin chain or other quantum lattice models with only nearest-neighbor interactions with periodic boundary conditions is

The two added sites, σ_1 and σ_2, are separated and put on the two sides of the right block. The σ_2 couples with the left block via the periodic boundary condition, although it looks separated from the left block.

However, the performance of DMRG in a periodic system is poorer than in an open boundary system if the same number of states is retained. This is because, in order to separate a closed chain into two parts, one has to cut two bonds. In contrast, an open chain is divided into two parts by cutting only one bond. This difference means that if D states are retained for an open boundary system, then D^2 states need to be retained to achieve roughly the same accuracy for the corresponding periodic system. This is manifested in the spectrum of the reduced density matrix, which decays much more slowly in a periodic system than in an open one. However, it should be pointed out that this is an artifact of the DMRG setup. The performance of the DMRG in a periodic system can be significantly improved by invoking the matrix product representation [39]. Nevertheless, the cost for determining all local matrices in a periodic MPS is very high compared to the DMRG calculation.

One can fold a periodic chain to convert it into an open chain. For example, a 14-site periodic chain

can be converted to a 2×7 ladder with open boundary conditions by folding the chain from the central bond

After folding, one can apply the DMRG algorithms previously introduced to this ladder system, but now each site contains two original lattice sites. From this, one can also understand why a periodic chain is more difficult to study than an open one.

7.6 Multiple Target States

In the above discussion, the basis states are optimized to represent the ground state accurately. If one wants to calculate other states, for example, the first or other low-lying excited states, one needs to use all these states as target states to build up the density matrix. In this case, the density matrix becomes

$$\rho = \sum_k w_k |\Psi_k\rangle \langle \Psi_k|, \qquad (7.44)$$

which is equivalent to the density matrix of a mixed state with weights w_k, whose sum is normalized to 1:

$$\sum_k w_k = 1. \qquad (7.45)$$

When multiple states are targeted, the number of basis states that need to be retained to maintain accuracy comparable to targeting a single state grows almost linearly with the number of target states. Best performance is, of course, obtained by running DMRG for each target state separately.

Sometimes one can calculate excited states in sectors with different symmetry quantum numbers from the ground state. In this case, there is no need to target multiple states simultaneously. For instance, to obtain the singlet-triplet excitation gap of the antiferromagnetic Heisenberg spin chain, one can do the DMRG calculation by targeting just the lowest energy state in sectors with well-defined total spin $S = 0$ and $S = 1$ separately. As a result, it can significantly reduce the truncation error compared to the calculation with multiple targeting states.

There is no optimal prior choice of weights. Finding a good combination of weights is a matter of trial and error. A rule of thumb is to weigh each target state according to its contribution to the final result [92]. Generally, we set all the weights equal.

7.7 Two-Dimensional Systems

A simple extension of DMRG to two dimensions would be to replace a single site added to a block with a column of sites. This approach might work if the number of sites in a column is small, for example, in a two-leg ladder. However, it is challenging to implement this approach if the number of sites in a column becomes large. First, it is computationally expensive to add a column to a block since the total dimension of the Hilbert space grows exponentially with the number of sites in a column. Second, the truncation error might become large since the number of basis states retained at each DMRG step in comparison with the total number of basis states drops exponentially with the number of sites in one column. Therefore, the two-dimensional algorithm should be designed such that only a single site is absorbed into a block at a time.

To carry out a DMRG calculation, one needs to map a higher-dimensional lattice onto a one-dimensional one by choosing a particular path to order all lattice sites. Two approaches are commonly adopted, referred to as the multichain and two-dimensional approaches, respectively [34]. These mappings break lattice symmetries and introduce long-range interactions.

7.7.1 Multichain Approach

The multichain approach is the most commonly adopted scheme used in the two-dimensional DMRG calculation. It is called the multichain approach because the length of the original lattice is unlimited, but the width is fixed.

The multichain approach maps a two-dimensional lattice onto a one-dimensional one by unzipping the lattice up and down along the *x*-axis, as illustrated in Fig. 7.6. This approach is simple to implement in the DMRG iteration. However, it does not preserve the symmetry of square lattices. The calculations on two successive square

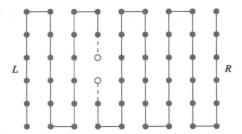

Figure 7.6 A one-dimensional mapping of a six-leg ladder with a typical configuration of the superblock. The left and right blocks are formed by dots connected
by solid lines. Hollowed dots are the sites added to the superblock.

lattices, such as the $(L-1) \times (L-1)$ and $L \times L$ lattices, are performed independently.
The information obtained from the iterations on the $(L-1) \times (L-1)$ lattice is not
used to prepare the initial blocks for the $L \times L$ lattice, which is a loss of efficiency.
It may also result in the loss of accuracy since the topological characteristics of
square lattices are not well manifested in the preparation of the initial block states
and the sweeping procedure of DMRG.

For a multichain system, if we divide the lattice into two parts, the entanglement entropy between them is proportional to the number of bonds (or links) cut,
according to the area law. The minimal number of bonds that need to be cut is just
the width of the lattice. Thus, the number of states needed to maintain accuracy
grows exponentially with the width of the lattice.

7.7.2 Square-Lattice Approach

The square-lattice approach orders the lattice sites along the diagonal direction and
is a bit more complicated to implement than the multichain one. However, it keeps
the topology of the square lattice during the mapping and allows us to build up the
initial blocks of the $L \times L$ lattice from the $(L-1) \times (L-1)$ lattice. Figure 7.7(a)
shows the order of sites on a 2×2 lattice.

This approach was proposed based on the observation that the initial superblocks
of an $L \times L$ lattice can always be built up using the blocks of an $(L-1) \times (L-1)$
lattice [34]. As an example, Fig. 7.7(b) shows how the superblock on a 3×3 lattice
is built up using the subblocks on a 2×2 lattice. The number labels the order of
the corresponding site in the mapped one-dimensional lattice. The initial left and
right blocks contain three and four sites linked by the solid line in the lower left
and upper right corners, respectively. However, the two added sites, 4 and 6, do
not lie next to each other. All the matrix elements for these initial blocks can be
obtained from the results previously obtained on the 2×2 lattice. After the DMRG

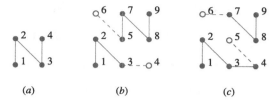

Figure 7.7 (a) A one-dimensional mapping of the 2×2 lattice. (b) The initial configuration of superblock for a 3×3 lattice system. The initial left block contains three sites linked by the solid line in the lower-left corner. The initial right block contains four sites linked by the solid line at the upper right corner. (c) Same as (b) for the next iteration. The number labels the order of the corresponding site in the unfolded one-dimensional lattice. Hollowed dots are the sites added to the left and right blocks.

calculation, the system, which contains the left block and site 4, is updated and taken as the new left block for the next iteration.

The superblock configuration for the next iteration is shown in Fig. 7.7(c). The left block contains four sites at the lower-left corner, and the right block contains three sites at the upper-right corner. The two added sites, sites 5 and 6, are now located next to each other. One can then perform the finite-lattice DMRG sweeping to optimize the ground-state wave function further.

Extending this procedure to an arbitrary $L \times L$ lattice is straightforward. As an example, Fig. 7.8 shows how to construct the first two superblocks on a 4×4 lattice using the subblocks of 3×3 lattice. In general, the initial superblocks on an $L \times L$ lattice can be formed based on the results of the left and right blocks on an $(L - 1) \times (L - 1)$ lattice. The first site at the lower-left corner is labeled as 1. The other sites are ordered similar to a folded zipper with unequal width along the diagonal. Initially, the two added sites have the coordinates $X_1 = (L - 1)L/2 + 1$ and $X_2 = L(L + 1)/2$ in the mapped one-dimensional lattice. We take the first $(L - 1)L/2$ sites in the lower-left corner as the initial left block, and all the sites in the upper right $(L-1) \times (L-1)$ square lattice not used by the left block as the initial right block. At the first few iterations of DMRG, the second site added is fixed at X_2, which is not next to X_1 in the mapped one-dimensional lattice. It continues until the two added sites become nearest neighbors in the mapped lattice. After that, the standard finite system sweeping is implemented as in a regular one-dimensional lattice.

This discussion applies not just to the square lattice. In fact, it can be readily extended to a two-dimensional lattice that can be topologically transformed to a square lattice by adding or removing some of the nearest- or next-nearest-neighbor interactions from the square lattice, such as triangular, hexagonal, and Kagome

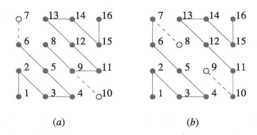

(a)   (b)

Figure 7.8 (a) The initial configuration of superblock on the 4×4 lattice, which is a superposition of two partial 3×3 lattices at the lower left and upper right corners, and two sites (i.e. sites 6 and 10). The number beside each lattice site defines the order in the mapped one-dimensional system. Solid lines link the sites in the left or right block. Hollowed dots are the sites added in. (b) Same as for (a) for the next iteration. The right block, sites 10–16, is a space reflective of the left block, sites 1–7, with respect to the center of the 4×4 lattice.

lattices. For example, a triangular lattice can be mapped onto a square lattice with extra next-nearest-neighbor connections by anticlockwise tilting the lattice:

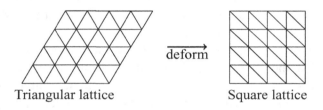

Triangular lattice   Square lattice

Other lattices can be similarly deformed onto square lattices.

8

Implementation of Symmetries

8.1 Symmetry Consideration

Implementation of symmetries or conservation laws is vital in practical calculations. It is realized by partitioning basis states in terms of the quantum numbers characterizing the symmetries of the system so that the Hamiltonian becomes block diagonal and operators obey specific selection rules between symmetry sectors. Hence, the usage of symmetries can significantly reduce the computational cost for computing and storing the matrix elements of these operators, allowing more basis states to be retained in the DMRG calculation. In particular, the quantum number that characterizes the symmetry of the superblock for a particular target state, for example, the ground state, is fixed, and the dimension of the Hilbert space for this subspace is generally smaller than the whole dimension of the Hilbert space. Thus, implementing symmetries allows us to obtain a more accurate eigenfunction with a smaller number of basis states.

Symmetries used in DMRG fall into two categories depending on whether the corresponding conservation laws are additive or nonadditive. The additive ones include the Abelian $U(1)$ and most of the discrete symmetries. The nonadditive ones include non-Abelian [34, 372, 373, 374] and permutation symmetries.

8.2 Continuous Abelian Symmetries

For the Heisenberg model, for example, the total spin along the z-axis S_z is conserved. The Hamiltonian can be block diagonalized according to the value of S_z. If this additive conservation law is considered, the basis states at each block should be labeled by two quantum numbers, (m, k), where m is the quantum number proportional to the eigenvalue of S_z:

$$S_z|m\rangle = m|m\rangle, \tag{8.1}$$

and k is the quantum number labeling the basis states in this $S_z = m$ subspace. If we use (m_l, k_l), (m_r, k_r), (m_1, k_1), and (m_2, k_2) to label, respectively, the basis states in the left block, right block, and two added sites, then the ground-state wave function can be represented as

$$|\Psi\rangle = \sum_{M,k} \Psi_{M,k} |M, k\rangle,$$ (8.2)

where $M = (m_l, m_1, m_2, m_r)$ and $k = (k_l, k_1, k_2, k_r)$. Then k is a vector of indices that labels the basis states in the corresponding quantum states. The four good quantum numbers in M are not independent. Their sum must equal the total spin m of the target state:

$$m_l + m_1 + m_2 + m_r = m.$$ (8.3)

Thus, the DMRG wave function is represented by seven independent quantum numbers: three from M and four from k. The target spin m is determined by the system to be studied. For the ground state of the spin-1/2 antiferromagnetic Heisenberg model with an even number of spins, for example, $m = 0$. However, for the same system with an odd number of spins, $m = 1/2$ or $-1/2$.

The product of a basis state in the left block and that in the left added site forms a basis state in the system:

$$|m_s, k_s\rangle = |m_l, k_l\rangle |m_1\rangle \delta_{m_1, m_s - m_1},$$ (8.4)

where k_s is an integer index that labels the basis state with the magnetic quantum number equal to m_s in the system. The delta-function is included in the expression to enforce the spin conservation, $m_s = m_l + m_1$. There is a one-to-one correspondence between (m_s, k_s) and (m_l, k_l, m_1). The total number of basis states for a given m_s, $D_s[m_s]$, is

$$D_s[m_s] = \sum_{m_1} D_l[m_s - m_1],$$ (8.5)

where $D_l[m_l]$ is the number of basis states with the magnetic quantum number equal to m_l in the left block.

From the conservation of S_z, it is simple to show that the reduced density matrix of the system, ρ^{sys}, is block diagonal according to the quantum number m_s:

$$\rho^{\text{sys}} = \sum_{m_s} \oplus \rho_{m_s}^{\text{sys}},$$ (8.6)

where $\rho_{m_s}^{\text{sys}}$ is the reduced density matrix in the sector whose total S_z is equal to m_s. Again, $\rho_{m_s}^{\text{sys}}$ is a Hermitian matrix. It can be diagonalized by a unitary matrix U_{m_s}:

$$\rho_{m_s}^{\text{sys}} = U_{m_s} \lambda_{m_s} U_{m_s}^\dagger = -\!\!\bigcirc\!\!-\!\!\diamond\!\!-\!\!\bigcirc\!\!- \atop {U_{m_s} \quad \lambda_{m_s} \quad U_{m_s}^\dagger}\,,$$ (8.7)

where λ_{m_s} is the diagonal eigenvalue matrix. The eigen-spectrum of ρ^{sys} is given by the direct sum of all λ_{m_s}:

$$\Lambda = \sum_{m_s} \oplus \lambda_{m_s}. \tag{8.8}$$

Keeping the D largest eigenvalues in Λ, we obtain the number of states retained at each spin sector. Correspondingly, the RG transformation matrix U is simply the direct sum of all U_{m_s}-matrices after truncation:

$$U = \sum_{m_s} \oplus U_{m_s}. \tag{8.9}$$

After truncation, the unitary matrix U_{m_s} becomes isometric. Using the basis states of the left block and the left added site, we can also write U_{m_s} as a three indexed tensor:

$$\left(U_{m_s}\right)_{k_s, k_s'} = U_{m_l k_l, m_s k_s'}[m_1], \tag{8.10}$$

or graphically:

$$m_s k_s -\!\!\square\!\!- m_s k_s' = m_l k_l \overset{m_1}{\to\!\square\!\to} m_s k_s' \quad . \tag{8.11}$$

Arrows are added to the dangling bonds to emphasize the spin conservation: the sum of ingoing spins equals that of outgoing ones, that is, $m_s = m_l + m_1$. $|m_s, k_s'\rangle$ is the basis state after transformation.

It is straightforward to generalize this discussion to include other additive symmetries. For example, for the Hubbard model, in addition to the spin conservation, one can also impose the charge conservation in the DMRG calculation. In this case, both the spin along the z-axis

$$S_z = \frac{1}{2} \sum_i \left(c_{i\uparrow}^\dagger c_{i\uparrow} - c_{i\downarrow}^\dagger c_{i\downarrow}\right) \tag{8.12}$$

and charge

$$N_c = \sum_i \left(c_{i\uparrow}^\dagger c_{i\uparrow} + c_{i\downarrow}^\dagger c_{i\downarrow}\right) \tag{8.13}$$

are conserved. Each basis state is labeled by three quantum numbers (m, n, k). Here m and n are, respectively, the eigenvalues of S_z and N_c, and k is an integer labeling a basis state in this symmetry sector. The Hamiltonian is now block diagonalized according to the quantum numbers (m, n).

8.3 Spin Reflection Symmetry

In many cases, in addition to the conservation of S_z and N_c, the Hamiltonian is also invariant under the spin reflection or particle-hole transformation. The spin reflection, for example, is to convert each basis state $|m, k\rangle$ with $S_z = m$ at each block into the state with opposite spin:

$$U_s|m, k\rangle = |-m, k\rangle, \tag{8.14}$$

where U_s is the spin-flip operator that satisfies $U_s^\dagger = U_s$ and $U_s^2 = I$. One can also impose this symmetry to further block diagonalize the Hamiltonian so that only the $m \geq 0$ basis states at each block are retained in the DMRG iteration. It can reduce the number of states retained at each block by almost a factor of 2.

However, extra caution should be paid to the $S_z = 0$ states at each building block. This is because by applying the spin flip operator U_s to an $S_z = 0$ state $|0, k\rangle$, $U_s|0\rangle$ is also an $S_z = 0$ state. If $|0, k\rangle$ is not an eigenstate of U_s, $U_s|0, k\rangle$ would be different from $|0, k\rangle$. In that case, we can use $|0, k\rangle$ and $U_s|0, k\rangle$ to generate a new basis state that is an even parity eigenstate of U_s:

$$|0_+, k\rangle = C\left(|0, k\rangle + U_s|0, k\rangle\right), \tag{8.15}$$

where C is the normalization constant. However, any two basis states thus generated may not be orthogonal to each other. In that case, we should use the Gram–Schmidt orthogonalization to reorthonormalize these basis states.

Similarly, we can define a set of odd-parity states from $|0, k\rangle$ and $U_s|0, k\rangle$:

$$|0_-, k\rangle = C\left(|0, k\rangle - U_s|0, k\rangle\right). \tag{8.16}$$

Again, we should use the Gram–Schmidt orthogonalization to ensure that these states are orthonormal to each other. The total number of these independent and orthonormalized even- and odd-parity states should be equal to the number of states originally retained for the $m = 0$ sector.

An operator O defined in a block can always be decoupled into a sum of two operators:

$$O = O_+ + O_-, \tag{8.17}$$

where

$$O_\pm = \frac{1}{2}\left(O \pm U_s^\dagger O U_s\right) \tag{8.18}$$

are the even and odd components of O under the spin reflection, respectively:

$$U_s^\dagger O_\pm U_s = \pm O_\pm. \tag{8.19}$$

In the DMRG calculation, only the matrix elements of $\langle m'k'|O_\pm|m,k\rangle$ with $m \geq 0$ need to be computed and stored. The matrix elements with $m < 0$ can be readily obtained using the parity properties of $O_\pm$:

$$\langle m', k'|O_\pm|m, k\rangle = \langle -m', k'|U_s^\dagger O_\pm U_s| - m, k\rangle = \pm\langle -m', k'|O_\pm| - m, k\rangle. \quad (8.20)$$

8.4 Spatial Reflection Symmetry

In addition to the charge, spin, or other good quantum numbers, one can also implement the spatial reflection symmetry in the DMRG calculation.

If the Hamiltonian is invariant under the spatial reflection transformation U_r,

$$U_r^\dagger H U_r = H, \quad (8.21)$$

then the wave function can be defined so that it is an eigenstate of U_r with definite parity:

$$|\Psi_\pm\rangle = \frac{1}{\sqrt{2}} (I \pm U_r) |\Psi\rangle. \quad (8.22)$$

As the system returns to itself by taking the spatial reflection transformation twice, $U_r^2 = I$, $|\Psi_\pm\rangle$ has even/odd parity under lattice reflection:

$$U_r|\Psi_\pm\rangle = \pm|\Psi_\pm\rangle. \quad (8.23)$$

This allows us to target a quantum state with a definite parity.

However, the spatial reflection swaps the lattice sites from left to right and from right to left. It could be implemented in the DMRG calculation only when the system and environment blocks have the same size, and for each basis state in the system, $|k, \sigma\rangle_{\text{sys}}$, there should be a corresponding basis state in the environment, $|\sigma, k\rangle_{\text{env}}$, such that

$$U_r|k, \sigma\rangle_{\text{sys}} = |\sigma, k\rangle_{\text{env}}, \qquad U_r|\sigma, k\rangle_{\text{env}} = |k, \sigma\rangle_{\text{sys}}. \quad (8.24)$$

From Eq. (8.23), it is simple to show that the wave function of the even-parity state $\Psi_{l\sigma_1\sigma_2r}^{(+)}$ is symmetric under the spatial reflection:

$$\Psi_{l\sigma_1\sigma_2r}^{(+)} = \Psi_{r\sigma_2\sigma_1l}^{(+)}. \quad (8.25)$$

Thus, only the wave functions with $(l, \sigma_1) \geq (r, \sigma_2)$ need to be evaluated and stored.

Similarly, the odd-parity state $\Psi_{l\sigma_1\sigma_2r}^{(-)}$ is antisymmetric under the spatial reflection:

$$\Psi_{l\sigma_1\sigma_2r}^{(-)} = -\Psi_{r\sigma_2\sigma_1l}^{(-)}. \quad (8.26)$$

Clearly, $\Psi^{(-)}_{l\sigma\sigma l} = 0$. In this case, only the wave functions with $(l, \sigma_1) > (r, \sigma_2)$ need to be evaluated and stored.

For either parity state, it is straightforward to show that the reduced density matrix of the system equals precisely that of the environment block:

$$\rho^{\text{sys}} = \rho^{\text{env}}. \tag{8.27}$$

This is, of course, a consequence of spatial reflection symmetry.

8.5 Non-Abelian Symmetries

The isotropic Heisenberg model possesses the spin SU(2) symmetry. The Hubbard model at half-filling possesses a higher SO(4) symmetry. The presence of non-Abelian symmetries offers a stronger simplification in practical calculations because many nonzero matrix elements are not independent of each other. Therefore, going beyond Abelian symmetries, non-Abelian symmetries allow compressing the nonzero blocks in terms of multiplet spaces significantly [34, 374].

Implementing SU(2) [34, 372, 374] or other non-Abelian symmetries [375] by eliminating some quantum numbers through the Clebsch–Gordan algebra and the Wigner–Eckart theorem is technically more involved. Nevertheless, it can significantly improve the performance once it is achieved. Generally, by extending the symmetry from U(1) to SU(2), the number of states retained can be reduced by a factor of 3 for achieving the same accuracy [34]. The gain in computational time is even higher.

In the following discussion, we take the one-dimensional Heisenberg model as an example to show how to implement SU(2) symmetry in the DMRG calculation. However, the idea can be readily extended to other models with non-Abelian symmetries. A more general discussion on non-Abelian symmetries in the framework of tensor network states could be found in Ref. [375].

To avoid the complication caused by using too many indices in labeling a quantum state in the standard four-block configuration of a superblock, we demonstrate the method by partitioning the system into two parts: a system and an environment. The Hamiltonian now contains the interactions within each block, H_s and H_e, and those between the blocks, $H_{s,e}$:

$$H = H_s + H_{s,e} + H_e. \tag{8.28}$$

For the one-dimensional Heisenberg model,

$$H_{s,e} = J\mathbf{S}_a \cdot \mathbf{S}_b, \tag{8.29}$$

where $\mathbf{S}_a$ is the spin operator on the rightmost lattice site of the system, and $\mathbf{S}_b$ is the spin operator on the leftmost lattice site of the environment.

8.5.1 Wave Function

In an SU(2)-invariant system, a basis state in the system is labeled by three quantum numbers $|j_s m_s k_s\rangle$, where j_s and m_s represent the total spin and the magnetic quantum number along the z-axis, respectively. k_s is the quantum number that labels the basis state in a given spin sector (j_s, m_s). If $\mathbf{S}_s$ is the total spin operator in the system, then

$$\mathbf{S}_s^2 |j_s m_s k_s\rangle = j_s(j_s + 1) |j_s m_s k_s\rangle, \qquad (8.30)$$

$$S_{s,z} |j_s m_s k_s\rangle = m_s |j_s m_s k_s\rangle. \qquad (8.31)$$

Similarly, we use (j_e, m_e, k_e) to label a basis state at the environment.

Given a spin j_s from the system and a spin j_e from the environment, their addition forms a tensor-product space that can be decomposed into a direct sum of total spins running from $|j_s - j_e|$ to $(j_s + j_e)$:

$$j_s \otimes j_e = (j_s + j_e) \oplus (j_s + j_e - 1) \oplus \cdots \oplus |j_s - j_e|. \qquad (8.32)$$

Assuming j to be the quantum number of the total spin, $\mathbf{S} = \mathbf{S}_s + \mathbf{S}_e$, and m its z-component of the whole system, the wave function of the target state can be represented as

$$|\Psi(jm)\rangle = \sum_{j_s k_s j_e k_e} \Psi_{j_s k_s j_e k_e} |jm; j_s k_s j_e k_e\rangle, \qquad (8.33)$$

where $|jm; j_s k_s j_e k_e\rangle$ is a basis state of total spin, (j, m), formed by the two spin states j_s and j_e:

$$|jm; j_s k_s j_e k_e\rangle = \sum_{m_s m_e} \langle j_s m_s \, j_e m_e | jm\rangle \, |j_s m_s k_s\rangle \, |j_e m_e k_e\rangle, \qquad (8.34)$$

and $\langle j_s m_s j_e m_e | jm\rangle$ is the Clebsch–Gordan coefficient which is nonzero only when the condition $|j_s - j_e| \leq j \leq (j_s + j_e)$ is satisfied. Equivalently, we can release this constraint in the summation just by setting $\Psi_{j_s k_s j_e k_e} = 0$ if $|j_s - j_e| > j$ or $j > (j_s + j_e)$.

The state $|jm; j_s k_s j_e k_e\rangle$ forms a complete basis set in the subspace whose total spin is equal to j and its z-component equal to m:

$$\sum_{j_s k_s j_e k_e} |jm; j_s k_s j_e k_e\rangle \langle jm; j_s k_s j_e k_e| = 1. \qquad (8.35)$$

From the orthonormalization condition of the Clebsch–Gordan coefficients, we find that the normalization condition of the wave function is determined by the equation

$$\langle \Psi(jm) | \Psi(jm)\rangle = \sum_{j_s k_s j_e k_e} \left| \Psi_{j_s k_s j_e k_e} \right|^2 = 1. \qquad (8.36)$$

8.5.2 Wigner–Eckart Theorem

The Hamiltonian we study is defined in the Hilbert space spanned by the basis states $|j_s m_s\, k_s\rangle$ in the system and $|j_e m_e k_e\rangle$ in the environment. To efficiently perform the DMRG calculation that preserves the SU(2) symmetry, we need to squeeze the basis space from $|j_s m_s\, k_s\rangle|j_e m_e k_e\rangle$ to $|j_s k_s\rangle|j_e k_e\rangle$ by eliminating the magnetic quantum numbers m_s and m_e. This is possible because the matrix element of a tensor operator can always be decomposed as a product of a Clebsch–Gordan coefficient and a reduced matrix element that does not depend on the magnetic quantum number according to the Wigner–Eckart theorem. We call $|j_s k_s\rangle|j_e k_e\rangle$ a reduced basis state.

The Wigner–Eckart theorem, derived by physicists Eugene Wigner and Carl Eckart, is a theorem about the matrix elements of irreducible tensor operators of rank-l, $T^{(l)}$, between two angular momentum eigenstates. $T^{(l)}$ is defined as a set of $2l + 1$ operators, $T_k^{(l)}$ ($k = -l, -l + 1, \ldots, l$), also called the spherical tensor operators, that fulfill the following defining commutation relations with the angular momentum $\mathbf{S} = (S_x, S_y, S_z)$:

$$[S_z, T_k^{(l)}] = k T_k^{(l)}, \tag{8.37}$$

$$[S_\pm, T_k^{(l)}] = \sqrt{l(l + 1) - k(k \pm 1)}\, T_{k\pm 1}^{(l)}, \tag{8.38}$$

where $S_\pm = S_x \pm i S_y$.

The Wigner–Eckart theorem states that in the eigenstate representation of the angular momentum, $\mathbf{S}^2$ and S_z, the matrix elements of the tensor operator $T_k^{(l)}$, between the state $|j_1 m_1 k_1\rangle$ and $|j_2 m_2 k_2\rangle$, can be reduced to a product of a term that depends on the magnetic quantum numbers (m_1, m_2, k) and a term that does not depend on the projection quantum numbers:

$$\langle j_2 m_2 k_2|\, T_k^{(l)}\, |j_1 m_1 k_1\rangle = \langle j_2 m_2|j_1 m_1 l k\rangle \langle j_2 k_2 \left\| T^{(l)} \right\| j_1 k_1\rangle, \tag{8.39}$$

where $\langle j_2 k_2 \left\| T^{(l)} \right\| j_1 k_1\rangle$ does not depend on the magnetic quantum numbers (m_1, m_2, k), and is referred to as the reduced matrix element of the tensor operator.

8.5.3 Reduced Matrix Elements of the Hamiltonian

In the reduced basis space, the matrix elements of the Hamiltonian are defined by

$$H_{j_s' k_s' j_e' k_e', j_s k_s j_e k_e} = \langle jm; j_s' k_s' j_e' k_e'| H |jm; j_s k_s j_e k_e\rangle, \tag{8.40}$$

and the matrix element of the Hamiltonian between any two states with the same target spin quantum numbers (j, m) can be expressed as

$$\langle \Phi(jm)| H |\Psi(jm)\rangle = \sum_{j_s' k_s' j_e' k_e'} \sum_{j_s k_s j_e k_e} \Phi^*_{j_s' k_s' j_e' k_e'} H_{j_s' k_s' j_e' k_e', j_s k_s j_e k_e} \Psi_{j_s k_s j_e k_e}. \tag{8.41}$$

To calculate the matrix elements of the Hamiltonian, we first consider the contribution of the system Hamiltonian H_s. As H_s is a rank-0 tensor (i.e. a scalar), we obtain from the Wigner–Eckart theorem that

$$\langle j'_s m_s {}'k'_s| H_s |j_s m_s k_s\rangle = \langle j'_s m_s {}'|j_s m_s 00\rangle \langle j'_s k'_s \|H_s\| j_s k_s\rangle$$
$$= \delta_{j'_s,j_s}\delta_{m_s{}',m_s} \langle j_s k'_s \|H_s\|j_s k_s\rangle \qquad (8.42)$$

and

$$\langle jm; j'_s k'_s j'_e k'_e| H_s |jm; j_s k_s j_e k_e\rangle$$
$$= \delta_{j'_s,j_s}\delta_{j'_e,j_e}\delta_{k'_e,k_e} \sum_{m_s\,m_e} \langle j_s m_s\, j_e m_e|jm\rangle \langle j'_s m_s\, j_e m_e|jm\rangle \langle j'_s m_s\, k'_s| H_s |j_s m_s k_s\rangle$$
$$= \delta_{j'_s,j_s}\delta_{j'_e,j_e}\delta_{k'_e,k_e} \langle j_s k'_s \|H_s\| j_s k_s\rangle \sum_{m_s\,m_e} \langle j_s m_s\, j_e m_e|jm\rangle \langle j_s m_s\, j_e m_e|jm\rangle$$
$$= \delta_{j'_s,j_s}\delta_{j'_e,j_e}\delta_{k'_e,k_e} \langle j_s k'_s \|H_s\| j_s k_s\rangle. \qquad (8.43)$$

Similarly, the contribution of the environment Hamiltonian H_e is

$$\langle jm; j'_s k'_s j'_e k'_e| H_e |jm; j_s k_s j_e k_e\rangle = \delta_{j'_s,j_s}\delta_{j'_e,j_e}\delta_{k'_s,k_s} \langle j_e k'_e \|H_e\| j_e k_e\rangle. \qquad (8.44)$$

We now calculate the matrix elements of $H_{s,e}$. It is straightforward to show that the spin operator

$$\mathbf{S} = (S_{-1}, S_0, S_1) = \left(\frac{S_x - iS_y}{\sqrt{2}}, S_z, \frac{-S_x - iS_y}{\sqrt{2}}\right) \qquad (8.45)$$

is a rank-1 tensor operator (or a vector). In this spherical tensor representation, $H_{s,e}$ becomes

$$H_{s,e} = J\left(S_{a,0}S_{b,0} - S_{a,1}S_{b,-1} - S_{a,-1}S_{b,1}\right) = J \sum_{k=-1,0,1} (-)^k S_{a,k}S_{b,-k}. \qquad (8.46)$$

Using the Wigner–Eckart theorem, we find that

$$\langle j'_e m'_e k'_e| \langle j'_s m_s {}'k'_s| H_{s,e} |j_s m_s k_s\rangle |j_e m_e k_e\rangle$$
$$= J \langle j'_s k'_s \|S_a\| j_s k_s\rangle \langle j'_e k'_e \|S_b\| j_e k_e\rangle$$
$$\sum_k (-)^k \langle j'_s m_s {}'|j_s m_s, 1, k\rangle \langle j'_e m'_e|j_e m_e, 1, (-k)\rangle. \qquad (8.47)$$

So the matrix element of $H_{s,e}$ is

$$\langle jm; j'_s k'_s j'_e k'_e| H_{s,e} |jm; j_s k_s j_e k_e\rangle$$
$$= J \langle j'_s k'_s \|S_a\| j_s k_s\rangle \langle j'_e k'_e \|S_b\| j_e k_e\rangle A_{jm}\left(j_s j'_s j_e j'_e\right), \qquad (8.48)$$

where

$$A_{jm}\left(j_s j_s' j_e j_e'\right) = \sum_{m_s m_e k} \sum_{m_s' m_e'} (-)^k \left\langle jm | j_s' m_s' j_e' m_e' \right\rangle \left\langle j_s m_s j_e m_e | jm \right\rangle$$
$$\left\langle j_s' m_s' | j_s m_s, 1, k \right\rangle \left\langle j_e' m_e' | j_e m_e, 1, (-k) \right\rangle. \tag{8.49}$$

Putting the above results together, we obtain the reduced matrix element of the Hamiltonian

$$H_{j_s' k_s' j_e' k_e' j_s k_s j_e k_e}$$
$$= \delta_{j_s' j_s} \delta_{j_e' j_e} \left(\delta_{k_e', k_e} \left\langle j_s k_s' \| H_s \| j_s k_s \right\rangle + \delta_{k_s', k_s} \left\langle j_e k_e' \| H_e \| j_e k_e \right\rangle \right)$$
$$+ J \left\langle j_s' k_s' \| S_a \| j_s k_s \right\rangle \left\langle j_e' k_e' \| S_b \| j_e k_e \right\rangle A_{jm}\left(j_s j_s' j_e j_e'\right). \tag{8.50}$$

8.5.4 Reduced Matrix Elements of the System Hamiltonian

In the DMRG calculation, the system consists of the left block and the left added site, and its Hamiltonian can be written as

$$H_s = H_l + J \mathbf{S}_{a-1} \cdot \mathbf{S}_a, \tag{8.51}$$

the first term on the right-hand side is the Hamiltonian of the left block and the second term is the interaction between the left block and the site added in. Similarly, a spin multiplet of the system, (j_s, k_s), is the product of a spin multiplet of the left block, (j_l, k_l), and the spin of the site added in, s, which is fixed and unique. Correspondingly, the basis state of the system can be expressed as

$$|j_s m_s k_s\rangle = \sum_{j_l m_l m_1} \langle j_l m_l s m_1 | j_s m_s \rangle |j_l m_l k_l\rangle |s m_1\rangle. \tag{8.52}$$

Given j_s, there is a one-to-one correspondence between k_s and (j_l, k_l). j_s, j_l, and s satisfy the triangle relation $|j_l - s| \le j_s \le j_l + s$.

As both the total spin and its third component are conserved, the Hamiltonian of the system should be diagonal for each given (j_s, m_s) and independent of the magnetic quantum number m_s:

$$\left\langle j_s' m_s' k_s' \right| H_s | j_s m_s k_s \rangle = \delta_{j_s' j_s} \delta_{m_s', m_s} \langle j_s k_s' \| H_s \| j_s k_s \rangle. \tag{8.53}$$

Similarly, H_l is diagonal in the basis space of the left block:

$$\langle j_l' m_l' k_l' | H_l \| j_l m_l k_l \rangle = \delta_{j_l' j_l} \delta_{m_l', m_l} \langle j_l k_l' \| H_l \| j_l k_l \rangle. \tag{8.54}$$

From this equation and the property of the Clebsch–Gordan coefficients, the matrix elements of H_l are reduced to

$$\langle j_s m_s k_s' | H_l | j_s m_s k_s \rangle = \sum_{j_l k_l k_l'} \langle j_l k_l' \| H_l \| j_l k_l \rangle. \tag{8.55}$$

The matrix elements of $J\mathbf{S}_{a-1} \cdot \mathbf{S}_a$ can be calculated similar to $H_{s,e}$. Making use of the properties of the rank-1 tensor operators, we find that

$$
\begin{aligned}
&\langle j'_l m'_l k'_l | \langle sm'_1 | J\mathbf{S}_{a-1} \cdot \mathbf{S}_a | j_l m_l k_l \rangle |sm_1 \rangle \\
&= J \sum_k (-)^k \langle j'_l k'_l \| S_{a-1} \| j_l k_l \rangle \langle s \| S_a \| s \rangle \langle j'_l m'_l | j_l m_l, 1, k \rangle \\
&\quad \langle sm'_1 | sm_1, 1, (-k) \rangle,
\end{aligned}
\tag{8.56}
$$

and

$$
\begin{aligned}
&\langle j_s m_s \, k'_s | J\mathbf{S}_{a-1} \cdot \mathbf{S}_a | j_s m_s \, k_s \rangle \\
&= \sum_{j'_l k'_l j_l k_l} J \langle j'_l k'_l \| S_{a-1} \| j_l k_l \rangle \langle s \| S_a \| s \rangle B_{j_s m_s} (j'_l, j_l),
\end{aligned}
\tag{8.57}
$$

where

$$
\begin{aligned}
B_{j_s m_s} (j'_l, j_l) &= \sum_{km'_l m'_1 m_l m_1} (-)^k \langle j_s m_s | j'_l m'_l sm'_1 \rangle \langle j_l m_l sm_1 | j_s m_s \rangle \\
&\quad \langle j'_l m'_l | j_l m_l, 1, k \rangle \langle sm'_1 | sm_1, 1, (-k) \rangle.
\end{aligned}
\tag{8.58}
$$

Putting these results together, we obtain the reduced matrix elements of H_s:

$$
\langle j_s k'_s \| H_s \| j_s k_s \rangle = \sum_{j'_l k'_l j_l k_l} \langle j'_l k'_l \| H_s \| j_l k_l \rangle,
\tag{8.59}
$$

where $\| j_s k_s \rangle = \| j_l k_l \rangle \otimes |s\rangle$ and $\| j_s k'_s \rangle = \| j'_l k'_l \rangle \otimes |s\rangle$, and

$$
\begin{aligned}
\langle j'_l k'_l \| H_s \| j_l k_l \rangle &= J \langle j'_l k'_l \| S_{a-1} \| j_l k_l \rangle \langle s \| S_a \| s \rangle B_{j_s j_s} (j'_l, j_l) \\
&\quad + \langle j_l k'_l \| H_l \| j_l k_l \rangle \delta_{j'_l j_l}.
\end{aligned}
\tag{8.60}
$$

8.5.5 Reduced Density Matrix

Solving the eigenequation, we can find the target eigenstate $|\Psi(jm)\rangle$. However, the density matrix $|\Psi(jm)\rangle \langle \Psi(jm)|$ of this state is not SU(2) invariant. An SU(2)-invariant density matrix is defined by

$$
\rho = \frac{1}{2j+1} \sum_m |\Psi(jm)\rangle \langle \Psi(jm)|.
\tag{8.61}
$$

It is simple to show that this density matrix commutes with the total spin operator:

$$
[\rho, \mathbf{S}] = 0.
\tag{8.62}
$$

The reduced density matrix of the system is defined by tracing over all environment variables in ρ:

$$\rho^{\text{sys}} = \text{Tr}_{\text{env}} \rho$$
$$= \sum_{j_s j_s' k_s k_s' j_e k_e} \Psi^*_{j_s' k_s' j_e k_e} \Psi_{j_s k_s j_e k_e} F\left(j, j_s, j_s', j_e\right) |j_s k_s\rangle \langle j_s' k_s'|, \qquad (8.63)$$

where

$$F\left(j, j_s, j_s', j_e\right)$$
$$= \frac{1}{2j+1} \sum_{m_s m_e m m_s'} \langle j_s m_s j_e m_e | jm\rangle \langle jm | j_s' m_s' j_e m_e\rangle |m_s\rangle \langle m_s'|$$
$$= \frac{1}{2j+1} \sum_{m_s m_e m} \langle j_s m_s j_e m_e | jm\rangle \langle jm | j_s' m_s j_e m_e\rangle |m_s\rangle \langle m_s|. \qquad (8.64)$$

Using the symmetry property of the Clebsch–Gordan coefficients

$$\langle jm | j_s m_s j_e m_e\rangle = (-)^{j_e + m_e} \sqrt{\frac{2j+1}{2j_s+1}} \langle j_s (-m_s) | j(-m) j_e m_e\rangle, \qquad (8.65)$$

we obtain

$$F\left(j, j_s, j_s', j_e\right) = \sum_{m_s} \frac{|m_s\rangle \langle m_s|}{\sqrt{2j_s+1}\sqrt{2j_s'+1}} \sum_{m_e m} \langle j_s'(-m_s) | j(-m) j_e m_e\rangle$$
$$\langle j(-m) j_e m_e | j_s(-m_s)\rangle$$
$$= \delta_{j_s' j_s}. \qquad (8.66)$$

Therefore, the reduced density matrix is

$$\rho^{\text{sys}} = \sum_{j_s k_s k_s' j_e k_e} \Psi^*_{j_s k_s' j_e k_e} \Psi_{j_s k_s j_e k_e} |j_s k_s\rangle \langle j_s k_s'|. \qquad (8.67)$$

As expected, ρ^{sys} is block diagonal according to the value of j_s. For each given j_s, the matrix elements of the corresponding reduced density matrix are given by

$$\rho^{\text{sys}}_{k_s, k_s'}(j_s) = \sum_{j_e k_e} \Psi^*_{j_s k_s' j_e k_e} \Psi_{j_s k_s j_e k_e}. \qquad (8.68)$$

8.5.6 Transformation Matrix

For a given j_s, we diagonalize the reduced density matrix with a unitary matrix:

$$\rho^{\text{sys}}(j_s) = U_{j_s} \lambda_{j_s} U_{j_s}^\dagger = \underset{U_{j_s} \quad \lambda_{j_s} \quad U_{j_s}^\dagger}{-\!\bigcirc\!-\!\diamond\!-\!\bigcirc\!-}. \qquad (8.69)$$

By grouping all the eigenvalues of the reduced density matrices and keeping the D largest eigenvectors, we determine which spin multiplets are retained in the system. After truncation, the unitary matrix U_{j_s} becomes isometric:

$$k_s \; \text{—O—} \; k_s^{\text{new}} \quad \rightarrow \quad k_s \; \text{—□—} \; k_s^{\text{new}}. \qquad (8.70)$$
$$\qquad U_{j_s} \qquad\qquad\qquad\qquad U_{j_s}$$

The transformation matrix U_{j_s} is to convert the quantum state from $|j_s k_s\rangle$ to $|j_s k_s^{\text{new}}\rangle$:

$$|j_s k_s^{\text{new}}\rangle = \sum_{k_s} U_{j_s k_s k_s^{\text{new}}} |j_s k_s\rangle. \qquad (8.71)$$

It can be represented as

$$k_s \; \text{—□—} \; k_s^{\text{new}} \;=\; |j_s k_s^{\text{new}}\rangle\langle j_s k_s|. \qquad (8.72)$$
$$\qquad U_{j_s}$$

Since the initial and final states have the same spin, U_{j_s} is a transformation operator whose net spin is zero. This is a manifestation of spin conservation.

Using this basis transformation matrix, we can update the matrix elements of tensor operators in the system. For example, the reduced matrix elements of H_s are updated according to the equation

$$\left\langle j_s \tilde{k}_s^{\text{new}} \,\|H_s\| \, j_s k_s^{\text{new}} \right\rangle = \sum_{k_s' k_s} U^*_{j_s k_s' \tilde{k}_s^{\text{new}}} \langle j_s k_s' \|H_s\| j_s k_s\rangle U_{j_s k_s k_s^{\text{new}}}. \qquad (8.73)$$

Other matrix elements of tensor operators in the system are similarly updated.

In the DMRG calculation, the system is composed of the left block and a spin added in from its right end, $|j_s k_s\rangle = |j_l k_l\rangle \otimes |s\rangle$. Their basis states are related to each other by the relation

$$|j_s m_s k_s\rangle = \sum_{m_l m_1} \langle j_l m_l s m_1 | j_s m_s\rangle |j_l m_l k_l\rangle |s m_1\rangle. \qquad (8.74)$$

If the spin multiplets of the left block, (j_l, k_l), and the spin added in, s, are used to represent the spin multiplets in the system, U_{j_s} is converted to a three-indexed tensor:

$$k_s \; \text{⇢□⇢} \; k_s^{\text{new}} \;=\; j_l k_l \; \text{⇢□⇢} \; j_s k_s^{\text{new}}. \qquad (8.75)$$
$$\qquad U_{j_s} \qquad\qquad\qquad\qquad U_{j_s}$$

Here an arrow is added to each dangling bond to indicate the flow of states in the DMRG calculation explicitly. Again, the total spin is conserved by this three-indexed tensor: the total spin of the incoming states equals that of the outgoing state.

9

DMRG with Nonlocal Basis States

9.1 General Consideration

The idea of DMRG holds generally. It works not only in real space, but also in momentum space, angular momentum space, and molecular orbital or other nonlocal basis space. In a nonlocal basis space, a lattice "site" is nothing but a single-particle basis state, and the "lattice" is a collection of all allowed or selected "sites." For example, a "site" can be a momentum point in momentum space, an angular momentum point in the angular momentum space, or an atomic or molecular orbital in an atom or molecule. By properly ordering these single-particle basis states, the whole lattice can be mapped onto a one-dimensional lattice for which DMRG is implemented.

In real space, the Hubbard interaction is diagonal and local. However, in momentum space, the Hubbard interaction becomes nonlocal and off-diagonal because it contains terms that link two or more momentum points. As a result, the two-particle interactions generally contain N^4 terms in a nonlocal basis space, where N is the total number of single-particle basis states. If the system is translation invariant, then the total number of interaction terms can be reduced to N^3 terms in momentum space using the momentum conservation.

If the matrix elements for all these N^4 terms (N^3 terms in a translation-invariant system) need to be evaluated and updated in the DMRG calculation, then the number of single-particle basis states N that can be handled with DMRG is small. A key issue in the extension of DMRG to a nonlocal basis space is to reduce the number of operators whose matrix elements should be evaluated and stored. This problem was solved by a regrouping technique of fermion operators first introduced by Xiang [56] for the Hubbard model. The idea of this regrouping technique holds generally. It can be applied to a system with arbitrary long-range interactions. By defining a set of complementary operators of fermions, one can reduce the number of operators whose matrix elements need to be evaluated and stored to the order of N^2 for a general potential. If the system is translation invariant with a factorizable interacting potential, for example, the Hubbard interaction, the number of complementary operators can be further reduced to the order of N in momentum space.

The many-body basis states of the left and right blocks are incomplete for a given superblock configuration. If the Hamiltonian contains terms with interactions between these two blocks, the matrix elements of these interaction terms will be less accurately approximated than other terms defined purely within a block. In a one-dimensional system with only nearest-neighbor interactions, there is no direct interaction between the left and right blocks. However, interactions between left and right blocks always exist in a nonlocal basis space, no matter how the basis states are ordered. Thus, if the same number of basis states is retained, results obtained by DMRG in nonlocal basis space are generally less accurate than those obtained in real space if a one-dimensional quantum lattice model with only nearest-neighbor interactions is studied [19].

9.2 Momentum-Space DMRG

In a translation-invariant system, momentum is conserved. The use of this symmetry is helpful to the DMRG calculation. It can be employed to block diagonalize the Hamiltonian. Not only can this save computer time, but also it allows us to keep many more states in the basis truncation since the number of nonzero matrix elements is now significantly reduced by the momentum conservation. Empirically, one can keep about ten times more many-body basis states in momentum space than in real space. Furthermore, the momentum-space DMRG treats the kinetic energy rigorously. Hence, this method works better in the weak coupling limit, in contrast to its real-space counterpart.

The momentum-space DMRG provides an alternative way to implement DMRG in two or higher dimensions with periodic boundary conditions. It keeps the translation symmetry during the DMRG iteration, which is an advantage compared to the real-space DMRG. However, the application of the momentum-space DMRG has its limitations. The entanglement entropy of the Hubbard model, for example, obeys the volume law in the momentum space [376]. Furthermore, it is difficult, if not impossible, to apply this method to a pure spin system, such as the Heisenberg model.

In the next section, we take the Hubbard model as an example to show how the DMRG method works in momentum space. An extension to other interacting fermion models will be discussed later.

9.2.1 Hubbard Model

The Hubbard model in an applied field is defined by the Hamiltonian

$$H = -t \sum_{\langle ij \rangle \sigma} c_{i\sigma}^{\dagger} c_{j\sigma} + U \sum_{i} n_{i\uparrow} n_{i\downarrow} - h \sum_{i} (n_{i\uparrow} - n_{i\downarrow}), \tag{9.1}$$

where $\langle ij \rangle$ denotes summation over nearest neighbors, h is an applied magnetic field, and $n_{i\sigma} = c_{i\sigma}^{\dagger} c_{i\sigma}$. In momentum space, this model reads

$$H = \sum_{k\sigma} \varepsilon_{k\sigma} c_{k\sigma}^{\dagger} c_{k\sigma} + \frac{U}{N} \sum_{k_1 k_2 k_3} c_{k_1 \uparrow}^{\dagger} c_{k_2 \uparrow} c_{k_3 \downarrow}^{\dagger} c_{k_1 - k_2 + k_3 \downarrow}, \tag{9.2}$$

where N is the lattice size. $\varepsilon_{k\sigma}$ is the energy dispersion of electrons:

$$\varepsilon_{k\sigma} = \begin{cases} -2t \cos k - \sigma h, & \text{one dimension.} \\ -2t(\cos k_x + \cos k_y) - \sigma h, & \text{two dimensions.} \end{cases} \tag{9.3}$$

Periodic boundary conditions are assumed.

Each momentum k point has four degrees of freedom; namely, four independent basis states: $|0\rangle, c_{k\uparrow}^{\dagger}|0\rangle, c_{k\downarrow}^{\dagger}|0\rangle, c_{k\uparrow}^{\dagger} c_{k\downarrow}^{\dagger}|0\rangle$. Generally, a k point with these four states can be regarded as a lattice site. However, in a practical calculation, one can even treat the spin degree of freedom as an extra coordinate and take a momentum-spin point (k, σ), which has only two degrees of freedom, as a lattice site. This definition of "site" can reduce the total dimension of the superblock and allow us to retain more states at each step of basis truncation.

There are many ways to order these momentum-spin points. In practical calculation, one should test a few possibilities of orders of k points and choose the one which gives the lowest ground-state energy. A natural choice is to order the momenta according to their distance from the Fermi momentum: the closer the distance, the closer its position to the middle of the mapped one-dimensional lattice. A more rigorous or optimized scheme is to order these momentum-spin points such that the entanglement entropy is minimized at each step of DMRG iterations. After a few sweeps, an optimal order of the momentum-spin points will be eventually obtained. A detailed discussion of this scheme is given in §9.4.

Again, a superblock is a collection of four building blocks: the left and right blocks plus two momentum-spin points. The system contains the left block with one momentum-spin point, and the environment contains the right block with another momentum-spin point. We use S and E to represent the collections of all system and environment momentum-spin points, respectively.

The Hubbard interaction contains terms that link two or more k points. This interaction is a summation of N^3 terms. If the matrix elements for all these N^3 terms need to be evaluated and updated in the DMRG iteration, then the lattice size N that can be handled is small. These N^3 interaction terms are not independent:

they can be grouped into the terms defined purely in each block and the terms containing interactions between two blocks [56].

To understand this, let us define the following operators:

$$a_{k\sigma} = c_{k\sigma}\delta_{(k\sigma)\in S}, \qquad b_{k\sigma} = c_{k\sigma}\delta_{(k\sigma)\in E}, \qquad (9.4)$$

where $a_{k\sigma}$ and $b_{k\sigma}$ are the annihilation operators of electrons in the system and environment, respectively. The electron operator can then be expressed as

$$c_{k\sigma} = a_{k\sigma} + b_{k\sigma}. \qquad (9.5)$$

The Hamiltonian of the superblock is a sum of three terms:

$$H = H_s + H_e + H_{se}, \qquad (9.6)$$

where H_s (H_e) contains all interactions among electrons in the system (environment). H_s is defined by the Hamiltonian (9.2) with $c_{k\sigma}$ replaced by $a_{k\sigma}$:

$$H_s = \sum_{k\sigma} \varepsilon_{k\sigma} a_{k\sigma}^\dagger a_{k\sigma} + \frac{U}{N} \sum_{k_1,k_2,k_3} a_{k_1\uparrow}^\dagger a_{k_2\uparrow} a_{k_3\downarrow}^\dagger a_{k_1-k_2+k_3\downarrow}. \qquad (9.7)$$

Similarly, H_e is defined by Eq. (9.2) with $c_{k\sigma}$ replaced by $b_{k\sigma}$.

H_{se} contains the interaction terms between the system and environment blocks. It can be represented as

$$H_{se} = \frac{U}{N} \sum_k \left\{ a_1(k\uparrow)b_1(-k\downarrow) - a_2^\dagger(k)b_2(k) + a_4^\dagger(k)b_4(k) \right.$$
$$\left. + \sum_\sigma \left[b_{k\sigma}^\dagger a_3(k\sigma) + h.c. \right] \right\} + (a \leftrightarrow b). \qquad (9.8)$$

(a_1, a_2, a_3, a_4) are newly introduced complementary operators defined in the system block:

$$a_1(k\sigma) = \sum_q a_{q\sigma}^\dagger a_{k+q\sigma},$$

$$a_2(k) = \sum_q a_{q\uparrow}^\dagger a_{k+q\downarrow}, \qquad (9.9)$$

$$a_3(k\sigma) = \sum_{q_1 q_2} a_{q_1\bar\sigma}^\dagger a_{q_2\bar\sigma} a_{k+q_1-q_2\sigma} \delta_{(k\sigma)\notin s},$$

$$a_4(k) = \sum_q a_{q\downarrow} a_{k-q\uparrow},$$

where $\bar\sigma = -\sigma$. The corresponding operators for the environment are similarly defined. They can be obtained simply by replacing (a, s) with (b, e). The total number of these complementary operators is $6N$ in each block. In the DMRG iterations, only the matrix elements of these complementary operators, together with those of H_S and H_E, need to be evaluated.

The total momentum, P, is conserved in momentum space. As the number of up spins $N_\uparrow$ and the number of down spins $N_\downarrow$ are also conserved, the basis states are classified by three quantum numbers $(N_\uparrow, N_\downarrow, P)$. All $(N_\uparrow, N_\downarrow, P)$ subspaces that a system or environment block can have are determined by the $(k\sigma)$ points in the block.

In momentum space, different lattices have different k points. Thus, the DMRG calculation in momentum space needs to fix the lattice size from the beginning. Given an order of momentum-spin points, the initial blocks can be built up using the infinite lattice algorithm of DMRG until all momentum-spin points are included. During this warm-up procedure, it is helpful to keep at least one to two basis states at each $(N_\uparrow, N_\downarrow, P)$ subspace in the truncation of Hilbert space. This can avoid the situation wherein the basis states retained are trapped in a few $(N_\uparrow, N_\downarrow, P)$ subspaces. Otherwise, some of the $(N_\uparrow, N_\downarrow, P)$ subspaces, which may substantially contribute to the ground state, may be completely discarded in the DMRG sweep.

After each step of iteration, the system is augmented. If $(k\sigma)$ is the site added to the system, the augmented system becomes $S' = S \oplus (k\sigma)$. $H_{s'}$ is related to H_s by the equation:

$$H_{s'} = H_s + n_{k\sigma} \varepsilon_{k\sigma} + \frac{U}{N} \left[n_{k\sigma} a_1(0, \bar{\sigma}) + c_{k\sigma}^\dagger a_3(k\sigma) + a_3^\dagger(k\sigma) c_{k\sigma} \right]. \tag{9.10}$$

Similarly, the complementary operators are updated by the formula:

$$\begin{aligned}
a_1'(p\sigma') &= a_1(p\sigma') + \delta_{\sigma,\sigma'} \left(a_{k-p\sigma}^\dagger c_{k\sigma} + c_{k\sigma}^\dagger a_{k+p\sigma} + n_{k\sigma} \delta_{p,0} \right), \\
a_2'(p) &= a_2(p) + \delta_{\sigma,\uparrow} c_{k\uparrow}^\dagger a_{k+p\downarrow} + \delta_{\sigma,\downarrow} a_{k-p,\uparrow}^\dagger c_{k\downarrow}, \\
a_3'(p\sigma') &= a_3(p\sigma') + \delta_{\sigma,\sigma'} a_1(p - k\bar{\sigma}) c_{k\sigma} \\
&\quad + \delta_{\sigma',\bar{\sigma}} \left[n_{k\sigma} a_{p\bar{\sigma}} - \sigma c_{k\sigma}^\dagger a_4(p+k) - c_{k\sigma} \tilde{a}_2(p-k) \right], \\
a_4'(p) &= a_4(p) + \sigma a_{p-k,\bar{\sigma}} c_{k\sigma},
\end{aligned} \tag{9.11}$$

where $\tilde{a}_2(p) = a_2(p)$ if $\sigma = \uparrow$ and $a_2^\dagger(-p)$ if $\sigma = \downarrow$.

9.2.2 Extension to Other Lattice Models

Extending the momentum-space DMRG for the Hubbard model to the periodic Anderson lattice or other interacting electron models with only on-site Coulomb interactions is straightforward. For these models, the complementary operators defined in Eq. (9.9) can be used without modification.

For other interacting fermion models, the complementary operators introduced in Eq. (9.9) need to be modified or redefined. In general, the number of complementary operators can be reduced to the order of N^2. If the potential $V(q - q')$ can be factorized into a sum of products of a function of q and that of q'; that is,

$$V(q - q') = \sum_l f_l(q)g_l(q'), \tag{9.12}$$

then the total number of complementary operators needed can be further reduced to the order of N. The Hubbard model is an example of a factorizable potential with $f_1 = g_1 = 1$. The nearest-neighbor Coulomb potential is also factorizable.

Below we take the one-dimensional spinless fermion model with the nearest-neighbor Coulomb interaction as an example to show how to construct the Hamiltonian for the superblock and define the complementary operators for a factorizable potential. The idea can be readily extended to higher dimensions and other interacting fermion models.

The one-dimensional nearest-neighbor Coulomb interaction of spinless fermions is defined by the Hamiltonian

$$H_V = V_0 \sum_i n_i n_{i+1}, \tag{9.13}$$

where V_0 is the coupling constant and $n_i = c_i^\dagger c_i$ is the fermion number operator. In momentum space, this potential reads

$$V = \frac{V_0}{N} \sum_{kqq'} \cos(q - q')c_q^\dagger c_{q'} c_{k+q'}^\dagger c_{k+q}. \tag{9.14}$$

This potential is factorizable because

$$\cos(q - q') = \cos q \cos q' + \sin q \sin q'. \tag{9.15}$$

A superblock consists of a system and an environment. Again, we use S and E to represent the system and environment and define the fermion operators in these blocks as

$$a_k = c_k \delta_{k \in S}, \qquad b_k = c_k \delta_{k \in E}. \tag{9.16}$$

The fermion operator can be written as

$$c_k = a_k + b_k. \tag{9.17}$$

Substituting this expression into Eq. (9.14) gives

$$V = V_s + V_e + V_{se}, \tag{9.18}$$

where V_s is the interacting potential of the system block

$$V_s = \frac{V_0}{N} \sum_{kqq'} \cos(q - q') a_q^\dagger a_{q'} a_{k+q'}^\dagger a_{k+q}. \tag{9.19}$$

V_e is similarly defined and can be obtained from V_s by replacing (a, s) with (b, e). The interaction between the system and environment is given by

$$V_{se} = \frac{V_0}{N} \sum_k \left[\cos k a_1 (-k) b_1 (k) + a_2^\dagger (k) b_2 (k) + a_4^\dagger (k) b_4 (k) \right]$$
$$+ \frac{V_0}{N} \sum_k \left[a_3^\dagger (k) b_k + h.c. \right] + (a \leftrightarrow b), \tag{9.20}$$

where the complementary operators in the system are defined by

$$a_1 (k) = \sum_q a_{k+q}^\dagger a_q,$$

$$a_2 (k) = \sum_q (\cos q, \sin q) \, a_q^\dagger a_{q+k}, \tag{9.21}$$

$$a_3 (k) = 2 \sum_{pq} \cos q a_p^\dagger a_{p+q} a_{k-q} \delta_{k \notin S},$$

$$a_4 (k) = \sum_q (\cos q, \sin q) \, a_q a_{k-q}.$$

Both $a_2(k)$ and $a_4(k)$ are 2-component vector operators. The corresponding operators in the environment are similarly defined.

9.3 DMRG in a General Basis Space

The idea introduced in the preceding section can be generalized to an arbitrary non-local basis space. This generalization is essential, especially in the DMRG study of the quantum Hall effect, quantum chemistry, nuclear physics, and other interacting fermion systems. In 1999, White and Martin [57] carried out the first DMRG calculation for the ground state energy of water molecule. Within a basis set of 25 Hartree–Fock orbitals, they found that the groundstate energy obtained by DMRG already converged to the exact result obtained from the exact diagonalization by just keeping 400 many-body basis states. In 2003, Chan and Gordon increased the number of Hartree–Fock orbitals to 41 and calculated the ground-state energy of water molecules by keeping up to 6,000 many-body basis states [66].

To perform DMRG in a nonlocal basis space, we need first to choose a proper representation of single-particle basis states on which the Hamiltonian is defined.

In addition, we also need to introduce some complementary operators to reduce the number of operators whose matrix elements are evaluated and stored.

For a system without momentum conservation, the Hamiltonian is generally defined by

$$H = \sum_{ij} t_{ij} c_i^\dagger c_j + \frac{1}{2} \sum_{ijkl} V_{ijkl} c_i^\dagger c_j c_k^\dagger c_l, \tag{9.22}$$

where the site index i represents a nonlocal basis state. For an interacting electron system, the spin degrees of freedom are implicitly included in the definition of i. In that case, i represents a state with both orbital and spin quantum numbers, similar to the momentum-spin state in the momentum space. The first term of H includes the contribution from the kinetic energy and the single-particle potential. The second term is the interaction potential. It is a sum of N^4 terms, where N is the total number of orbitals.

This Hamiltonian is Hermitian, which requires that the hopping integral t_{ij} is Hermitian under the swap of the two subscripts:

$$t_{ij} = t_{ji}^*. \tag{9.23}$$

The two-body interaction potential is symmetric under the exchange of two particles. Hence, the potential tensor V_{ijkl} satisfies

$$V_{ijkl} = V_{klij} = V_{lkji} = V_{jilk}. \tag{9.24}$$

The Hamiltonian conserves the total number of fermions:

$$N_e = \sum_i c_i^\dagger c_i. \tag{9.25}$$

Using this property, we can rewrite the first term in Eq. (9.22) as

$$\sum_{ij} t_{ij} c_i^\dagger c_j = \frac{1}{N_e} \sum_{ijk} t_{ij} c_i^\dagger c_j c_k^\dagger c_k$$

$$= \frac{1}{2N_e} \sum_{ijkl} \left(t_{ij}\delta_{kl} + t_{kl}\delta_{ij} \right) c_i^\dagger c_j c_k^\dagger c_l, \tag{9.26}$$

which formally becomes a sum of four-fermion interaction terms, like the second term in the Hamiltonian.

Now we can define a new potential tensor:

$$V_{ijkl}' = V_{ijkl} + \frac{1}{N_e} \left(t_{ij}\delta_{kl} + t_{kl}\delta_{ij} \right), \tag{9.27}$$

and write the Hamiltonian as

$$H = \frac{1}{2} \sum_{ijkl} V_{ijkl}' c_i^\dagger c_j c_k^\dagger c_l. \tag{9.28}$$

Using Eqs. (9.23) and (9.24), it is simple to show that V' has the same symmtry as V:

$$V'_{ijkl} = V'_{klij} = V'_{lkji} = V'_{jilk}. \tag{9.29}$$

In a quantum chemistry calculation (similarly in other cases), the single-particle basis state created by $c_i^\dagger$ is generally taken as a molecular orbital obtained by the Hartree–Fock approximation. The DMRG calculation depends on how these single-particle basis states are defined and ordered. One can optimize the DMRG results by defining a new set of single-particle basis states, d_i, from a linear superposition of the initial basis set. These two sets of basis states are linked by a unitary matrix U:

$$d_i = \sum_{i'} U_{i'i}^* c_{i'}, \qquad c_i = \sum_{i'} U_{ii'} d_{i'}. \tag{9.30}$$

In this new set of basis states, the Hamiltonian becomes

$$H = \frac{1}{2} \sum_{ijkl} W_{ijkl} d_i^\dagger d_j d_k^\dagger d_l, \tag{9.31}$$

$$W_{ijkl} = \sum_{i'j'k'l'} V'_{i'j'k'l'} U_{i'i} U_{j'j}^* U_{k'k} U_{l'l}^*. \tag{9.32}$$

To perform DMRG calculations, we separate the fermion operators in the system and environment blocks into two sets of operators:

$$a_i = d_i \delta_{i \in S}, \qquad b_i = d_i \delta_{i \in E}. \tag{9.33}$$

Using these operators, the Hamiltonian is separated into three terms:

$$H = H_s + H_e + H_{se}. \tag{9.34}$$

H_s contains all the terms in the system block:

$$H_s = \frac{1}{2} \sum_{ij} \sum_{k \in E} W_{ikkj} a_i^\dagger a_j + \frac{1}{2} \sum_{ijkl} W_{ijkl} a_i^\dagger a_j a_k^\dagger a_l. \tag{9.35}$$

Similarly, H_e contains all the interactions within the environment block, which can be obtained from the above expression by replacing (a, e) with (b, s). H_{se} describes the interactions between the two blocks:

$$H_{se} = \sum_{ij} \sum_{k \in S} W_{ikkl} a_i^\dagger b_j + \sum_{ijkl} \left(W_{ijkl} a_i^\dagger a_k^\dagger a_l b_j + h.c \right)$$

$$+ \frac{1}{2} \sum_{ijkl} W_{ijkl} a_i^\dagger a_j b_k^\dagger b_l + \frac{1}{2} \sum_{ijkl} W_{ijkl} a_i^\dagger a_k^\dagger b_l b_j$$

$$- \frac{1}{2} \sum_{ijkl} W_{ijkl} a_i^\dagger a_l b_k^\dagger b_j + (a \leftrightarrow b, S \to E). \tag{9.36}$$

To define the complementary operators, we take the following SVDs for the potential tensor:

$$W_{ijkl} = W_{ij,kl}^{(1)} = \sum_m U_{ij,m}^{(1)} \lambda_m^{(1)} V_{kl,m}^{(1)},$$

$$W_{ijkl} = W_{ik,jl}^{(2)} = \sum_m U_{ik,m}^{(2)} \lambda_m^{(2)} V_{jl,m}^{(2)}, \tag{9.37}$$

$$W_{ijkl} = W_{il,kj}^{(3)} = \sum_m U_{il,m}^{(3)} \lambda_m^{(3)} V_{kj,m}^{(3)},$$

where the singular values of $\lambda^{(\alpha)}$ are descending ordered. Substituting (9.37) into (9.36), we obtain

$$H_{se} = \sum_m \left\{ a_1^\dagger (m) \, b_m + \left[a_2^\dagger (m) \, b_m + h.c \right] \right\} + \frac{1}{2} \sum_m \lambda_m^{(1)} a_3 (m) \, b_3 (m)$$

$$+ \frac{1}{2} \sum_m \left[\lambda_m^{(2)} a_4^\dagger (m) \, b_4 (m) - \lambda_m^{(3)} a_5 (m) \, b_5 (m) \right] + (a \leftrightarrow b), \tag{9.38}$$

where the complementary operators of the system are defined by

$$a_1^\dagger (m) = \sum_i \sum_{k \in S} W_{ikkm} a_i^\dagger \delta_{m \notin S}, \tag{9.39}$$

$$a_2^\dagger (m) = \sum_{ikl} W_{imkl} a_i^\dagger a_k^\dagger a_l \delta_{m \notin S}, \tag{9.40}$$

$$a_3 (m) = \sum_{ij} U_{ij,m}^{(1)} a_i^\dagger a_j, \tag{9.41}$$

$$a_4^\dagger (m) = \sum_{ik} U_{ik,m}^{(2)} a_i^\dagger a_k^\dagger, \tag{9.42}$$

$$a_5 (m) = \sum_{il} U_{il,m}^{(3)} a_i^\dagger a_l. \tag{9.43}$$

The corresponding complementary operators of the environment are defined by replacing $(S, a, U^{(\alpha)})$ in the above expressions with $(E, b, V^{(\alpha)})$.

The total number of independent a_1 and a_2 operators is of order N. In general, the rank of $\lambda_m^{(\alpha)}$ ($\alpha = 1, 2, 3$) is of order N^2. Thus, the total number of all complementary operators is of order N^2. In practical calculations, we often find that the singular value $\lambda_m^{(\alpha)}$ decays very quickly with increasing m. In this case, we can lower the rank of $\lambda^{(\alpha)}$ by discarding all small singular values, which, of course, will introduce an extra error in addition to the basis truncation error. If we assume $N^{(\alpha)}$ to be the number of singular values of $\lambda^{(\alpha)}$ retained, then the corresponding truncation error is given by

$$\varepsilon^{(\alpha)} = \frac{1}{\text{Tr}\lambda^{(\alpha)}} \sum_{m > N^{(\alpha)}} \lambda_m^{(\alpha)}. \tag{9.44}$$

This approximation is acceptable if and only if all the three truncation errors $\varepsilon^{(\alpha)}$ are sufficiently small.

In the DMRG iteration, the matrix elements of a_1 and a_2 operators can be updated from their matrix elements stored in the previous iteration. However, to evaluate the matrix elements of a_3, a_4, and a_5, we cannot use the matrix elements of these operators obtained in the previous iteration. In fact, the matrix elements of a_3, a_4, and a_5 do not need to be stored. Instead, what really needs to be stored and updated are the matrix elements of $a_{i\sigma}^\dagger a_{j\sigma'}$ and $a_{i\sigma}^\dagger a_{j\sigma'}^\dagger$. The matrix elements of a_3, a_4, and a_5 can be obtained from these operators using Eqs. (9.41)–(9.43).

9.4 Optimization of Single-Particle Basis States

DMRG is a minimal entanglement approach. It tends to minimize the entanglement entropy between the system and environment blocks at each iteration step. This goal, however, may not always be optimally achieved in the DMRG calculation when the basis states become nonlocal. In fact, for a given basis set, two problems should be solved in order to optimize a DMRG calculation:

(i) Ordering of basis states: unlike in real space, there is no natural lattice structure or geometry to order these basis states. For example, one can order the basis states according to their occupation numbers. However, it may not be an optimal scheme to implement.

(ii) Basis optimization: the DMRG calculation depends on the choice of the single-particle basis states since the interaction integrals are different in different basis representations. Different basis sets are related to each other by unitary transformations. Therefore, it is desirable to find an optimized transformation matrix that can reduce the entanglement entropy but does not increase the computational complexities.

In a DMRG calculation, it is impossible to solve these two problems from the beginning or just in one step. However, one can improve the basis states and, at the same time, reorder them by variationally optimizing the basis transformation matrix such that the entanglement entropy is minimized at each step of DMRG iterations [259]. To understand this, let us consider the wave function generated by DMRG:

$$|\Psi\rangle = \sum_{l,\sigma_i,\sigma_{i+1},r} \Psi\left(l,\sigma_i,\sigma_{i+1},r\right)|l,\sigma_i,\sigma_{i+1},r\rangle, \tag{9.45}$$

where l and r are the quantum numbers of basis states for the left and right blocks, respectively. If $a_i^\dagger$ and $a_{i+1}^\dagger$ are the fermion creation operators at sites i and $i+1$, respectively, then

$$|\sigma_i\rangle = \left(a_i^\dagger\right)^{\sigma_i}|0\rangle, \qquad |\sigma_{i+1}\rangle = \left(a_{i+1}^\dagger\right)^{\sigma_{i+1}}|0\rangle. \tag{9.46}$$

Both l and r are many-body basis states. It is difficult to manipulate them to optimize the single-particle basis states. However, σ_i and σ_{i+1} are single-particle states. They can be readily mixed to form two new orthonormal single-particle basis states, α_i and α_{i+1}, by taking a unitary transformation:

$$\begin{pmatrix} \alpha_i^\dagger \\ \alpha_{i+1}^\dagger \end{pmatrix} = U(\theta) \begin{pmatrix} a_i^\dagger \\ a_{i+1}^\dagger \end{pmatrix}, \tag{9.47}$$

where $\alpha_i^\dagger$ and $\alpha_{i+1}^\dagger$ are the fermion creation operators for the new basis states at these two sites. $U(\theta)$ is a 2×2 orthogonal matrix:

$$U(\theta) = \begin{pmatrix} \cos\theta & \sin\theta \\ -\sin\theta & \cos\theta \end{pmatrix}, \tag{9.48}$$

and θ is the mixed angle between the two basis states. The transformation at $\theta = \pi/2$ is to swap the two basis states. Here we only consider an orthogonal transformation since the Hamiltonian is real.

In this new basis representation, the ground-state wave function becomes θ-dependent:

$$|\Psi(\theta)\rangle = \sum_{l,\alpha_i,\alpha_{i+1},r} \Psi_\theta(l, \alpha_i, \alpha_{i+1}, r) |l, \alpha_i, \alpha_{i+1}, r\rangle. \tag{9.49}$$

Ψ_θ is related to Ψ by a unitary transformation:

$$\Psi_\theta(l, 0, 0, r) = \Psi(l, 0, 0, r), \tag{9.50}$$

$$\Psi_\theta(l, 1, 1, r) = \Psi(l, 1, 1, r), \tag{9.51}$$

$$\begin{pmatrix} \Psi_\theta(l, 0, 1, r) \\ \Psi_\theta(l, 1, 0, r) \end{pmatrix} = U^\dagger(\theta) \begin{pmatrix} \Psi(l, 0, 1, r) \\ \Psi(l, 1, 0, r) \end{pmatrix}. \tag{9.52}$$

Equation (9.51) is valid because $\alpha_i^\dagger \alpha_{i+1}^\dagger = a_i^\dagger a_{i+1}^\dagger$.

To optimize the basis states, we group (l, α_i) as the system and (α_{i+1}, r) as the environment, and evaluate the entanglement Renyi entropy between them in the ground state [259]:

$$S_\mu(\theta) = \frac{1}{1-\mu} \ln \mathrm{Tr}(\rho_\theta)^\mu, \tag{9.53}$$

where ρ_θ is the reduced density matrix defined by

$$\rho_\theta(l'\alpha_i', l\alpha_i) = \sum_{r\alpha_{i+1}} \Psi_\theta^*(l', \alpha_i', \alpha_{i+1}, r)\, \Psi_\theta(l, \alpha_i, \alpha_{i+1}, r). \tag{9.54}$$

Ψ_θ is normalized to 1. In the limit $\mu \to 1$, $S_\mu(\theta)$ reduces to the von Neumann entropy:

$$S(\theta) = -\mathrm{Tr}\rho_\theta \ln \rho_\theta. \tag{9.55}$$

The optimized θ is determined by minimizing the Renyi entropy. The exponent μ does not change the spectrum of the reduced density matrix, but it changes the weight of each eigenvalue of ρ_θ in the Renyi entropy. In principle, μ can be used as a variational parameter to minimize the entanglement entropy. In practical calculations, we find it is sufficient to take $\mu \rightarrow 1$; that is, taking the von Neumann entropy, or $\mu = 1/2$. The results obtained with these two kinds of criteria are not qualitatively different. However, the calculation with $\mu = 1/2$ converges faster in many cases.

If a conservation law, for example, the momentum conservation, is imposed, the two single-basis states cannot be mixed with an arbitrary angle θ. In this case, θ can take just two values, 0 and $\pi/2$. $\theta = \pi/2$ is to swap the two basis states.

Once the optimized unitary transformation angle θ is determined, the basis space is updated. The unitary basis transformation matrix defined in Eq. (9.30) is changed to

$$U = U_0 \cdot \left(I_l \otimes U_\theta^\dagger \otimes I_r \right), \tag{9.56}$$

where I_l and I_r are the identity matrices for the L and R blocks, respectively. U_0 is the basis transformation matrix obtained in the preceding step.

9.4.1 Ordering of Basis States

The above transformation provides an efficient scheme to optimize the single-particle basis states. It also provides a practical approach to reorder the basis states in the DMRG iteration. However, this basis transformation is implemented between two neighboring sites at each step of DMRG iterations. It underestimates the correlation between two basis sites that are strongly correlated but initially separated far from each other. This strong correlation happens, for example, when a square lattice is mapped onto a one-dimensional one by a snake path on which two neighboring sites on the original lattice may become well separated. Due to the truncation of many-body states, the DMRG iteration might be stuck at local minima. In this case, the correlation between these two sites may not be properly or efficiently recovered by the successive single-particle basis transformations.

A simple solution to this problem is to randomly mix or reorder some of the basis states. Alternatively, one can reorder all the basis sites according to a criterion based on the mutual information [377, 378]. The mutual information between two basis states, σ_1 and σ_2, is defined by

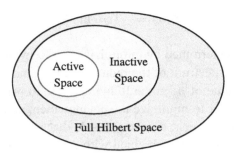

Figure 9.1 Schematic illustration of the active basis space used in the DMRG calculation, the inactive basis space, and the full Hilbert space. The sum of the active and inactive basis space is the whole basis space generated by the self-consistent Hartree–Fock or other kinds of approximations.

$$I_{\sigma_1,\sigma_2} = S_{\sigma_1} + S_{\sigma_2} - S_{\sigma_1\sigma_2} \geq 0, \qquad (9.57)$$

where S_{σ_1}, S_{σ_2}, $S_{\sigma_1\sigma_2}$ are the entanglement entropy of σ_1, σ_2, and $\sigma_1\sigma_2$, respectively. It is a measure of the mutual entanglement between σ_1 and σ_2. The larger is $I_{\sigma_1\sigma_2}$, the stronger is the correlation between σ_1 and σ_2. If they are completely independent, then $I_{\sigma_1\sigma_2} = 0$.

We order the basis sites by requiring that any two strongly correlated basis states are arranged as closely as possible. It can be achieved by minimizing the entanglement distance:

$$f = \sum_{ij} I_{\sigma_i,\sigma_j} \cdot |i-j|, \qquad (9.58)$$

where $|i-j|$ is the distance between the two basis sites at i and j.

After reordering all basis states, we rebuild all system and environment subblocks needed for carrying out the finite-lattice sweep. We then carry out the basis transformations introduced above to further optimize and reorder the basis states.

9.5 Optimizing Active Basis Space

To improve the accuracy of DMRG, one can increase the size of the single-particle basis space and the number of many-body states retained in the DMRG calculation. However, this demands a dramatic increase in computing resources. In a molecular system, the dimension of the Hilbert space is infinite. The single-particle basis states (or orbitals) used in the DMRG calculation, as illustrated in Fig. 9.1, are only a small portion of the basis states obtained with the self-consistent Hartree–Fock or other kinds of approximations. These basis states are obtained without fully considering the correlation effect of all electrons. As other basis states not included in the DMRG calculation may contribute significantly to the ground state, it suggests that the active basis set directly obtained with the self-consistent Hartree–Fock or other single-particle approximation may not be the optimal one.

One can use DMRG to optimize the active basis states [67]. For this purpose, we first generate a bigger orbital set using a mean-field approach, such as the self-consistent Hartree–Fock theory and then divide them into two groups. The first group contains active orbitals used in the DMRG calculation. The second group contains all other orbitals, taken as a reservoir. We define the Hamiltonian in the whole basis space but carry out the DMRG calculations only using the active orbitals.

Initially, the orbitals in the reservoir are either fully occupied or not occupied if their energies are below or above the Fermi level. By freezing the orbitals in the reservoir, one can rewrite the Hamiltonian as

$$H = E_0 + \sum_{ij \in \mathcal{A}} t_{ij}^a c_i^\dagger c_j + \frac{1}{2} \sum_{ijkl \in \mathcal{A}} V_{ijkl} c_i^\dagger c_j^\dagger c_l c_k, \qquad (9.59)$$

where E_0 is the energy contributed by the orbitals in the reservoir:

$$E_0 = \sum_{ij \in \mathcal{I}} \left(t_{ii} \delta_{ij} - V_{ijji} \langle c_j^\dagger c_j \rangle + V_{ijij} \langle c_j^\dagger c_j \rangle \right) \langle c_i^\dagger c_i \rangle. \qquad (9.60)$$

Here $\langle c_i^\dagger c_i \rangle = 1$ if orbital i is below the Fermi level or 0 otherwise. $\mathcal{A}$ and $\mathcal{I}$ represent the active and inactive basis space, respectively. $E_0 = 0$ if all the orbitals below the Fermi level are included in the active space. The coupling constant t_{ij}^a is the hopping integral between two active orbitals. It includes the contribution from the potential energy between active and inactive orbitals:

$$t_{ij}^a = t_{ij} + \sum_{k \in \mathcal{I}'} \left(V_{ikkj} - V_{ikjk} \right) \langle c_k^\dagger c_k \rangle. \qquad$$

By solving the Hamiltonian with DMRG in the active orbital space, one can evaluate the one-particle density matrix from the ground-state wave function $|\Psi\rangle$:

$$\rho_{ij}^{(1)} = \langle \Psi | c_i^\dagger c_j | \Psi \rangle. \qquad (9.61)$$

The eigenvectors of $\rho^{(1)}$ define a new set of orthogonal one-electron basis states, called natural orbitals. The eigenvalue of $\rho^{(1)}$ is the occupation number of the corresponding natural orbital in the ground state. The orbitals with the highest or lowest occupation number have the least contribution to the exchange and correlation energy. They are less important compared to other orbitals. Thus by diagonalizing the single-particle density matrix, one can optimize the basis states by freezing 5% to 10% of less important active natural orbitals and activating the same number of inactive orbitals in the reservoir. This update of the basis states defines a new set

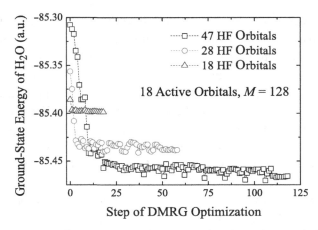

Figure 9.2 The ground-state energy of H_2O as a function of the optimization step for the three sets of Pople-type basis states, which contain 47, 28, and 18 Hartree–Fock orbitals, respectively. There are 18 orbitals in the active space. In the DMRG calculation, 128 many-body basis states are retained. (Adapted from Ref. [67].) Copyright by the American Physical Society.

of active orbital space. Again these orbitals can be orthogonalized and optimized with the orbitals in the inactive space using DMRG. If all the orbitals in the reservoir are activated once by repeating the above procedure, we define it as a cycle of optimization. Generally, a few cycles are needed to obtain the most optimized orbitals.

Figure 9.2 shows how the ground-state energy of water molecule varies with the step of optimization using 18 active orbitals obtained with DMRG. Three sets of Pople-type Hartree–Fock bases, 6-311++G(2d, 2p), 6-311+G* and 6-31G*, are used. They contain 47, 28, and 18 Hartree–Fock orbitals, respectively. For the third basis set, all 18 orbitals are used. In this case, the orbital optimization is to orthogonalize the molecular orbitals through the unitary transformation defined by the single-particle density matrix. There are orbital exchanges between active and reservoir spaces for the other two cases shown in Fig. 9.2. At each time, the three least-occupied orbitals in the active space are swapped with the orbitals in the reservoir. A cycle of optimization needs 4 and 10 times of swapping for the systems with 28 and 47 Hartree–Fock orbitals, respectively. Figure 9.2 shows the results for 10 cycles of optimizations and additional 10 instances of the DMRG sweeping in the optimized active space.

For all three cases shown in Fig. 9.2, the orbital optimization significantly improves the DMRG results. The improvement is most striking in the first cycle of orbital optimization. After that, the improvement becomes relatively small, which suggests that two to three cycles of orbital optimization are sufficient in practical applications.

Another feature revealed by Fig. 9.2 is that the greater the number of Hartree–Fock orbitals, the lower (hence better) the ground-state energy. This feature is a natural consequence of a larger basis set involving more correlations not considered in the Hartree–Fock approximation. Thus to improve the result, one should include as many orbitals as possible in the whole basis space once the memory space for storing the matrix elements of V_{ijkl} is allowed.

10

Matrix Product States

10.1 The DMRG Wave Function

In DMRG, the wave function is represented using the basis states of the system and environment blocks. Using the local basis states, one can also represent the system and environment basis states through the basis transformation matrices obtained in the DMRG iterations. We can use these transformations to represent the DMRG wave function as a *matrix product state* (MPS). An MPS, as pointed out by Östlund and Rommer [20], can also be regarded as a variational wave function determined by variationally minimizing the ground-state energy.

To generate an MPS, let us consider a DMRG wave function:

$$|\Psi\rangle = \sum_{\sigma_i \sigma_{i+1}} \sum_{l_{i-1} r_{i+2}} \Psi(l_{i-1}, \sigma_i, \sigma_{i+1}, r_{i+2})|l_{i-1}, \sigma_i, \sigma_{i+1}, r_{i+2}\rangle. \qquad (10.1)$$

We first use SVD to decompose this wave function into the form

$$\Psi(l_{i-1}, \sigma_i, \sigma_{i+1}, r_{i+2}) = \quad \begin{matrix} l_{i-1}, \sigma_i & \sigma_{i+1}, r_{i+2} \\ \mid & \mid \\ \bigcirc\!\!-\!\!\diamond\!\!-\!\!\bigcirc \\ U_i \quad \lambda_i \quad V_{i+1} \end{matrix} \qquad . \qquad (10.2)$$

The singular value matrix λ_i is just the square root of the eigenvalue matrix of the reduced density matrix. U_i and V_{i+1} are the basis transformation matrices. U_i can be represented as a three-leg tensor:

$$\begin{matrix} l, \sigma \\ \mid \\ \boxed{U_i}\!\!-\!m \end{matrix} \quad = \quad \begin{matrix} \sigma \\ \mid \\ l -\!\boxed{U_i}\!\!-\!m \end{matrix} \quad \equiv \quad (U_i)_{l,m}[\sigma]. \qquad (10.3)$$

Similarly, V_{i+1} can also be written as a three-leg tensor:

$$m \overset{\sigma,r}{\underset{V_{i+1}}{-\bigcirc-}} \; = \; m \overset{\sigma}{\underset{V_{i+1}}{-\bigcirc-}} r \equiv (V_{i+1})_{m,r}[\sigma]. \tag{10.4}$$

In the above expressions, the bond dimension of m equals the product of the bond dimension of l (or r) and the dimension of σ. Hence, if the bond dimension of l or r is D, and that of σ is d, then m has dimension dD. After the DMRG truncation, the dimension of m reduces to D, and U_i and V_{i+1} become isometric:

$$(U_i)_{l,m}[\sigma] = l \overset{\sigma}{\underset{U_i}{-\bigtriangleup-}} m \quad, \qquad (V_{i+1})_{m,r}[\sigma] = m \overset{\sigma}{\underset{V_{i+1}}{-\bigtriangleup-}} r. \tag{10.5}$$

They are left and right canonicalized, respectively:

$$\overset{U_i^*}{\underset{U_i}{\boxminus}} \; = I, \qquad \overset{V_{i+1}^*}{\underset{V_{i+1}}{\boxminus}} \; = I. \tag{10.6}$$

Here we use the same symbols to represent the tensors before and after truncation. Below it is the tensors after truncation that are used.

The local basis states in the left and right blocks are labelled by l_{i-1} and r_{i+2}, respectively. By adding a site into the left or right block, their basis states are transformed according to the equations:

$$|l_i\rangle = \sum_{l_{i-1}\sigma_i} (U_i)_{l_{i-1},l_i}[\sigma_i] |l_{i-1}\sigma_i\rangle, \tag{10.7}$$

$$|r_{i+1}\rangle = \sum_{r_{i+2}\sigma_{i+1}} (V_{i+1})_{r_{i+1},r_{i+2}}[\sigma_{i+1}] |\sigma_{i+1}r_{i+2}\rangle. \tag{10.8}$$

U_i and V_{i+1} are just the basis transformation matrices obtained in the DMRG calculation. Since both U_i and V_{i+1} are isometric, $|l_i\rangle$ and $|r_i\rangle$ are biorthonormal basis states in the truncated basis space:

$$\langle l_i | l_i'\rangle = \delta_{l_i,l_i'}, \qquad \langle r_{i+1} | r_{i+1}'\rangle = \delta_{r_{i+1},r_{i+1}'}. \tag{10.9}$$

Using the basis transformation equations (10.7) and (10.8) recursively, we can eventually express the DMRG ground-state wave function as an MPS:

$$|\Psi\rangle = \sum_{\sigma_1\cdots\sigma_N} \Psi_{\sigma_1\cdots\sigma_N} |\sigma_1\cdots\sigma_N\rangle, \tag{10.10}$$

$$\Psi_{\sigma_1\cdots\sigma_N} = \begin{array}{cccc} \sigma_1 & \sigma_i & \sigma_{i+1} & \sigma_N \\ & & & \\ \square & \cdots & \square \diamond \square & \cdots & \square \\ U_1 & U_i & \lambda_i \; V_{i+1} & V_N \end{array} , \qquad (10.11)$$

where U-matrices are left canonical and V-matrices are right canonical. U_1 and V_N are $1 \times D$ and $D \times 1$ matrices, respectively.

10.2 Canonical Representations

Equation (10.11) shows that the DMRG wave function is an MPS. It can be shown that any given wave function can always be represented as an MPS if there is no restriction to the bond dimension. However, the MPS representation of a wave function is not unique and possesses much redundancy. For example, the wave function

$$\Psi_{\sigma_1\cdots\sigma_N} = \begin{array}{ccc} \sigma_1 & \sigma_2 & \sigma_N \\ & & \\ (A_1) (A_2) & \cdots & (A_N) \end{array} \qquad (10.12)$$

is invariant under arbitrary local gauge transformations for the local matrices

$$A_i[\sigma_i] \to P_{i-1} A_i[\sigma_i] P_i^{-1}, \qquad (10.13)$$

where P_i with $P_0 = P_N = 1$ is an arbitrary set of invertible matrices. These redundant degrees of freedom can be partially fixed by taking gauge transformations so that the basis states thus generated are orthonormalized or canonicalized, and the corresponding MPS is in a canonical form.

By the canonical form, we mean that matrix products over a part of the lattice are orthogonal eigenvectors of the corresponding reduced density matrix. There are many ways to convert an MPS into a canonical form. Consequently, there are also different kinds of canonical representations of MPS. The most commonly used canonical forms include:

(i) Right canonical form
   An MPS is right canonicalized if all local matrices satisfy the right canonical conditions:

$$\Lambda_{i-1} \begin{array}{c} \langle A_i^* | \\ \\ \langle A_i | \end{array} = \diamond \Lambda_i , \qquad \begin{array}{c} \langle A_i^* | \\ \\ \langle A_i | \end{array} = I, \qquad (10.14)$$

where Λ_i is the diagonal eigenvalue matrix of the reduced density matrix

$$\rho_i = \mathrm{Tr}_{\sigma_{i+1}\cdots\sigma_N} |\Psi\rangle \langle\Psi|. \qquad (10.15)$$

The left canonical representation of the wave function is defined similarly, with all local matrices being left canonicalized.

(ii) $(\lambda - \Gamma)$-canonical form

In Eq. (10.12), if we decouple A_i into a product of a site tensor Γ_i and a diagonal bond matrix $\lambda_i = \sqrt{\Lambda_i}$:

$$-\!\!\left(A_i\right)\!\!- \;\; = \;\; -\!\!\left(\Gamma_i\right)\!\!-\!\!\left\langle\lambda_i\right\rangle\!\!- \;, \tag{10.16}$$

we can rewrite the wave function as

$$\Psi_{\sigma_1\ldots\sigma_N} = \underset{\lambda_0\;\;\Gamma_1\;\;\lambda_1\;\;\Gamma_2\;\;\lambda_2}{\overset{\sigma_1\qquad\sigma_2\qquad\qquad\sigma_N}{\diamond\!\!-\!\!\bigcirc\!\!-\!\!\diamond\!\!-\!\!\bigcirc\!\!-\!\!\diamond\cdots-\!\!\bigcirc\!\!-\!\!\diamond}}\;\underset{\Gamma_N\;\;\lambda_N}{}. \tag{10.17}$$

In this expression, $\lambda_0 = \lambda_N = 1$. If the wave function is canonicalized, then Γ_i and λ_i satisfy the equations:

$$\Lambda_{i-1}\!\left\langle\diamond\right|\begin{matrix}\overset{\frown}{\left(\Gamma_i^*\right)}\\[4pt]\underset{\smile}{\left(\Gamma_i\right)}\end{matrix} \;\; = I, \qquad \begin{matrix}-\!\!\left(\Gamma_i^*\right)\overset{\frown}{}\\[4pt]-\!\!\left(\Gamma_i\right)\underset{\smile}{}\end{matrix}\!\left|\diamond\right\rangle\!\Lambda_i \;\; = I. \tag{10.18}$$

(iii) Left-right canonical form

The left-right canonical representation is generally more convenient to use in practical calculations. For example, the wave function generated by DMRG, Eq. (10.11), is a realization of this representation. In Eq. (10.17), if we set

$$-\!\!\left(V_i\right)\!\!- \;\; = \;\; -\!\!\left(\Gamma_i\right)\!\!-\!\!\left\langle\lambda_i\right\rangle\!\!- \;, \tag{10.19}$$

$$-\!\!\left(U_i\right)\!\!- \;\; = \;\; -\!\!\left\langle\lambda_{i-1}\right\rangle\!\!-\!\!\left(\Gamma_i\right)\!\!- \;, \tag{10.20}$$

then $|\Psi\rangle$ is just the wave function given in Eq. (10.11). This MPS is in the left-right canonical representation because V_i is right canonicalized, satisfying the right canonical equation (10.14), and U_i is left canonicalized:

$$\begin{matrix}\overset{\frown}{-\!\!\left(U_i^*\right)}\\[4pt]\underset{\smile}{-\!\!\left(U_i\right)}\end{matrix} \;\; = I, \qquad \begin{matrix}-\!\!\left(U_i^*\right)\overset{\frown}{}\\[4pt]-\!\!\left(U_i\right)\underset{\smile}{}\end{matrix}\!\left|\diamond\right\rangle\!\Lambda_i \;\; = \left|\diamond\right\rangle\!\Lambda_{i-1} \;. \tag{10.21}$$

The canonical representation removes redundant virtual degrees of freedom. It provides an efficient way to reveal the physical properties of MPS, especially in the following three aspects:

(i) It facilitates the computation of expectation values of local operators. For instance, in the left–right canonical representation, the expectation value of a local operator O_i is determined purely by the local tensor at that site:

$$\langle \Psi | O_i | \Psi \rangle =$$

$$= \qquad (10.22)$$

Similarly, the expression for a two-point correlation function $\langle \Psi | O_i O_j | \Psi \rangle$ involves only the local matrices between i and j.

(ii) It simplifies the truncation of bond indices. If an MPS is in a left–right canonical form, one can retain the largest D diagonal matrix elements of Λ_i. This truncation scheme is optimal. It is also how DMRG works.

(iii) A canonicalized MPS remains canonical upon any local unitary basis transformation.

10.3 Canonical Transformation

An arbitrarily given MPS may not be canonical. However, it can be canonicalized by taking a series of local transformations. For example, a right canonicalized MPS can be realized in two steps.

(i) Step I: Right canonicalization

Let us start from an MPS defined by Eq. (10.12) whose local tensors do not satisfy the canonical conditions. The first step is to find a set of matrices V_i from A_i that are right isometric. It can be done by taking LQ decompositions from the right end of the chain.

We first take a LQ decomposition for A_N:

$$
\alpha \underset{}{\longrightarrow} \overset{\sigma}{\underset{A_N}{\bigcirc}} \underset{}{\longrightarrow} \beta \quad = \quad \alpha \underset{}{\longrightarrow} \overset{}{\underset{L_N}{\bigcirc}} \underset{}{\longrightarrow} \overset{\sigma}{\underset{V_N}{\square}} \underset{}{\longrightarrow} \beta \tag{10.23}
$$

where the bond dimension of β equals 1. V_N satisfies the right isometric condition:

$$
\begin{matrix} \boxed{V_N^*} \\ | \\ \boxed{V_N} \end{matrix} = I. \tag{10.24}
$$

We then absorb L_N into A_{N-1} and take another LQ decomposition:

$$
\underset{A_{N-1}}{\bigcirc} \underset{L_N}{\bigcirc} = \underset{L_{N-1}}{\bigcirc} \underset{V_{N-1}}{\square} \tag{10.25}
$$

Again V_{N-1} is right isometric.

Repeating these steps, each time we absorb the L-matrix obtained in the previous step and perform an LQ decomposition for this combined tensor to generate a new right isometric matrix, V_i, and a lower triangular matrix, L_i:

$$
\underset{A_i}{\bigcirc} \underset{L_{i+1}}{\bigcirc} = \underset{L_i}{\bigcirc} \underset{V_i}{\square} . \tag{10.26}
$$

Once the iteration reaches the left end of the chain, we define

$$
\alpha \underset{}{\longrightarrow} \overset{\sigma}{\underset{V_1}{\square}} \underset{}{\longrightarrow} \beta \quad = \quad \alpha \underset{}{\longrightarrow} \overset{\sigma}{\underset{A_1}{\bigcirc}} \underset{}{\longrightarrow} \underset{L_2}{\bigcirc} \underset{}{\longrightarrow} \beta, \tag{10.27}
$$

where the bond dimension of α is 1. V_1 is also right isometric. The right isometric condition of V_1 is just the normalization condition of the wave function:

$$
\langle \Psi | \Psi \rangle = \alpha \begin{bmatrix} \boxed{V_1^*} \\ | \\ \boxed{V_1} \end{bmatrix} = \begin{matrix} \boxed{V_1^*} \\ | \\ \boxed{V_1} \end{matrix} = 1. \tag{10.28}
$$

The second equality holds because the dimension of α equals 1.

(ii) Step II: Left canonicalization

Now we start from the left end of the chain. We first diagonalize the following Hermitian matrix by a unitary matrix X_1:

$$
\sum_\sigma V_1^\dagger[\sigma]V_1[\sigma] = X_1^\dagger \Lambda_1 X_1 \tag{10.29}
$$

or graphically:

where Λ_1 is the diagonal eigenvalue matrix. A_1 is then reset to

$$A_1[\sigma] = V_1[\sigma]X_1^\dagger. \qquad (10.30)$$

It is simple to show that A_1 satisfies the two right canonical equations:

$$\qquad\qquad\qquad\qquad\qquad\qquad\qquad\qquad\qquad (10.31)$$

with $\Lambda_0 = 1$.
By diagonalizing the matrix

$$\sum_\sigma V_2^\dagger[\sigma]X_1^\dagger\Lambda_1 X_1 V_2[\sigma] = X_2^\dagger\Lambda_2 X_2 \qquad (10.32)$$

with a unitary matrix X_2, we reset the local tensor at site 2 as

$$A_2[\sigma] = X_1 V_2[\sigma]X_2^\dagger. \qquad (10.33)$$

Again A_2 is right canonical:

$$\qquad\qquad\qquad\qquad\qquad\qquad\qquad\qquad\qquad (10.34)$$

To keep on with this procedure to the right end of the lattice, we obtain the last canonicalized local tensor:

$$A_N[\sigma] = X_{N-1}V_N[\sigma]. \qquad (10.35)$$

From the normalization condition, it is simple to verify that the two right canonical conditions are satisfied:

$$\qquad\qquad\qquad\qquad\qquad\qquad\qquad\qquad\qquad (10.36)$$

$$\begin{array}{c} \boxed{A_N^*} \\ \\ \boxed{A_N} \end{array} = I. \tag{10.37}$$

From these two canonicalization steps, we can also show that the canonical form is uniquely determined up to permutations and degeneracies in the diagonal eigenspectra Λ_i.

10.4 Implementation of Symmetries

In case the system is invariant under certain symmetry transformations, the Hilbert space is divided into several symmetry sectors in which the Hamiltonian is block diagonalized. Targeting a state in a specific symmetry sector can save both computational time and memory space significantly.

A symmetry group can be classified as either Abelian or non-Abelian. An Abelian group, also called a commutative group, is a group in which the result of applying the group operation to two group elements does not depend on their order. Abelian groups generalize the arithmetic of the addition of integers. A non-Abelian group is noncommutative. It is a group in which at least two elements do not commute with each other.

An essential step in implementing symmetry is constructing an MPS with this symmetry. In the following, we will take the SU(2) Heisenberg model as an example to show how to implement the Abelian U(1) and non-Abelian SU(2) symmetries in the MPS calculation. It is simple to extend it to a lattice model with other symmetries. In particular, the discussion on the U(1) symmetry can be readily generalized to a system with other additive quantum numbers, such as a Z_N-symmetric system. On the other hand, the discussion on SU(2) symmetry can be generalized to a system with other non-Abelian symmetries, such as the SU(3) symmetry.

10.4.1 Abelian Symmetries

Let us first consider an MPS with open boundary conditions:

$$|\Psi\rangle = \sum_{\sigma_1 \cdots \sigma_N} A_1[\sigma_1] \cdots A_N[\sigma_N] |\sigma_1 \cdots \sigma_N\rangle, \tag{10.38}$$

where A_1 is a $1 \times D$ matrix and A_N is a $D \times 1$ matrix. To understand the symmetry properties of MPS, we note that the local tensor A_i is to map the states

$$|\psi(\alpha_{i-1})\rangle = \sum_{\sigma_1 \cdots \sigma_{i-1}} (A_1[\sigma_1] \cdots A_{i-1}[\sigma_{i-1}])_{\alpha_{i-1}} |\sigma_1 \cdots \sigma_{i-1}\rangle, \tag{10.39}$$

defined on the subsystem spanned by the first $(i-1)$ sites to the states

$$|\psi(\alpha_i)\rangle = \sum_{\alpha_{i-1}\sigma_i} (A_i)_{\alpha_{i-1}\alpha_i}[\sigma_i]|\psi(\alpha_{i-1})\rangle, \tag{10.40}$$

on the subsystem spanned by the first i sites. Therefore, the local tensor A_i behaves like an operator that maps incoming states $|\psi(\alpha_{i-1})\rangle$ to a system enlarged by one site. The spin conservation implies that the total spin of the input states should be that of the output states.

For the Heisenberg model, the total spin along the z-axis, S_z, is conserved. If we use $[\alpha_i]$ to denote the magnetic quantum number of the state $|\psi(\alpha_i)\rangle$, then the spin conservation can be imposed by simply requiring the local tensor A_i to satisfy the conservation law:

$$[\alpha_i] - [\alpha_{i-1}] = \sigma_i. \tag{10.41}$$

Graphically, the local spin conservation law can be represented as

$$[\alpha_{i-1}] \longrightarrow \!\!\!\!\!\overset{\overset{\textstyle\sigma_i}{\downarrow}}{(A_i)}\!\!\!\!\!\longrightarrow [\alpha_i]. \tag{10.42}$$

The U(1) symmetry is imposed by requiring that the sum of the incoming spins (with the arrows pointing towards the tensor) equals that of the outgoing one (with the arrow pointing away from the tensor).

The sequence of local tensors builds up the target state site by site from the vacuum state. Consequently, the quantum numbers appearing in the MPS tensors on opposite ends are the vacuum and target states, respectively. This could be understood more clearly by summing over all physical spins using Eq. (10.41):

$$\sum_i \sigma_i = [\alpha_{N+1}] - [\alpha_0], \tag{10.43}$$

where $[\alpha_{N+1}]$ and $[\alpha_0]$ are the virtual spins of the rightmost and leftmost bonds, respectively. In practical calculations, we can always set the leftmost bond as the vacuum state so that $[\alpha_0] = 0$. In this case, the above equation becomes

$$\sum_i \sigma_i = [\alpha_{N+1}], \tag{10.44}$$

hence the virtual spin at the right end of the lattice, $[\alpha_{N+1}]$, is the spin of the target state.

In a periodic system, both A_1 and A_N are $D \times D$ matrices, and the wave function becomes

$$|\Psi\rangle = \sum_{\sigma_1\cdots\sigma_N} \mathrm{Tr}A_1[\sigma_1]\cdots A_N[\sigma_N]|\sigma_1\cdots\sigma_N\rangle. \tag{10.45}$$

As the left bond of A_1 is linked to the right bond of A_N, we have $[\alpha_N] = [\alpha_0]$. In this case, Eq. (10.43) becomes

$$\sum_i \sigma_i = 0, \tag{10.46}$$

corresponding to a target state with total spin zero.

To target a state of total spin $S_z = m$, we should add $(-m)$ to the physical spin at one of the lattice sites, say σ_N, so that the local tensor at that site becomes

$$\tag{10.47}$$

and the local tensors at other sites are unchanged. Now the physical spin at site N becomes $\sigma_N - m$, and

$$\sum_i \sigma_i = m. \tag{10.48}$$

The corresponding MPS becomes

$$|\Psi\rangle = \sum_{\sigma_1 \cdots \sigma_N} \mathrm{Tr}A_1[\sigma_1] \cdots A_{N-1}[\sigma_{N-1}]A_N[\sigma_N - m]|\sigma_1 \cdots \sigma_N\rangle. \tag{10.49}$$

Alternatively, we can target a finite spin state by applying the spin-flip operators,

$$S^\pm = \sum_i S_i^\pm, \tag{10.50}$$

to a spin-zero MPS. For example, the MPS of a total spin $S_z = 1$ state could be set as

$$|\Psi\rangle = S^+ \sum_{\sigma_1 \cdots \sigma_N} \mathrm{Tr}A_1[\sigma_1] \cdots A_N[\sigma_N]|\sigma_1 \cdots \sigma_N\rangle. \tag{10.51}$$

This expression is useful if the MPS is translation invariant with site-independent local tensors:

$$|\Psi\rangle = \sum_{\sigma_1 \cdots \sigma_N} \mathrm{Tr}A[\sigma_1] \cdots A[\sigma_N]S^+|\sigma_1 \cdots \sigma_N\rangle. \tag{10.52}$$

It is straightforward to generalize these discussions to systems of interacting fermions. For example, for a spinless fermion system, we can set σ_i equal to 1 if site i is occupied or -1 otherwise. At half-filling, the sum of local fermion numbers is

$$\sum_i \sigma_i = 0. \tag{10.53}$$

The m in Eq. (10.49) measures the number of doped electrons if $m > 0$ or doped holes if $m < 0$ with respect to half-filling. To study a state with only one fermion or hole excitation away from the half-filling, we can also use creation or annihilation operators to add an electron or hole to a half-filled MPS.

10.4.2 Non-Abelian Symmetries

In an SU(2) symmetric system, a basis state is labeled by the spin quantum number s that uniquely links an irreducible representation of SU(2) and its projection along the z-axis, or the magnetic quantum number. On the other hand, one can reduce the wave function so that it does not explicitly depend on the magnetic quantum number in the SU(2) representation, as discussed in §8.5. This reduction in the wave function implies that the corresponding local tensors could also be represented with the reduced quantum numbers that do not depend on the magnetic quantum number. Therefore, the MPS in an open boundary system should take the form

$$|\Psi\rangle = \sum_{s_1 \cdots s_N} A_1[s_1] \cdots A_N[s_N] |s_1 \cdots s_N\rangle, \tag{10.54}$$

where s_i is the spin quantum number (not the magnetic quantum number) at site i. As the spin at each site is fixed (i.e. $s_i = s$, for the spin-s Heisenberg model), the above summation contains just one term. In principle, we can remove the symbol of summation in the above expression simply by setting all local spins s_i to s. However, in other quantum models, such as the Hubbard model, the physical basis states s_i at each lattice site may appear in more than one irreducible representation of SU(2) or other non-Abelian groups. Therefore, we keep the convention to include the physical spin quantum number explicitly at each local tensor and allow the summation to run over all possible spin configurations, although there is only one configuration for the SU(2) Heisenberg model.

Similarly, to preserve the SU(2) symmetry, the basis states at each virtual bond should be in certain irreducible representations of SU(2) as well. If we use (j_i, α_i) and (j_{i-1}, α_{i-1}) to label, respectively, the virtual basis states on the right and left bonds, the local tensor can then be represented as

$$
(A_i)_{j_{i-1}\alpha_{i-1}j_i\alpha_i}[s_i] = j_{i-1}\alpha_{i-1} \rightarrow \!\!\!\overset{\displaystyle s_i}{\underset{\displaystyle \uparrow}{(A_i)}}\!\!\! \rightarrow j_i\alpha_i. \tag{10.55}
$$

Here j_i and j_{i-1} refer to the spin quantum numbers on the right and left bonds, respectively. α_i is an integer that labels the basis state in the subspace in which the spin quantum number is j_i; α_{i-1} is similarly defined.

A_i thus defined is in fact a reduced local tensor since it does not depend on the magnetic quantum numbers of (j_{i-1}, s, j_i). As indicated by the directions of arrows shown in Eq. (10.55), it maps the incoming states $\|j_{i-1}, \alpha_{i-1}, s\rangle$ to the outgoing state $\|j_i, \alpha_i\rangle$:

$$A_i = \sum_{j_{i-1}\alpha_{i-1}j_i\alpha_i} (A_i)_{j_{i-1}\alpha_{i-1}j_i\alpha_i}[s] \|j_i\alpha_i\rangle\langle j_{i-1}\alpha_{i-1}s\|. \tag{10.56}$$

Here we have used $\| \cdot \cdot \rangle$ to represent a reduced basis state that does not depend on the magnetic quantum numbers. $\| j_i \alpha_i \rangle \langle j_{i-1} \alpha_{i-1} s \|$ is related to the true basis states of angular momenta by the equation

$$
\begin{aligned}
&\| j_i \alpha_i \rangle \langle j_{i-1} \alpha_{i-1} s \| \\
&= \sum_{m_1 m_{i-1} m_i} \begin{pmatrix} j_{i-1} & s & j_i \\ m_{i-1} & m_1 & -m_i \end{pmatrix} | j_i m_i \alpha_i \rangle \langle j_{i-1} m_{i-1} \alpha_{i-1} | \langle s m_1 |,
\end{aligned} \quad (10.57)
$$

where the prefactor is the Wigner 3j-symbol. Since the superposition of the two incoming spins equals the outgoing spin; that is,

$$
j_i = j_{i-1} \otimes_{\mathrm{SU}(2)} s_i, \quad (10.58)
$$

the spin is conserved. $\otimes_{\mathrm{SU}(2)}$ is the addition operator of SU(2). In other words, the net spin carried by the local tensor is zero. This could be readily verified from the rotational symmetry of the 3j-symbol.

If the leftmost bond of the MPS represents a vacuum state with $j_0 = 0$, Eq. (10.58) implies that the spin quantum number at the rightmost bond, j_N, is the targeted spin quantum number:

$$
\prod_i \otimes_{\mathrm{SU}(2)} s_i = j_N. \quad (10.59)
$$

In a periodic system, the MPS becomes

$$
|\Psi\rangle = \sum_{s_1 \cdots s_N} \mathrm{Tr} A_1[s_1] \cdots A_N[s_N] | s_1 \cdots s_N\rangle. \quad (10.60)
$$

Since $j_N = j_0$ in a periodic system, the total spin of this MPS is zero:

$$
\prod_i \otimes_{\mathrm{SU}(2)} s_i = 0. \quad (10.61)
$$

To target a state with total spin q, again, we need to add a q-spin to one of the local tensors, say A_N, so that it becomes

$$
\begin{array}{c}
s_N \otimes_{\mathrm{SU}(2)} q \\
\downarrow \\
j_{N-1}\alpha_{N-1} \longrightarrow \!\!\boxed{A_N}\!\!\longrightarrow j_N \alpha_N.
\end{array} \quad (10.62)
$$

In this case, the MPS becomes

$$
|\Psi\rangle = \sum_{s_1 \cdots s_N} \mathrm{Tr} A_1[s_1] \cdots A_{N-1}[s_{N-1}] A_N[s_N \otimes_{\mathrm{SU}(2)} q] | s_1 \cdots s_N\rangle. \quad (10.63)
$$

11

Infinite Matrix Product States

11.1 Translation Invariant MPS

In a system with periodic boundary conditions, a matrix product state can be expressed as

$$\Psi_{\sigma_1 \cdots \sigma_N} = \text{[diagram: } A_1 - A_2 - \cdots - A_N \text{ with legs } \sigma_1, \sigma_2, \sigma_N\text{]} \qquad (11.1)$$

In a translation-invariant system, $A_i = A$ becomes site-independent, and the above wave function becomes

$$\Psi_{\sigma_1 \cdots \sigma_N} = \text{[diagram: } A - A - \cdots - A \text{ with legs } \sigma_1, \sigma_2, \sigma_N\text{]} .$$

This wave function is defined on a finite lattice system of N sites with periodic boundary conditions. In the thermodynamic limit, $N \to \infty$, it becomes

$$\Psi_{\cdots \sigma_1 \cdots \sigma_i \cdots} = \text{Tr}\left(\cdots \; A - A - \cdots - A - A \; \cdots \right). \qquad (11.2)$$

This MPS is generally not canonical. Nevertheless, there are many ways to transform it into a canonical form. A simple one is to canonicalize the local tensors by taking a series of QR and LQ decompositions. This scheme is computationally stable because it does not involve the calculation of the inverse of a matrix which is often unstable if some eigenvalues of this matrix are very small.

The left canonicalization is to find a bond matrix R so that the local tensor defined by

$$-\boxed{A_L}- \quad = \quad -(R)-(A)-(R^{-1})- \tag{11.3}$$

satisfies the equation

$$\boxed{\begin{array}{c} (A_L^*) \\ | \\ (A_L) \end{array}} = I. \tag{11.4}$$

Equation (11.3) can also be expressed as

$$-(R)-(A)- \quad = \quad -(A_L)-(R)- . \tag{11.5}$$

This equation defines an iterative scheme to determine R and A_L. Take a QR decomposition for the matrix $M[\sigma] = RA[\sigma]$ starting from an arbitrary initial R:

$$-(M)- \underset{\text{QR}}{=\!=} -(A_L)-(R')- , \tag{11.6}$$

where A_L is just the Q-matrix obtained from the QR decomposition and R' is an upper triangular matrix. According to the property of QR decomposition, A_L thus obtained is left canonical. Repeating these QR decompositions by setting $R = R'$, the converged R is just the desired result.

Similarly, A can be right canonicalized by a gauge bond matrix L:

$$-\boxed{A_R}- \quad = \quad -(L^{-1})-(A)-(L)- . \tag{11.7}$$

A_R is a right canonical tensor:

$$\boxed{\begin{array}{c} (A_R^*) \\ | \\ (A_R) \end{array}} = I. \tag{11.8}$$

Again, L can be determined by iteratively solving the following equation with LQ decomposition:

$$-(A)-(L)- \quad = \quad -(L)-\boxed{A_R}- . \tag{11.9}$$

At each step of decomposition, $A[\sigma]L$ is LQ decomposed, and L is then updated with the lower triangular matrix obtained from the decomposition. The converged A_R, or the Q-matrix obtained from the LQ decomposition, is just the right canonical tensor to find.

After the left–right canonicalization, the wave function can now be represented as

$$\Psi = \mathrm{Tr}\left(\cdots \boxed{A_L}\boxed{A_L}\!-\!\!\bigcirc\!\!-\!\langle A_R|\,\langle A_R|\, \cdots \right), \tag{11.10}$$

where

$$C = RL. \tag{11.11}$$

From the orthonormal conditions of the left and right canonical tensors, it is simple to show that the singular spectrum of C; that is, λ in the expression

$$-\!\bigcirc\!\!-\;=\;-\!\boxed{U}\!-\!\langle\lambda\rangle\!-\!\boxed{V^\dagger}\!- \tag{11.12}$$

is just the square root of the eigenspectrum of the reduced density matrix. Taking the following gauge transformations

$$B_L[\sigma] = U^\dagger A_L[\sigma]U, \tag{11.13}$$
$$B_R[\sigma] = V^\dagger A_R[\sigma]V, \tag{11.14}$$

we can then convert the MPS into the standard left–right canonical form

$$\Psi = \mathrm{Tr}\left(\cdots \boxed{B_L}\boxed{B_L}\!-\!\langle\lambda\rangle\!-\!\langle B_R|\,\langle B_R|\, \cdots \right). \tag{11.15}$$

This expression suggests that the renormalization effect of the environment on the system is effectively measured by λ up to a unitary gauge transformation. As the system and environment sublattices can be disconnected by cutting just one bond in one dimension, it implies that we can take λ as a bond field to mimic the environment contribution. This provides a robust scheme, generally referred to as the simple update approach, to determine an MPS in one dimension.

The simple update uses a diagonal matrix to mimic the contribution of the environment. This approach applies quasi-exactly to an arbitrary lattice without loops, but not for a lattice that cannot be separated into two parts by cutting just one bond, such as the square lattice. In that case, it is no longer accurate to mimic the environment's contribution by a diagonal matrix for each bond connecting the system and environment sublattices. Nevertheless, it might still be a good approximation in accounting for the renormalization effect of the environment.

It is simple to show that B_L is left canonical and satisfies the equations (Ψ is assumed to be normalized):

$$= I, \qquad \lambda^2 = \lambda^2. \tag{11.16}$$

Similarly, B_R satisfies the right-canonical equations:

$$= I, \qquad \lambda^2 = \lambda^2. \tag{11.17}$$

Furthermore, by setting

$$-B_R- = -\Gamma-\lambda-, \tag{11.18}$$

we can also represent Ψ as

$$\Psi = \ldots -\Gamma-\lambda-\Gamma-\lambda-\Gamma-\lambda-\ldots. \tag{11.19}$$

In this $\lambda - \Gamma$-representation, the above canonical equations become

$$\lambda^2 = I, \qquad \lambda^2 = I. \tag{11.20}$$

As an example, let us consider the MPS representation of the AKLT state. The local tensors, as defined in Eq. (5.38), are

$$A[1] = \begin{pmatrix} 0 & 1 \\ 0 & 0 \end{pmatrix}, \quad A[0] = \frac{1}{\sqrt{2}} \begin{pmatrix} -1 & 0 \\ 0 & 1 \end{pmatrix}, \quad A[-1] = \begin{pmatrix} 0 & 0 \\ -1 & 0 \end{pmatrix}. \tag{11.21}$$

These tensors are not canonicalized. However, it is simple to show that

$$B[\sigma] = \sqrt{\frac{2}{3}} A[\sigma] \tag{11.22}$$

are both left and right canonicalized, satisfying the equation

$$= I \tag{11.23}$$

with $\lambda = I$.

11.2 Transfer Matrix and Canonical Transformation

There is more than one way to canonicalize an infinite MPS. In addition to the approach previously introduced, one can also determine the R and L matrices for canonicalizing the MPS by taking a canonical transformation. This canonicalization is achieved by diagonalizing the transfer matrix defined by

$$\mathcal{E}_{\alpha'\alpha,\beta'\beta} = \sum_\sigma A^*_{\alpha'\beta'}[\sigma]A_{\alpha\beta}[\sigma] = \qquad \begin{array}{c} \alpha' -\!\!\boxed{A^*}\!\!- \beta' \\ \big| \\ \alpha -\!\!\boxed{A}\!\!- \beta \end{array} \tag{11.24}$$

in two steps:

(i) Left canonicalization:

   To left canonicalize A, we first diagonalize the transfer matrix to find its dominant eigenvalue η and the corresponding left eigenvector $\langle l|$:

$$\begin{array}{c} \boxed{A^*} \\ \big| \\ \big(l\big) \\ \big| \\ \boxed{A} \end{array} \quad = \quad \eta \ \big(l\big) \ . \tag{11.25}$$

   If the MPS wave function is normalized, it is simple to show that $\eta = 1$. As l is Hermitian and nonnegative, one can diagonalize it with a unitary matrix U:

$$-\!\!\big(l\big)\!\!- \quad = \quad -\!\!\boxed{U^\dagger}\!\!-\!\!\langle\kappa\rangle\!\!-\!\!\boxed{U}\!\!- \ , \tag{11.26}$$

   where κ is a positive diagonal matrix. The matrix that left canonicalizes $A[\sigma]$ is then found to be

$$-\!\!\big(R\big)\!\!- \quad = \quad -\!\!\underset{\kappa^{1/2}}{\langle\rangle}\!\!-\!\!\boxed{U}\!\!- \ . \tag{11.27}$$

(ii) Right canonicalization:

   We can also determine the right canonical tensor A_R by diagonalizing the transfer matrix. In this case, one needs to find the right eigenvector $|r\rangle$ corresponding to the dominant eigenvalue η:

$$\begin{array}{c} -\!\!\boxed{A^*} \\ \big| \quad \big(r\big) \\ -\!\!\boxed{A} \end{array} \quad = \quad \eta \ \big(r\big) \ . \tag{11.28}$$

   Again, $\eta = 1$ if the wave function is normalized. r is a positive definite and Hermitian matrix. By diagonalizing it with a unitary matrix V,

$$-\!\!\!\bigcirc\!\!\!r\!\!\!-\!\!\!- \quad = \quad -\!\!\!\boxed{V}\!\!\!-\!\!\!\diamondsuit\!\!\!\kappa\!\!\!-\!\!\!\boxed{V^\dagger}\!\!\!-\!\!\!- \quad , \tag{11.29}$$

we then find the matrix that right canonicalizes $A[\sigma]$:

$$-\!\!\!\bigcirc\!\!\!L\!\!\!-\!\!\!- \quad = \quad -\!\!\!\boxed{V}\!\!\!-\!\!\!\diamondsuit\!\!\!- \atop \kappa^{1/2} \quad . \tag{11.30}$$

11.3 Expectation Values of Physical Observables

We first evaluate the expectation value of a local observable O_i in an MPS:

$$\langle O_i \rangle = \frac{\langle \Psi | O_i | \Psi \rangle}{\langle \Psi | \Psi \rangle}. \tag{11.31}$$

The denominator $\langle \Psi | \Psi \rangle$ now can be expressed as a product of transfer matrices:

$$\langle \Psi | \Psi \rangle = \mathrm{Tr} \mathcal{E}^N, \tag{11.32}$$

where N is the lattice size. In the thermodynamic limit, $N \to \infty$, $\langle \Psi | \Psi \rangle$ is determined purely by the maximal eigenvalue of $\mathcal{E}$; that is, η:

$$\langle \Psi | \Psi \rangle = \eta^N. \tag{11.33}$$

Here $\eta = 1$ if the wave function is normalized.

Similarly, one can also express $\langle \Psi | O_i | \Psi \rangle$ as a trace of a set of transfer matrices:

$$\langle \Psi | O_i | \Psi \rangle = \mathrm{Tr} \left(\mathcal{E}[O_i] \mathcal{E}^{N-1} \right), \tag{11.34}$$

where

$$\mathcal{E}_{\alpha'\alpha,\beta'\beta}[O_i] = \begin{array}{c} \alpha' -\!\!\!\boxed{A^*}\!\!\!- \beta' \\ \boxed{O_i} \\ \alpha -\!\!\!\boxed{A}\!\!\!- \beta \end{array} \quad . \tag{11.35}$$

In the thermodynamic limit,

$$\langle \Psi | O_i | \Psi \rangle = \eta^{N-1} \; \boxed{l} \; \boxed{O_i} \; \boxed{r} \; . \tag{11.36}$$

Here l and r are the left and right dominant eigenvectors of $\mathcal{E}$, respectively.

Thus, the expectation value of O_i is

$$\langle O_i \rangle = \frac{1}{\eta}\; \begin{array}{c} \text{[diagram]} \end{array} .\tag{11.37}$$

The correlation function of local operators can be similarly evaluated. The two-site correlation function of O_i is defined by

$$C_x = \langle \delta O_{i+x} \delta O_i \rangle,\tag{11.38}$$

where $\delta O_i = O_i - \langle O_i \rangle$. In the thermodynamic limit,

$$C_x = \frac{1}{\eta^{x+1}} \langle l | \mathcal{E}[\delta O_{i+x}] \mathcal{E}^{x-1} \mathcal{E}[\delta O_i] | r \rangle.\tag{11.39}$$

It can also be written as

$$C_x = \sum_{k \geq 2} \left(\frac{\eta_k}{\eta} \right)^{x-1} \frac{\langle l | \mathcal{E}[\delta O_{i+x}] | r_k \rangle \langle l_k | \mathcal{E}[\delta O_i] | r \rangle}{\eta^2},\tag{11.40}$$

where η_k is the eigenvalue of $\mathcal{E}$. l_k and r_k are the corresponding left and right eigenvectors

$$\begin{array}{c} \text{[diagram]} \end{array} = \eta_k \; \begin{array}{c} \text{[diagram]} \end{array} ,\tag{11.41}$$

$$\begin{array}{c} \text{[diagram]} \end{array} = \begin{array}{c} \text{[diagram]} \end{array} \eta_k .\tag{11.42}$$

Here $\eta_1 = \eta$ is the dominant eigenvalue. In the summation for k, the maximal eigenvalue is excluded because $\langle l | \mathcal{E}[\delta O_i] | r \rangle = 0$.

In the limit $x \to \infty$, the summation in the expression is dominated by the first leading eigenvalue η_n, on which $\langle l | \mathcal{E}[\delta O_{i+r}] | r_n \rangle \langle l_n | \mathcal{E}[\delta O_i] | r \rangle \neq 0$. In this case,

$$C_x = \left(\frac{\eta_n}{\eta} \right)^{x-1} \frac{\langle l | \mathcal{E}[\delta O_{i+r}] | r_n \rangle \langle l_n | \mathcal{E}[\delta O_i] | r \rangle}{\eta^2}.\tag{11.43}$$

Usually, η_n is the second largest eigenvalue of $\mathcal{E}$. C_x decays exponentially with the distance between the two operators:

$$C_x \sim e^{-x/\xi},\tag{11.44}$$

where ξ is the correlation length, determined by the ratio between η and η_n:

$$\xi = \ln^{-1} \frac{\eta}{|\eta_n|}. \tag{11.45}$$

Since $|\eta_n| < \eta$ (assuming the system is injective), this means that MPS is always finite correlated [33] and not suited for describing critical states where the correlation length diverges.

11.4 String Order Parameter

The discussion in §11.3 can be readily extended to evaluate the expectation value of a string product of operators O_i in the thermodynamic limit

$$\left\langle \prod_{i=1}^{n} O_i \right\rangle = \frac{1}{\eta^n} \text{Tr} \left(l \prod_{i=1}^{n} \mathcal{E}[O_i] r \right). \tag{11.46}$$

If $\mathcal{E}[O_i]$ is site independent, then in the limit $n \to \infty$, the expectation becomes

$$\left\langle \prod_{i=1}^{n} O_i \right\rangle = \left(\frac{\gamma}{\eta} \right)^n \text{Tr} \left(l r_\gamma \right) \text{Tr} \left(l_\gamma r \right), \tag{11.47}$$

where γ is the maximal eigenvalue of $\mathcal{E}[O_i]$. If $|\gamma| < \eta$, this expectation value decays exponentially with n. l_γ and r_γ are the corresponding left and right eigenvectors:

$$\tag{11.48}$$

In Eq. (11.47), if $O_i = U_i \neq I$ is a unitary operator, this equation then defines a string order parameter that is finite if and only if $|\gamma| = \eta$ in the limit $n \to \infty$.

11.4.1 Maximal Eigenvalue of $\mathcal{E}[U_i]$

For a string operator whose transfer matrix $\mathcal{E}[U_i]$ is site-independent, our discussion suggests that the string order exists only when the absolute value of the maximal eigenvalue $|\gamma|$ equals η. This condition sets a strong constraint on the unitary operator U_i.

In fact, for an arbitrary given unitary operator U, it can be shown that the maximal eigenvalue of $\mathcal{E}[U]$; that is, γ, cannot be larger than η in its absolute value [379]. Otherwise, the string order parameter would diverge in the limit $n \to \infty$.

To prove $|\gamma| \leq \eta$, let us consider a translation-invariant MPS whose local tensor A is left canonicalized:

$$(11.49)$$

Here we assume that the wave function is normalized to 1. Hence $\eta = 1$. The maximal eigenvalue γ and the corresponding eigenvector W of $\mathcal{E}[U]$ satisfie the eigenequation

$$(11.50)$$

A unitary matrix can always be written as an exponential of an anti-Hermitian matrix. In the representation at which this anti-Hermitian matrix is diagonal, the unitary matrix can be expressed as

$$U = \sum_i e^{i\theta_i} |i\rangle \langle i|. \tag{11.51}$$

Substituting it into Eq. (11.50), the eigenequation becomes

$$(11.52)$$

where

$$A_i = \sum_\sigma \langle i|\sigma \rangle A[\sigma]. \tag{11.53}$$

Like $A[\sigma]$, A_i is also left canonical, satisfying the equation

$$(11.54)$$

Multiplying $\lambda^2 W^\dagger$ from the right to Eq. (11.50) and taking trace, we obtain

$$\sum_i \mathrm{Tr}\left(e^{i\theta_i} A_i^\dagger W A_i \lambda^2 W^\dagger\right) = \sum_i \mathrm{Tr}\left(X_i^\dagger Y_i\right) = \gamma \, \mathrm{Tr}\left(W\lambda^2 W^\dagger\right), \tag{11.55}$$

where

$$X_i = A_i W \lambda \tag{11.56}$$

$$Y_i = e^{i\theta_i} W A_i \lambda, \tag{11.57}$$

Using the Cauchy–Schwarz inequality

$$\mathrm{Tr}\left(X_i^\dagger Y_i\right) \mathrm{Tr}\left(X_i Y_i^\dagger\right) \leq \mathrm{Tr}\left(X_i^\dagger X_i\right) \mathrm{Tr}\left(Y_i^\dagger Y_i\right), \tag{11.58}$$

we obtain

$$|\gamma| \left|\mathrm{Tr}\left(W\lambda^2 W^\dagger\right)\right| \leq \sum_i \left|\mathrm{Tr}\left(X_i^\dagger Y_i\right)\right| \leq \sum_i \left[\mathrm{Tr}\left(X_i^\dagger X_i\right) \mathrm{Tr}\left(Y_i^\dagger Y_i\right)\right]^{1/2}. \tag{11.59}$$

From the canonical conditions of A_i, it is simple to show that

$$\mathrm{Tr}\left(X_i^\dagger X_i\right) = \mathrm{Tr}\left(Y_i^\dagger Y_i\right) = \mathrm{Tr}\left(W\lambda^2 W^\dagger\right). \tag{11.60}$$

We therefore have

$$|\gamma| \left|\mathrm{Tr}\left(W\lambda^2 W^\dagger\right)\right| \leq \left|\mathrm{Tr}\left(W\lambda^2 W^\dagger\right)\right|, \tag{11.61}$$

which implies

$$|\gamma| \leq 1. \tag{11.62}$$

11.4.2 Condition for the Existence of String Order

If $|\gamma| = 1$, the Cauchy–Schwarz inequality becomes an equality. This is valid only when X_i is proportional to Y_i; that is,

$$\alpha A_i W \lambda = e^{i\theta_i} W A_i \lambda. \tag{11.63}$$

Multiplying by the adjoint and taking trace gives

$$|\alpha|^2 \mathrm{Tr}\left(W^\dagger A_i^\dagger A_i W \lambda^2\right) = \mathrm{Tr}\left(W A_i \lambda^2 A_i^\dagger W^\dagger\right). \tag{11.64}$$

By summing in i and using the canonical conditions, we obtain $|\alpha| = 1$; that is, $\alpha = e^{i\theta}$ and

$$W A_i = e^{i(\theta - \theta_i)} A_i W. \tag{11.65}$$

Using this equation, it is simple to show that

$$\sum_i A_i^\dagger W^\dagger W A_i = W^\dagger \sum_i A_i^\dagger A_i W = W^\dagger W. \tag{11.66}$$

Thus, $W^\dagger W$ is an eigenvector of $\sum_i A_i^\dagger A_i$ with eigenvalue 1. Since the unit matrix is the only eigenvector of $\sum_i A_i^\dagger A_i$ with eigenvalue 1, we have

$$W^\dagger W = I. \tag{11.67}$$

Hence W is unitary.

From the canonical property of A_i, we can show that

$$\lambda^2 = W \sum_i A_i \lambda^2 A_i^\dagger W^\dagger = \sum_i A_i W \lambda^2 W^\dagger A_i^\dagger = W \lambda^2 W^\dagger. \tag{11.68}$$

Thus, W commutes with λ^2:

$$W \lambda^2 = \lambda^2 W. \tag{11.69}$$

This indicates that at least one pair of eigenvalues of λ^2 must be degenerate, or $W = I$ otherwise. Thus, the entanglement spectrum degeneracy is a must for the existence of topological string order [379].

11.4.3 String Order in the AKLT State

The ground state of the AKLT model discussed in §5.3 actually possesses a nontrivial string order [363]. The corresponding string order parameter is

$$O_{\text{string}}^z = -\lim_{l \to \infty} \left\langle S_0^z \prod_{i=1}^{l} U_i S_{l+1}^z \right\rangle, \qquad U_i = e^{i\pi S_i^z}. \tag{11.70}$$

To demonstrate the existence of this topological string order, let us consider the ground state of this model. As shown in §5.3, the AKLT ground-state wave function is a translation-invariant MPS whose local tensors $A[\sigma]$ are defined in Eq. (5.38). By normalizing the wave function to 1, the local tensors become

$$A[1] = \sqrt{\frac{2}{3}} \begin{pmatrix} 0 & 1 \\ 0 & 0 \end{pmatrix},$$

$$A[0] = \frac{1}{\sqrt{3}} \begin{pmatrix} -1 & 0 \\ 0 & 1 \end{pmatrix}, \tag{11.71}$$

$$A[-1] = \sqrt{\frac{2}{3}} \begin{pmatrix} 0 & 0 \\ -1 & 0 \end{pmatrix}.$$

The transfer matrix corresponding to this MPS is a 4×4 matrix:

$$\mathcal{E} = \sum_\sigma A^*[\sigma] \otimes A[\sigma] = \begin{pmatrix} 1/3 & 0 & 0 & 2/3 \\ 0 & -1/3 & 0 & 0 \\ 0 & 0 & -1/3 & 0 \\ 2/3 & 0 & 0 & 1/3 \end{pmatrix}. \tag{11.72}$$

Its four eigenvalues are $(1, -1/3, -1/3, -1/3)$. The maximal eigenvalue $\eta = 1$, and the corresponding eigenvector $l = r = (1/\sqrt{2}, 0, 0, 1/\sqrt{2})$.

Similarly, the transfer matrix corresponding to the unitary matrix, $\mathcal{E}[U_i]$, is

$$
\mathcal{E}[U_i] = \begin{pmatrix} 1/3 & 0 & 0 & -2/3 \\ 0 & -1/3 & 0 & 0 \\ 0 & 0 & -1/3 & 0 \\ -2/3 & 0 & 0 & 1/3 \end{pmatrix}.
\tag{11.73}
$$

It has precisely the same four eigenvalues as $\mathcal{E}$. The maximal eigenvalue γ is also equal to 1, and the corresponding eigenvector $l_\gamma = r_\gamma = (1/\sqrt{2}, 0, 0, -1/\sqrt{2})$. As $\gamma = \eta$, the AKLT state should have a finite topological string order generated by the unitary matrix U_i.

Furthermore, by canonicalizing the local tensors, it is simple to show that the bond matrix λ of the AKLT state is proportional to the identity matrix, which has two degenerate eigenvalues $(1, 1)$. This confirms that the AKLT state does have a string order.

For the AKLT state, the string order parameter can be explicitly calculated. Using the expression

$$
\mathcal{E}[S^z] = \begin{pmatrix} 0 & 0 & 0 & 2/3 \\ 0 & 0 & 0 & 0 \\ 0 & 0 & 0 & 0 \\ -2/3 & 0 & 0 & 0 \end{pmatrix},
\tag{11.74}
$$

we find that

$$
O^z_{\text{string}} = - \lim_{l \to \infty} \left\langle S_0^z \prod_{i=1}^{l} U_i S_{l+1}^z \right\rangle = \left(l \mathcal{E}[S^z] l_\gamma \right)^2 = \frac{4}{9}.
\tag{11.75}
$$

11.5 MPS with a Finite Unit Cell

MPS can also be canonicalized for a system that is invariant under translations of more than one site. We will take an MPS, which is translation-invariant by two sites, as an example to illustrate the canonical transformation method. The method, however, works generally. It can be readily extended to an MPS with a unit cell of more than two sites.

A two-site translation-invariant MPS is a successive product of two alternating matrices:

$$
\Psi_{\dots \sigma_1 \sigma_2 \dots} = \cdots \overset{\sigma_1}{\underset{}{\textcircled{A}}} - \overset{\sigma_2}{\underset{}{\textcircled{B}}} - \overset{\sigma_3}{\underset{}{\textcircled{A}}} - \overset{\sigma_4}{\underset{}{\textcircled{B}}} - \cdots.
\tag{11.76}
$$

This wave function can be converted into a translation-invariant form by considering the superposition of this MPS and its translation by one lattice site:

$$\Psi_{\cdots\sigma_1\sigma_2\cdots} = \mathrm{Tr}\left(\cdots \begin{array}{cccc} \sigma_1 & \sigma_2 & \sigma_3 & \sigma_4 \\ \!\!\!\overset{|}{-}\!\!A\!\!-\!\!B\!\!-\!\!A\!\!-\!\!B\!\!-\cdots \end{array}\right.$$

$$\left. + \cdots \begin{array}{cccc} \sigma_1 & \sigma_2 & \sigma_3 & \sigma_4 \\ \!\!\!\overset{|}{-}\!\!B\!\!-\!\!A\!\!-\!\!B\!\!-\!\!A\!\!-\cdots \end{array}\right). \tag{11.77}$$

This wave function can be further expressed as a single MPS with higher bond dimensions:

$$\Psi_{\cdots\sigma_1\sigma_2\cdots} = \mathrm{Tr}\left(\cdots C[\sigma_1]C[\sigma_2]C[\sigma_3]C[\sigma_4]\cdots\right), \tag{11.78}$$

where

$$C[\sigma] = \begin{pmatrix} 0 & A[\sigma] \\ B[\sigma] & 0 \end{pmatrix}. \tag{11.79}$$

If $A[\sigma]$ $(B[\sigma])$ is a $D_1 \times D_2$ $(D_2 \times D_1)$ matrix, then $C[\sigma]$ is a square matrix of dimension $(D_1 + D_2)$.

11.5.1 Majumdar–Ghosh State

As an example, let us consider the ground state of the so-called spin-1/2 Majumdar–Ghosh model, defined by the Hamiltonian

$$H = \sum_i \left(\mathbf{S}_i \cdot \mathbf{S}_{i+1} + \frac{1}{2}\mathbf{S}_i \cdot \mathbf{S}_{i+2} \right). \tag{11.80}$$

It is simple to show that the ground state of this model is dimerized and doubly degenerate. One of the dimerized states is

$$\Psi = \cdots \underset{1 \quad 2}{\overset{}{\bigcirc\!\!-\!\!\bullet}}\ \underset{3 \quad 4}{\overset{}{\bigcirc\!\!-\!\!\bullet}}\ \underset{5 \quad 6}{\overset{}{\bigcirc\!\!-\!\!\bullet}}\ \underset{7 \quad 8}{\overset{}{\bigcirc\!\!-\!\!\bullet}}\cdots , \tag{11.81}$$

where each dot represents an $S=1/2$ spin and two dots connected by a line form a spin singlet:

$$\underset{}{\overset{}{\bigcirc\!\!-\!\!\bullet}} = |\!\uparrow\downarrow - \downarrow\uparrow\rangle. \tag{11.82}$$

This wave function can be written as an MPS as expressed by Eq. (11.76) with $A[\sigma]$ a 1×2 matrix and $B[\sigma]$ a 2×1 matrix:

$$A[\uparrow] = \begin{pmatrix} 0, & 1 \end{pmatrix}, \quad A[\downarrow] = \begin{pmatrix} 1, & 0 \end{pmatrix}, \tag{11.83}$$

$$B[\uparrow] = \begin{pmatrix} -1 \\ 0 \end{pmatrix}, \quad B[\downarrow] = \begin{pmatrix} 0 \\ 1 \end{pmatrix}. \tag{11.84}$$

It can also be written as a translation-invariant MPS, defined by Eq. (11.78). The local tensor is now a 3×3 matrix and defined by

$$C[\uparrow] = \begin{pmatrix} 0 & 0 & 1 \\ -1 & 0 & 0 \\ 0 & 0 & 0 \end{pmatrix}, \qquad C[\downarrow] = \begin{pmatrix} 0 & 1 & 0 \\ 0 & 0 & 0 \\ 1 & 0 & 0 \end{pmatrix}. \tag{11.85}$$

11.5.2 Canonical Representation

The MPS defined by Eq. (11.76) is right canonical if $A[\sigma]$ and $B[\sigma]$ satisfy the following equations:

$$\tag{11.86}$$

$$\tag{11.87}$$

where λ_1 and λ_2 are the entanglement spectra on the right bonds of A and B tensors, respectively.

If A and B are not canonicalized, there are many ways to transform them into the canonical form. One commonly adopted approach is to canonicalize them by iteratively singular-value decomposing the products of the tensors on the two neighboring sites. This canonicalization is done in the following steps:

(i) From the product of A and B,

$$\tag{11.88}$$

we define a bond-entanglement weighted matrix

$$\tag{11.89}$$

Initially, λ_2 is set as an arbitrary positive diagonal matrix. The rank of T is upper bounded by the right bond dimension of A, say D_2. We then diagonalize

T using two unitary matrices U and V:

$$T_{\alpha\sigma_1,\beta\sigma_2} \;=\; \alpha -\!\boxed{U}\!-\!\langle\lambda_1\rangle\!-\!\boxed{V}\!- \beta \quad . \tag{11.90}$$

λ_1 is the semi-positive diagonal singular matrix. The number of positive singular values of λ_1 equals the rank of T, which is less than or equal to D_2. In general, λ_1 has a number of zero singular values, which can be discarded.

After the truncation, both U and V become isometric. We then update the local tensors by the formula

$$-\!\langle B\rangle\!- \;=\; -\!\langle V\rangle\!- \;,\qquad -\!\langle A\rangle\!- \;=\; -\!\langle M\rangle\!-\!\boxed{V^\dagger}\!- \quad . \tag{11.91}$$

It is straightforward to show that these updated tensors, A and B, satisfy the equations

$$\boxed{B^*}\;\Big]\;=I,\qquad \lambda_2^2\,\langle\;\begin{array}{c}\overline{A^*}\\ \\ \overline{A}\end{array}\;=\;\langle\;\rangle\,\lambda_1^2 \quad . \tag{11.92}$$

Finally, λ_1 is normalized so that $\mathrm{Tr}\lambda_1^2 = 1$.

(ii) Repeat the calculations in (i), by swapping A with B and λ_2 with λ_1. Now M becomes

$$\begin{array}{c}\sigma_2 \quad\quad \sigma_3\\ \alpha -\!\langle M\rangle\!- \beta\end{array} \;=\; \begin{array}{c}\sigma_2 \quad \sigma_3\\ \alpha -\!\langle B\rangle\!-\!\langle A\rangle\!- \beta\end{array} \;, \tag{11.93}$$

and T becomes

$$T_{\alpha\sigma_2,\beta\sigma_3} \;=\; \begin{array}{c}\sigma_2 \quad\quad\quad \sigma_3\\ \alpha -\!\langle\lambda_1\rangle\!-\!\langle M\rangle\!- \beta\end{array} \quad . \tag{11.94}$$

Again, T is diagonalized by two unitary matrices U and V:

$$T_{\alpha\sigma_2,\beta\sigma_3} \;=\; \begin{array}{c}\sigma_2 \quad\quad \sigma_3\\ \alpha -\!\boxed{U}\!-\!\langle\lambda_2\rangle\!-\!\boxed{V}\!- \beta\end{array} \quad . \tag{11.95}$$

λ_2 is the singular bond matrix between B and A. We then discard all zero singular values in λ_2 and update the local tensors by the formula

$$
\overset{|}{A} = \overset{|}{V} , \quad \overset{|}{B} = \overset{|}{M}{-}\overset{|}{V^\dagger} . \tag{11.96}
$$

These local tensors satisfy the equations

$$
\boxed{A^*} \atop \boxed{A} = I, \quad \lambda_1^2 \overset{\boxed{B^*}}{\underset{\boxed{B}}{\diamond}} = \diamond \lambda_2^2 . \tag{11.97}
$$

Finally, we normalize λ_2 so that $\mathrm{Tr}\lambda_2^2 = 1$.

(iii) Repeat these two steps until the entanglement spectra λ_1 and λ_2 become completely converged. After that, the local tensors A and B become canonicalized, satisfying Eqs. (11.92) and (11.97) simultaneously.

12

Determination of MPS

12.1 Variational Optimization

There are many ways to determine the local tensors in the MPS representation of the ground state for a given Hamiltonian. DMRG is one example. In the DMRG calculation, the Hamiltonian is updated, and its dimension is truncated due to the basis truncation. In the MPS representation, on the other hand, the truncation could be done just for the local tensors without affecting the Hamiltonian. Thus, if we calculate the ground state directly in the MPS representation, we can, in principle, obtain a slightly more accurate result for the ground-state energy than that obtained with DMRG.

Variational optimization is a simple approach commonly used in evaluating the ground-state MPS [39]. It may not be the most efficient, but it can be used to calculate an MPS in a finite lattice for an arbitrary Hamiltonian with either open or periodic boundary conditions. The interaction is not necessarily short-ranged. Furthermore, this method can be implemented to a tensor product state in a finite or infinite lattice in higher dimensions.

This method takes all tensor elements of local tensors in an MPS as variational parameters and determines them by iteratively minimizing the ground-state energy. A generalized eigenequation is solved each time to optimize one or a few local tensors. Here we take a system with open boundary conditions as an example to demonstrate this method. It is straightforward to extend this method to a periodic system.

Let us consider an MPS defined by Eq. (10.12). In the variational optimization, the local tensors of the ground state are determined by variationally minimizing the expectation value of the Hamiltonian:

$$\langle H \rangle = \frac{\langle \Psi | H | \Psi \rangle}{\langle \Psi | \Psi \rangle}. \tag{12.1}$$

At the minimum, the local tensors are determined by the extreme equation: that is, the Ritz–Rayleigh equation:

$$\frac{\partial \langle H \rangle}{\partial A_i} = \frac{1}{\langle \Psi | \Psi \rangle} \left[\frac{\partial \langle \Psi | H | \Psi \rangle}{\partial A_i} - \lambda \frac{\partial \langle \Psi | \Psi \rangle}{\partial A_i} \right] = 0, \tag{12.2}$$

where

$$\lambda = \frac{\langle \Psi | H | \Psi \rangle}{\langle \Psi | \Psi \rangle} \tag{12.3}$$

is just the ground-state energy at the solution point.

It is difficult to determine all the local tensors just by solving Eq. (12.2) in one run. However, these local tensors can be determined iteratively. At each iteration, we take one local tensor, for example, $A^{(i)}$, as a free tensor variable and keep all other local tensors fixed. In this case, both $\langle \Psi | H | \Psi \rangle$ and $\langle \Psi | \Psi \rangle$ are quadratic functions of A_i, which can be represented as

$$\langle \Psi | H | \Psi \rangle = \quad \boxed{A_i^\dagger} \!-\! \boxed{\mathcal{H}_i} \!-\! \boxed{A_i} \quad , \tag{12.4}$$

$$\langle \Psi | \Psi \rangle = \quad \boxed{A_i^\dagger} \!-\! \boxed{\mathcal{N}_i} \!-\! \boxed{A_i} \quad . \tag{12.5}$$

Here the $D \times D \times d$-dimensional local tensor A_i is vectorized. $\mathcal{H}_i$ and $\mathcal{N}_i$ are the effective Hamiltonian and norm matrices, respectively. They can be determined by tracing out all local tensors excluding A_i and $A_i^\dagger$ in $\langle \Psi | H | \Psi \rangle$ and $\langle \Psi | \Psi \rangle$, respectively.

Substituting the above expressions into Eq. (12.2), the Ritz–Rayleigh equation becomes

$$\boxed{\mathcal{H}_i} \!-\! \boxed{A_i} \;\; = \lambda \;\; \boxed{\mathcal{N}_i} \!-\! \boxed{A_i} \quad . \tag{12.6}$$

This is a generalized eigenequation for determining A_i. We can obtain a converged ground-state energy λ by iteratively solving this equation for all the local tensors.

The generalized eigenequation can be efficiently solved if the condition number, which is basically determined by the ratio between the smallest and largest singular values of $\mathcal{N}_i$, is not too small. For an open chain system, one can always take a canonical transformation to normalize $\mathcal{N}_i$ so that it equals the identity matrix. This canonical transformation is implicitly implemented in the DMRG calculation. For a periodic MPS, however, there is no simple way to turn $\mathcal{N}_i$ into an identity.

In the variational iteration, the energy will, in principle, go down. However, the converging speed depends strongly on the initial guess of local tensors. The performance starting from a random initial MPS is generally poor. It is also not guaranteed that the global minimum, instead of a local minimum, can be reached.

In a system with open boundary conditions, the initial MPS can be determined using DMRG. This initial MPS may not be the true solution, but it is undoubtedly better than a random one. One can also improve the initial guess by starting with a smaller D and gradually enlarging D once all local tensors converge.

It is simple to show that both $\mathcal{H}_i$ and $\mathcal{N}_i$ are Hermitian. In an open boundary system, the cost for evaluating $\mathcal{N}_i$ and $\mathcal{H}_i$ is of the order of ND^3. In a periodic system, the cost for evaluating $\mathcal{H}_i$ and $\mathcal{N}_i$ is significantly higher.

12.1.1 Norm Matrix

To evaluate the norm matrix, we start with the left-right canonicalized MPS:

$$\Psi = \underset{U_1}{\overline{\Box}}\cdots\underset{U_{i-1}}{\overline{\Box}}\underset{A_i}{\bigcirc}\underset{V_{i+1}}{\overline{\Box}}\cdots\underset{V_N}{\overline{\Box}} \ . \tag{12.7}$$

A_i is the variational tensor to be determined. All other tensors are either left or right canonicalized. In this representation, the norm of the wave function is

$$\langle\Psi|\Psi\rangle = \ \begin{array}{ccccc} U_1^* & U_{i-1}^* & A_i^* & V_{i+1}^* & V_N^* \\ \end{array} \ . \tag{12.8}$$
$$\begin{array}{ccccc} U_1 & U_{i-1} & A_i & V_{i+1} & V_N \end{array}$$

This expression can be reduced using the defining properties of the left and right canonical tensors to

$$\langle\Psi|\Psi\rangle = \ \overset{A_i^*}{\underset{A_i}{\Box}} \ = \ A_i^\dagger \ \boxed{I} \ A_i \ . \tag{12.9}$$

Thus, the norm matrix $\mathcal{N}_i$ is simply an identity matrix. In this case, the Ritz–Rayleigh equation is well conditioned and becomes the conventional eigenequation:

$$\boxed{\mathcal{H}_i} \ A_i = \lambda \ A_i \ . \tag{12.10}$$

12.1.2 Effective Hamiltonian Matrix

The effective Hamiltonian matrix $\mathcal{H}_i$ is determined by both the Hamiltonian itself and the local tensors and is slightly more difficult to calculate. For simplicity, let us consider a system whose Hamiltonian contains just the nearest-neighbor interactions:

$$H = \sum_j h_{j,j+1}, \tag{12.11}$$

where $h_{j,j+1}$ is the interaction between sites j and $j+1$. The discussion, nevertheless, can be readily extended to an arbitrary interacting Hamiltonian.

To evaluate $\mathcal{H}_i$, we denote all the lattice sites to the left of site i as the left block and those to the right of i as the right block. The Hamiltonian can then be written as a sum of four terms:

$$H = H_{<i} + h_{i-1,i} + h_{i,i+1} + H_{>i}, \tag{12.12}$$

where $H_{<i}$ ($H_{>i}$) is the Hamiltonian for the left (right) block. We now consider the contribution of these terms to $\mathcal{H}_i$ separately.

(i) Contribution of $h_{i-1,i} + h_{i,i+1}$:

Both $h_{i-1,i}$ and $h_{i,i+1}$ interact directly with A_i. Their contribution to the expectation value of the Hamiltonian can be readily evaluated using the properties of the left and right canonical tensors. For example,

$$\langle \Psi | h_{i-1,i} | \Psi \rangle =$$

$$= \tag{12.13}$$

Similarly, the matrix element of $h_{i,i+1}$ is

$$\langle \Psi | h_{i,i+1} | \Psi \rangle = \tag{12.14}$$

(ii) Contribution of $H_{<i}$:

$H_{<i}$ contains all the terms in the left block. As it does not interact directly with A_i, we take it as a single operator. Its matrix element in the ground state is

$$\langle \Psi | H_{<i} | \Psi \rangle = \quad\quad\quad\quad\quad\quad\quad\quad\quad\quad\quad\quad\quad\quad\quad\quad$$

$$= \quad\quad\quad\quad\quad\quad\quad\quad\quad\quad\quad\quad\quad\quad\quad\quad . \quad\quad (12.15)$$

In this expression, the matrix element of $H_{<i}$ is defined by

$$\quad\quad\quad\quad\quad\quad\quad\quad\quad\quad\quad\quad\quad\quad\quad\quad\quad\quad\quad (12.16)$$

From the recursive relation of the left block Hamiltonian:

$$H_{<i} = H_{<i-1} + h_{i-2,i-1}, \quad\quad\quad\quad (12.17)$$

we can further express the matrix element of $H_{<i}$ as

$$\quad\quad\quad\quad\quad\quad\quad\quad\quad\quad\quad\quad\quad\quad\quad\quad\quad\quad\quad$$

$$\quad\quad\quad\quad\quad\quad\quad\quad\quad\quad\quad\quad\quad\quad\quad\quad\quad\quad\quad . \quad\quad (12.18)$$

This sets a recursive formula for evaluating all the matrix elements of the Hamiltonian, $H_{<l}$ with $l \leq i$, in the left block, starting from $H_{<2}$.

(iii) Contribution of $H_{>i}$:

Following the derivation given in step (ii), it is simple to show that the matrix element of $H_{>i}$ is

$$\langle \Psi | H_{>i} | \Psi \rangle = \quad \text{(12.19)}$$

where the matrix element of $H_{>i}$ is defined by

$$H_{>i} \quad = \quad \text{(12.20)}$$

Again, $H_{>i}$ is related to $H_{>i+1}$ by the formula

$$H_{>i} = H_{>i+1} + h_{i+1,i+2}, \quad \text{(12.21)}$$

and their matrix elements are determined by the equation

$$H_{>i} \quad = \quad H_{>i+1} \quad + \quad h_{i+1,i+2} \quad . \quad \text{(12.22)}$$

Again, this equation sets up a recursive formula to calculate all the matrix elements of the Hamiltonian in the right block, from right to left, starting from the right end of the chain.

By removing the two A_i tensors from the above matrix elements of the Hamiltonian, we immediately obtain the matrix elements of $\mathcal{H}_i$:

$$(\mathcal{H}_i)_{\alpha\beta} = \quad h_{i-1,i} \quad + \quad h_{i,i+1}$$

$$+ \quad \boxed{H_{<i}} \overset{\alpha}{\underset{\beta}{\,}} \quad + \quad \overset{\alpha}{\underset{\beta}{\,}} \boxed{H_{>i}} \; . \tag{12.23}$$

12.1.3 Update of Local Tensors

$\mathcal{H}_i$ is a $dD^2 \times dD^2$ matrix. An iterative eigensolver, such as the Lanczos or Jacobi–Davidson large sparse matrix solver, can be used to find its lowest eigenvalue and eigenvector. The computational cost is roughly proportional to $d^2 D^4$ times a large prefactor determined by the number of multiplications of $\mathcal{H}_i$ with a vector of dimension dD^2 needed in the diagonalization of a sparse matrix.

Once A_i is determined by solving the eigenequation (12.10), we move to one of the neighboring sites of i, say site $i+1$, and optimize the local tensor at that site. Before doing this, we first left canonicalize A_i by taking a QR decomposition:

$$-\!\!\overset{|}{(A_i)}\!\!- \;=\; -\!\!\overset{|}{\boxed{U_i}}\!\!-\!\!(R)\!\!- \; . \tag{12.24}$$

U_i is left isometric. Now we set U_i as the local tensor at site i and redefine the local tensor at site $i+1$ as

$$-\!\!\overset{|}{\underset{A_{i+1}}{\bigcirc}}\!\!- \;=\; -\!\!(R)\!\!-\!\!\overset{|}{\underset{V_{i+1}}{\diamond}}\!\!- \; . \tag{12.25}$$

The ground-state wave function then becomes

$$\Psi \;=\; \overset{|}{\underset{U_1}{\triangleright}}\!\cdots\!\overset{|}{\underset{U_i}{\triangleright}}\!\overset{|}{\underset{A_{i+1}}{\bigcirc}}\!\overset{|}{\underset{V_{i+2}}{\triangleleft}}\!\cdots\!\overset{|}{\underset{V_N}{\triangleleft}} \; . \tag{12.26}$$

It has the desired left-right canonical form. A_{i+1} is the local tensor that will be variationally optimized in the next iteration.

On the other hand, we can move the variation tensor from right to left, namely to the site $i-1$. In this case, we should first take an LQ decomposition for A_i and then optimize the local tensor at site $i-1$, A_{i-1}, defined by absorbing the resulting L-matrix into U_{i-1}.

12.2 Excited States

Two approaches can be used to evaluate the low-lying excited states with MPS. One is the variational minimization, which determines a set of low-energy excited states using one MPS. The other is a modified Hamiltonian approach. The first excited state, for example, can be determined by minimizing the energy of a modified Hamiltonian defined by lifting the ground-state energy above the energy of the first excited one. The variational minimization allows many excited states to be determined in one run. It is an accurate approach for determining the energy gap between different energy states. On the other hand, the modified Hamiltonian approach is extendable to higher dimensions. Below we give an introduction to these two approaches.

12.2.1 Variational Minimization

This approach is based on a generalized Rayleigh–Ritz variational principle that works not only for the ground state but also the low-lying excited states [380]. It states that for a set of orthonormal states $\{|\Psi_1\rangle, |\Psi_2\rangle, \ldots, |\Psi_m\rangle\}$,

$$|\Psi_\alpha\rangle = \sum_{\sigma_1,\ldots,\sigma_N} \Psi_\alpha(\sigma_1,\ldots,\sigma_N)|\sigma_1,\ldots,\sigma_N\rangle, \qquad (\alpha = 1,\ldots,m), \qquad (12.27)$$

the sum of their expectation values of the Hamiltonian is not less than the sum of the lowest m eigenvalues of the Hamiltonian:

$$\sum_{\alpha=1}^{m} \langle\Psi_\alpha|H|\Psi_\alpha\rangle \geq \sum_{\alpha=1}^{m} E_\alpha, \qquad (12.28)$$

where $E_1 \leq E_2 \leq \cdots \leq E_m$ are the m lowest eigenenergies. The inequality (12.28) suggests that one can optimize these orthonormal states by variationally minimizing the sum of their energy expectation values.

To carry out the variational minimization, we first introduce the MPS wave function for these orthonormal states $\{|\Psi_\alpha\rangle, \alpha = 1,\ldots,m\}$. We divide the lattice into two parts and assume that the local tensors on the left part of the lattice are left canonicalized, and the local tensors on the right part are right canonicalized. In addition to these left and right canonicalized tensors, we introduce a three-leg tensor C on the bond linking the left and right parts of the lattice. Hence, the MPS has the form

$$\Psi_\alpha(\sigma_1,\ldots,\sigma_N) = \underset{A_1}{\overset{\sigma_1}{\square}} \cdots \underset{A_i}{\overset{\sigma_i}{\square}} \underset{\underset{\alpha}{C}}{} \underset{B_{i+1}}{\overset{\sigma_{i+1}}{\square}} \cdots \underset{B_N}{\overset{\sigma_N}{\square}} . \qquad (12.29)$$

It is similar to the MPS wave function for the ground state, except that tensor C now carries an external index α. If $\alpha = m = 1$, the above wave function reduces to the conventional representation of the ground state.

Tensor C is introduced to orthonormalize these quantum states. To see this more clearly, let us evaluate the overlaps of these states:

$$(12.30)$$

In obtaining the second equality, the properties of the left and right canonical tensors are used. Thus, the orthonormalization condition, $\langle \Psi_\beta | \Psi_\alpha \rangle = \delta_{\alpha\beta}$, is fulfilled if the local tensor C satisfies the equation:

$$(12.31)$$

Variational determination of all the local tensors can be done in the same way as for the ground state. In this case, it is the sum of the expectation values:

$$f(A, B, C) = \sum_\alpha \langle \Psi_\alpha | H | \Psi_\alpha \rangle, \tag{12.32}$$

instead of the ground-state energy, that is minimized with respect to A, B, and C. At each step of optimization, the orthonormal conditions should be preserved, and the A and B tensors should be optimized within the left and right canonical subspaces [381], respectively.

Once all the local tensors are optimized, we are ready to determine the m-low energy eigenvalues and the corresponding eigenstates. Since it is the total energy of these m-states that is minimized, the wave function Ψ_α thus obtained is not automatically an eigenfunction of the Hamiltonian. We should diagonalize the Hamiltonian within the subspace spanned by these m-orthonormal basis states to find the eigenstates. For doing this, we should first evaluate the matrix elements of the Hamiltonian:

$$H_{\alpha\beta} = \langle \Psi_\alpha | H | \Psi_\beta \rangle. \tag{12.33}$$

By diagonalization, we obtain the lowest m-eigenstates and the corresponding eigenenergies.

12.2.2 Modified Hamiltonian Approach

Given the ground-state wave function, $|\Psi_0\rangle$, one can find the first excited state of H by projecting out the ground state from H. This is achieved by defining a new Hamiltonian H' to lift the ground-state energy to a higher energy level so that the first excited state of H becomes the ground state of H'. This modified Hamiltonian is defined by

$$H' = H + w|\Psi_0\rangle\langle\Psi_0|, \tag{12.34}$$

where w is an arbitrary positive constant larger than the first excited energy gap. $|\Psi_0\rangle$ is assumed normalized.

The first excited state of H, $|\Psi\rangle$, now can be obtained by minimizing the energy of H'. This, as discussed in §12.1, is equivalent to solving the following generalized eigenequation:

$$\mathcal{H}'_i A_i = \lambda \mathcal{N}_i A_i, \tag{12.35}$$

where $\mathcal{N}_i$ is the norm matrix and

$$\mathcal{H}'_i = \mathcal{H}_i + \mathcal{W}_i \tag{12.36}$$

is the effective Hamiltonian matrix of H'. The formula for evaluating the matrix elements of $\mathcal{H}_i$ is the same as that discussed in §12.1.2. $\mathcal{W}_i$ is defined by the equation

$$A_i^\dagger \mathcal{W}_i A_i = w \langle\Psi|\Psi_0\rangle\langle\Psi_0|\Psi\rangle. \tag{12.37}$$

The overlap between $|\Psi_0\rangle$ and $|\Psi\rangle$ can be calculated by directly contracting all the tensors except A_i.

The above generalized eigenequation reduces to the simple eigenequation

$$\mathcal{H}'_i A_i = \lambda A_i \tag{12.38}$$

in the left-right canonical representation of Ψ. In this representation, all the tensors on the left-hand side of i are left canonicalized, and those on the right-hand side of i are right canonicalized.

This variational scheme can also be extended to successively find the second or even higher excited states. At each step, the wave functions of the lower energy states should be accurately evaluated. The errors in the lower energy states will be amplified in the calculation of higher excited states. Thus, this scheme cannot be used to find too many excited states.

12.3 Imaginary Time Evolution

If the Hamiltonian contains only the nearest-neighbor interactions, the ground state can be efficiently determined by imaginary time evolution. This is an advantage of RG formulation in terms of MPS. Another advantage of determining the MPS wave function using the imaginary time evolution is that the truncation error is not accumulated in the iteration. In principle, the wave function will converge to the true ground state no matter how poor the initial guess of the wave function is, provided that it is not orthogonal to the ground state.

The imaginary time evolution is to use the evolution operator $e^{-\beta H}$ to act successively on an arbitrary initial state. This corresponds to a successive cooling down in temperature. In the limit $\beta \to \infty$, the temperature becomes zero, and this evolution will project out the ground state. More specifically, starting from an arbitrary initial state $|\Psi\rangle$, the state will become

$$|\Psi(\beta)\rangle = e^{-\beta H} |\Psi\rangle = \sum_n e^{-\beta E_n} |n\rangle \langle n|\Psi\rangle \tag{12.39}$$

after the evolution. E_n is the eigenenergy of the Hamiltonian H and $|n\rangle$ the corresponding eigenvector. Assuming $|0\rangle$ to be the ground state, the above equation can also be written as

$$|\Psi(\beta)\rangle = e^{-\beta E_0} \left[|0\rangle \langle 0|\Psi\rangle + \sum_{n\neq 0} e^{-\beta(E_n-E_0)} |n\rangle \langle n|\Psi\rangle \right]. \tag{12.40}$$

This equation suggests that as long as the ground state is not degenerate and β is larger than the inverse of the minimum excitation gap of H, the state $|\Psi(\beta)\rangle$ will converge exponentially to the ground state in the limit $\beta \to \infty$.

As an example, let us consider a one-dimensional quantum lattice model with nearest-neighbor interactions, $h_{i,i+1}$. These interaction terms generally do not commute with each other, thus one cannot evaluate the imaginary time evolution in one step. Instead, one can divide β into many small steps, $\beta = M\tau$ and

$$|\Psi(\beta)\rangle = \left(e^{-\tau H}\right)^M |\Psi\rangle, \tag{12.41}$$

and evaluate the time evolution step by step. At each step, the Trotter–Suzuki decomposition is used to decouple $e^{-\tau H}$ into a product of two terms:

$$e^{-\tau H} = e^{-\tau H_{\text{odd}}} e^{-\tau H_{\text{even}}} + o(\tau^2), \tag{12.42}$$

where

$$H_{\text{odd}} = \sum_{i=\text{odd}} h_{i,i+1}, \quad H_{\text{even}} = \sum_{i=\text{even}} h_{i,i+1}. \tag{12.43}$$

The Trotter error does not depend on the system size and can be reduced by using a smaller Trotter step τ or a higher-order Trotter–Suzuki decomposition formula. The second-order Trotter–Suzuki decomposition formula reads

$$e^{-\tau H} = e^{-\tau H_{\text{odd}}/2} e^{-\tau H_{\text{even}}} e^{-\tau H_{\text{odd}}/2} + o(\tau^3). \tag{12.44}$$

All the bond interaction terms in H_{odd} (similarly in H_{even}) commute with each other. The time evolutions of these terms can be carried out simultaneously. By applying $\exp(-\tau H_{\mathrm{odd}})$ to the wave function, the state becomes

$$|\Psi'\rangle = e^{-\tau H_{\mathrm{odd}}}|\Psi\rangle$$

$$= \sum_{\sigma_1 \cdots \sigma_N} e^{-\tau H_{\mathrm{odd}}} \quad |\sigma_1 \dots \sigma_N\rangle$$

$$= \sum_{\sigma_1 \cdots \sigma_N} \qquad \cdots \qquad |\sigma_1 \dots \sigma_N\rangle, \qquad (12.45)$$

where $B_{i,i+1}$ is defined on two neighboring sites,

$$\qquad (12.46)$$

and

$$= \langle \sigma_i \sigma_{i+1} | e^{-\tau h_{i,i+1}} | \sigma_i' \sigma_{i+1}' \rangle. \qquad (12.47)$$

One can decompose $B_{i,i+1}$, for example, using SVD, into a product of two matrices,

$$\qquad (12.48)$$

B_i and B_{i+1} are three-leg tensors. The thicker bond has a higher dimension, which equals the product of the dimension of α-bond and that of σ-bond; that is, Dd. If we update the local tensors A_i by B_i and A_{i+1} by B_{i+1}, Ψ' becomes

$$\Psi' = \qquad . \qquad (12.49)$$

It has the same form as Ψ, except that the virtual dimension on the odd-even bonds equals Dd. After truncating this bond dimension to D, Ψ' is reset as the initial MPS for the next evolution step. We obtain the ground-state wave function by repeating these steps until the energy converges.

Several schemes are used to truncate the thick bonds in this MPS. Canonicalization, full update, and simple update are the three commonly used schemes. Their computational cost and converging speeds are different. The canonicalization is an optimized truncation scheme. The full update minimizes the truncation error at each step of the iteration. The simple update is not optimal at each truncation, but it is the most efficient scheme.

12.3.1 Canonicalization

The wave function obtained in (12.49) can be canonicalized using the gauge transformation introduced in §10.3 so that the wave function is perfectly conditioned, as in DMRG. It provides an optimal scheme of truncation. The optimal local tensors are obtained by keeping the largest D singular values on each thick bond.

Canonicalization is an accurate and robust method for optimizing the truncation scheme at each step of imaginary time evolution. It can be extended to an infinite lattice system using the canonical transformation described in §11.2, or to a quantum system defined on a Bethe lattice or Cayley tree. This method can also be used to evaluate the leading eigenvectors of a transfer matrix or other nonunitary projection operators in the framework of MPS [382]. Hence, it can be used to evaluate the thermodynamic properties of two-dimensional classical models introduced in Chapter 14 or one-dimensional quantum lattice models in Chapter 16. In two or higher dimensions, it is impossible to divide a system into two separate parts by just cutting one bond. This kind of canonicalization scheme cannot be used.

12.3.2 Full Update

The variational optimization is also a useful approach for accurately determining a renormalized wave function in the imaginary time evolution. This approach can be applied not just to a finite lattice system with open boundary conditions but also to an infinite lattice, a periodic chain with or without translation invariance, or even to a quantum lattice system in two or higher dimensions.

Given Ψ', represented by Eq. (12.49), the goal of this approach is to find another MPS:

$$\Psi = \underset{B_1}{-} \underset{B_2}{-} \underset{B_3}{-} \cdots \underset{B_N}{-}, \qquad (12.50)$$

which gives the best approximation to Ψ'. This is equivalent to variationally opti-mizing all local tensors B_i so that the difference between these two wave functions; that is,

$$F = \big\| \, |\Psi'\rangle - |\Psi\rangle \, \big\|^2 = \langle\Psi'|\Psi'\rangle + \langle\Psi|\Psi\rangle - \langle\Psi|\Psi'\rangle - \langle\Psi'|\Psi\rangle, \qquad (12.51)$$

is minimized. F is a quadratic and concave function of local tensors of Ψ. The first term on the right-hand side does not contain any variational parameters. If we fix all the local tensors of Ψ except one of them, say B_i, F is quadratic in B_i. In this case, $\langle\Psi|\Psi\rangle$ and $\langle\Psi|\Psi'\rangle$ can be represented, respectively, as

$$\langle\Psi|\Psi\rangle = \qquad (12.52)$$

$$\langle\Psi|\Psi'\rangle = \qquad (12.53)$$

$\mathcal{N}_i$ is the norm matrix of Ψ. $\mathcal{M}_i$ is a three-leg vector defined by

$$(\mathcal{M}_i)_\alpha = \qquad (12.54)$$

Taking the derivative of F with respect to $B_i^\dagger$, we find the equation for optimizing B_i:

$$\qquad (12.55)$$

This is a set of linear equations that could be solved with the standard numerical recipes.

In the left-right canonical representation of Ψ, where the tensors on the left- and right-hand sides of B_i are left and right canonicalized, the norm matrix is an identity matrix. In this case, the solution is

$$B_i = \mathcal{M}_i. \qquad (12.56)$$

This one-site optimization is repeated, from left to right and then from right to left, until convergence is reached. To optimize the local tensor at a new site, one should reorthonormalize Ψ so that it is left-right canonicalized about that point.

12.3.3 Simple Update

In the imaginary time evolution, if the time step τ is small, then the evolution operator $\exp(-\tau H_{i,i+1})$ is nearly unitary. In this case, one can take a mean-field approach to convert a global optimization problem into a local one [382]. The mean field parameters are the bond vectors λ_i defined in Eq. (10.17). In the canonical representation, λ_i measures the entanglement between its left and right blocks. It can also be regarded as an effective field acting on that bond.

In the time evolution [35, 38], the evolution operator is unitary, and the contribution from the environment connecting by this bond is described by the entanglement field λ_i. In the imaginary time evolution, as the evolution operator is not unitary, the MPS is not canonical after each step of evolution. In this case, λ_i is not equal to the entanglement spectrum. Nevertheless, it can still be regarded as an approximate entanglement field acting on that bond. This approximation allows us to evaluate the environment contribution using local parameters without taking time-consuming canonical transformation. It yields a fast-converging and easily implemented algorithm for determining the ground-state wave function, provided that the imaginary time evolution operator is sufficiently close to a unitary one [383].

Vidal and coworkers used this mean-field approximation [382, 383] as an extension of the TEBD in one dimension. An extension to higher dimensions with the interpretation of λ_i as an effective entanglement field was proposed by Jiang *et al.* [116], leading to an approximate but efficient method, which is dubbed as *simple update*, for evaluating ground-state wave functions in two or higher dimensions.

It is more convenient to implement this mean-field approach, starting from the $\lambda - \Gamma$ representation of MPS defined by Eq. (10.17). By applying $\exp(-\tau H_{\mathrm{odd}})$ to this state, the wave function becomes

$$
\Psi' = \quad \overset{\overset{\displaystyle \sigma_1\,\sigma_2}{\smile}}{\underset{\lambda_0 \quad B_{1,2} \quad \lambda_2}{\diamondsuit\!-\!\bigcirc\!-\!\diamondsuit}} \overset{\overset{\displaystyle \sigma_3\,\sigma_4}{\smile}}{\underset{B_{3,4} \quad \lambda_4}{\bigcirc\!-\!\diamondsuit}} \cdots \overset{\overset{\displaystyle \sigma_{N-1}\sigma_N}{\smile}}{\underset{B_{N-1,N}\;\lambda_N}{\bigcirc\!-\!\diamondsuit}} \,, \tag{12.57}
$$

where $B_{i,i+1}$ is defined by

$$
\overset{\overset{\displaystyle \sigma_i\,\sigma_{i+1}}{\smile}}{\underset{B_{i,i+1}}{-\!\bigcirc\!-}} \;=\; \overset{\displaystyle \sigma_i \qquad \sigma_{i+1}}{\underset{\Gamma_i \quad \lambda_i \quad \Gamma_{i+1}}{\boxed{v^{i,i+1}}}} \,. \tag{12.58}
$$

If the basis states of the left and right blocks connected to matrix $B_{i,i+1}$ in the above formula are orthonormal, then the environment contribution to $B_{i,i+1}$ is proportional to λ_{i-1} from the left side and λ_{i+1} from the right side. In the mean-field calculation, however, the basis states on both the left and right blocks are not orthogonalized. In this case, λ_{i-1} and λ_{i+1} are just an approximate measure of the entanglement spectra. Nevertheless, they provide an effective measure of the environment contribution. In particular, the environment renormalized $B_{i,i+1}$-tensor is

$$\left(\mathcal{B}_{i,i+1}\right)_{\alpha\sigma_i,\sigma_{i+1}\beta} = \alpha \overset{\displaystyle \sigma_i \; \sigma_{i+1}}{\underset{\lambda_{i-1} \; B_{i,i+1} \; \lambda_{i+1}}{\diamond{-}\bigcirc{-}\diamond}} \beta. \tag{12.59}$$

Here we have grouped α and σ_i as the row index, and β and σ_{i+1} as the column index of $\mathcal{B}_{i,i+1}$. $\mathcal{B}_{i,i+1}$ is a Dd-dimensional matrix. By decoupling it with SVD

$$\mathcal{B}_{i,i+1} = \underset{}{-\!\boxed{U}\!-\!\diamond\!-\!\boxed{V}\!-} , \tag{12.60}$$

and keeping its D largest singular values and the corresponding basis vectors, we obtain

$$\mathcal{B}_{i,i+1} \approx \underset{}{-\!\boxed{U}\!-\!\langle\diamond\rangle\!-\!\boxed{V}\!-} . \tag{12.61}$$

Here we have used the same symbols to represent the unitary matrices, U and V in Eq. (12.60), and those after truncation in Eq. (12.61).

After this, we update the local tensors by the formula

$$-\!\bigcirc\!- = \underset{\lambda_{i-1}^{-1}}{-\!\diamond\!-\!\boxed{U}\!-} , \qquad -\!\bigcirc\!- = \underset{\Gamma_{i+1}}{-\!\boxed{V}\!-\!\diamond\!-} , \qquad \lambda_{i+1}^{-1} \tag{12.62}$$

and $\lambda_i = \lambda$. The diagonal bond matrices λ_{i-1} and λ_{i+1} remain unchanged. This completes one step of imaginary time evolution with H_{odd}. Similarly, one can carry out the imaginary time evolution with H_{even} and update all site tensors and bond matrices. The wave function will eventually converge by repeating these steps.

In this simple update procedure, the truncation error is larger than that obtained by the full update at each step of evolution. However, as the truncation error is not accumulated with the iterations, minimizing the truncation error at every step of iteration is no longer needed. In the limit $\beta \to \infty$, the results thus obtained can reach the same accuracy as those obtained by the canonical transformation [130]. The truncation error of the converged result is controlled purely by the bond dimension D, rather than the mean-field approximation of the entanglement spectra. It can be reduced by increasing the bond dimension.

Figure 12.1 compares the difference in the ground state energy obtained by taking the canonical transformation, E_{can}, with that obtained by the simple update, E_{SU}, starting from a randomly generated wave function for the one-dimensional

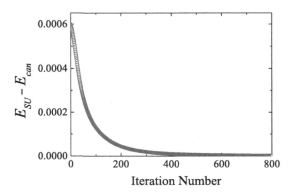

Figure 12.1 Difference in the ground-state energy obtained by the simple update, E_{SU}, with that by the canonical transformation, E_{can}, for the one-dimensional spin-1/2 Heisenberg model with $D = 20$.

Heisenberg model with $D = 20$. Both E_{can} and E_{SU} converge to the true ground-state energy quickly. As expected, E_{SU} is always higher than E_{can}. However, the difference between these two energies, $E_{SU} - E_{can}$, approaches zero with the increase of the iteration number, indicating that the wave function obtained using the simple update converges to the true ground-state wave function. The simple update does not need to perform the variational calculation at each projection step. It is much more efficient than the full update in evaluating the ground-state wave function.

It should be emphasized that the simple update approach should not be directly applied to the calculation of time-dependent or thermodynamic quantities wherein both the Trotter and truncation errors are accumulated with the iterations, and the long-time (or imaginary time in the calculation of thermodynamic quantities) results may not converge. In this case, a more rigorous treatment of the renormalization effect from the environment should be considered to minimize both the Trotter and truncation errors.

12.4 Purification

In a mixed state, to determine the expectation value of an observable $\hat{O}$, we generally need to evaluate the ensemble average of this observable from a set of orthonormal basis states $|n\rangle$ with a probability distribution w_n. The density matrix for this mixed state is defined by

$$\rho = \sum_{n=1}^{m} w_n |n\rangle \langle n|. \tag{12.63}$$

The expectation value of $\hat{O}$ is then given by

$$\bar{O} = \text{Tr}\left(\hat{O}\rho\right) = \sum_n w_n \langle n|\hat{O}|n\rangle. \tag{12.64}$$

To evaluate this sum from the expectation value for each basis state $|n\rangle$ is difficult. This difficulty, however, can be resolved by introducing a set of auxiliary basis states $|n_a\rangle$ to turn the density matrix into a partial trace over a Schmidt decomposition of a pure state [78, 384]. The auxiliary state space can be taken as a copy of the original one. The original and auxiliary states form an augmented Hilbert space.

By introducing a pure state in the augmented Hilbert space,

$$|\Psi\rangle = \sum_{n=1}^m \sqrt{w_n}|n\rangle \otimes |n_a\rangle, \tag{12.65}$$

where $|n_a\rangle$ is simply a copy of $|n\rangle$, defined in an auxiliary subspace. The density matrix can be expressed as a partial trace of the pure state $|\Psi\rangle$ over the auxiliary states:

$$\rho = \text{Tr}_{n_a}|\Psi\rangle\langle\Psi|. \tag{12.66}$$

Using this purified state, we can write the expectation value defined by Eq. (12.64) as

$$\bar{O} = \frac{\langle\Psi|\hat{O}|\Psi\rangle}{\langle\Psi|\Psi\rangle}. \tag{12.67}$$

This purification scheme is useful. It can be used not just in one dimension but also in higher dimensions. In the following sections, we apply it to evaluate the partition function of a quantum system and the dynamical evolution of the ground state in a quenched disorder system in one dimension. The quantum interactions are assumed to be local.

12.4.1 Thermodynamics

Finite temperature calculations can be carried out based on the purification of thermal density matrix [78]:

$$\rho = e^{-\beta H} = \sum_\sigma e^{-\beta H/2}|\sigma\rangle\langle\sigma|e^{-\beta H/2}, \tag{12.68}$$

where $|\sigma\rangle = |\sigma_1, \ldots, \sigma_N\rangle$ is a set of complete basis states and $|\sigma_i\rangle$ is the basis state at site i . By introducing a copy of the basis state $|\sigma_{i,a}\rangle$ for a given physical basis state $|\sigma_i\rangle$, we can form a maximally entangled basis state:

$$|\Psi_0\rangle = \prod_i \sum_{\sigma_i} |\sigma_i\rangle \otimes |\sigma_{i,a}\rangle. \tag{12.69}$$

This is also the purified thermodynamic state at infinite temperature. It is an MPS of bond dimension 1, if we take the basis sets $|\sigma_i\rangle$ and $|\sigma_{i,a}\rangle$ as an enlarged basis set at site i.

The thermal density matrix can be expressed as a partial trace on an auxiliary pure state $|\Psi(\beta)\rangle$ in the enlarged Hilbert space:

$$\rho = \mathrm{Tr}_a |\Psi(\beta)\rangle\langle\Psi(\beta)|, \tag{12.70}$$

where

$$|\Psi(\beta)\rangle = e^{-\beta H/2} |\Psi_0\rangle. \tag{12.71}$$

The thermal average of an observable $\hat{O}$ is given by

$$\bar{O} = \frac{\langle\Psi(\beta)|\hat{O}|\Psi(\beta)\rangle}{\langle\Psi(\beta)|\Psi(\beta)\rangle}. \tag{12.72}$$

Here $|\Psi(\beta)\rangle$ can be found by performing an imaginary time evolution using the canonicalization or full update method introduced in §12.3, starting from the initial wave function $|\Psi_0\rangle$. What we need to do is to carry out imaginary time evolution up to $\beta/2$, and then calculate expectation values as for a pure state. This scheme can also be extended to evaluate the real-time evolution at finite temperature [385].

12.4.2 Parallelizing Quenched Disorders

For a disordered quantum many-body system, it is a difficult task to evaluate the quenched average of a physical observable $\hat{O}$:

$$\bar{O} = \sum_q \rho(q) \frac{\mathrm{Tr}\left[\hat{O}e^{-\beta H(q)}\right]}{Z_q}, \tag{12.73}$$

where $q = (q_1, q_2, \ldots, q_N)$ are random variables that take values within a finite discrete set, $q_i \in (s_1, \ldots, s_m)$, with a probability distribution given by $\rho(q)$. $H(q)$ is the Hamiltonian for a given random configuration q, and Z_q is the corresponding partition function:

$$Z_q = \mathrm{Tr}e^{-\beta H(q)}. \tag{12.74}$$

In order to evaluate precisely the random average, one would need to perform m^N simulations, one for each possible realization of random variables.

One generally adopts a replica trick to solve this complex summation problem of random configurations. This trick is to represent the free energy as a product of n copies of the partition function Z_q in a given random configuration. The crux of this trick is to convert $\ln Z_q$ into a power of Z_q from which the disorder average can be more easily performed. By positing that n is no longer constrained to be an integer and that an analytic continuation to the limit $n \to 0$ can be safely taken from the solution of Z_q^n for all positive integers n, the free energy can then be calculated using the formula

$$\ln Z_q = \lim_{n \to 0} \frac{Z_q^n - 1}{n}. \tag{12.75}$$

Similarly, the expectation value of an observable can be evaluated using an effective Hamiltonian of n replicas, $H_{\text{eff}}^{(n)}$, in the limit $n \to 0$:

$$\overline{O} = \lim_{n \to 0} \frac{\text{Tr}\hat{O}e^{-\beta H_{\text{eff}}^{(n)}}}{\text{Tr}e^{-\beta H_{\text{eff}}^{(n)}}}, \tag{12.76}$$

where $H_{\text{eff}}^{(n)}$ is defined by

$$e^{-\beta H_{\text{eff}}^{(n)}} = \sum_q \rho(q) Z_q^n. \tag{12.77}$$

However, the above limit $n \to 0$ is not well defined. It poses many mathematical questions and introduces many subtleties or uncertainties in the extrapolation to the limit $n \to 0$ from the results obtained with positive integer n.

The difficulty in evaluating the random average of an observable is that before taking the random average, one has first to take the thermal average. In other words, both the numerator and the denominator in Eq. (12.73) depend on the random distribution. However, for a pure state under the real-time evolution, this random average can be done efficiently by introducing auxiliary quantum fields to simulate quenched disorders in parallel. This algorithm was first proposed by Paredes *et al.* [384].

To understand how the algorithm works, let us consider the random average of an observable in a time-dependent pure state:

$$\overline{O}(t) = \sum_q \rho(q) \frac{\langle \psi_q(t) | \hat{O} | \psi_q(t) \rangle}{\langle \psi_q(t) | \psi_q(t) \rangle} = \sum_q \rho(q) \langle \psi_q(t) | \hat{O} | \psi_q(t) \rangle. \tag{12.78}$$

In Eq. (12.78), the second equality holds because $\langle \psi_q(t) | \psi_q(t) \rangle = 1$ is valid at any time under a unitary evolution. $|\psi_q(t)\rangle$ is the evolved state in a given random configuration with a time-dependent Hamiltonian $H(q, t)$:

$$|\psi_q(t)\rangle = \exp\left[-i \int_0^t dt H(q, t)\right] |\psi_0\rangle. \tag{12.79}$$

$|\psi_0\rangle$ is the initial state at $t = 0$.

In this case, one can represent all possible random configurations as a superposition state of an auxiliary quantum system and map a quantum random system onto a dimension-enlarged random-free system. To do this, one can introduce a quantum operator Q_i and assume the random variable q_i at site i to be its eigenvalue

$$Q_i |q_i\rangle = q_i |q_i\rangle, \qquad (12.80)$$

and $|q_i\rangle$ the corresponding eigenvector. The auxiliary state $|q\rangle = |q_1, q_2, \ldots, q_N\rangle$ is in one-to-one correspondence to the random configuration q, with a weight equal to the probability $\rho(q)$. Since q are classical variables, Q_i should commute with other Q operators and with all quantum operators defined in H_q. Using these ancilla states, Eq. (12.78) can then be expressed as an expectation value on a pure state defined on a dimension-enlarged Hilbert space:

$$\overline{O}(t) = \langle \Psi(t)|\hat{O}|\Psi(t)\rangle, \qquad (12.81)$$

where

$$|\Psi(t)\rangle = \sum_q |\psi_q(t)\rangle \otimes \sqrt{\rho(q)}|q\rangle. \qquad (12.82)$$

This wave function can also be obtained from the time evolution of an effective Hamiltonian $H(Q, t)$, which is obtained by replacing $q = (q_1, q_2, \ldots, q_N)$ in $H(q, t)$ by $Q = (Q_1, Q_2, \ldots, Q_N)$, from an initial state defined by the superposition of $|\psi_0\rangle$ and an auxiliary ancilla state $|\psi_a\rangle$:

$$|\Psi_0\rangle = |\psi_0\rangle \otimes |\psi_a\rangle, \qquad (12.83)$$

where

$$|\psi_a\rangle = \sum_q \sqrt{\rho(q)} \, |q\rangle. \qquad (12.84)$$

The time evolution of $|\Psi_0\rangle$ then gives the desired wave function:

$$|\Psi(t)\rangle = \exp\left[-i \int_0^t dt H(Q, t)\right]|\Psi_0\rangle. \qquad (12.85)$$

This means that the random average can be encoded and efficiently simulated as a purified quantum state.

As a particular case of application, the above algorithm can be used to simulate in parallel the random average of all ground states. To do this, one needs to prepare the initial state as a superposition of the ground state without disorder at $t = 0$, $|\psi_0\rangle$, and the ancilla state $|\psi_a\rangle$. One then switches on the random potentials from 0 at $t = 0$ to r at $t = T$ adiabatically. The effective Hamiltonian is now defined by

$$H(Q, t) = H(g(t)Q), \qquad (12.86)$$

where $g(t)$ is a slowly varying function of time with $g(0) = 0$ and $g(T) = 1$. In the limit $T \to \infty$, the time-evolved state should converge to a superposition state of the ground state of $H(q)$ with a distribution given by $\rho(q)$. This leads to a parallel simulation of all possible ground states for a random quantum system. Here we assume that there are no level crossings during the whole period of evolution. If there is a level crossing, the initial and final states will be in different phases, and the above algorithm does not simulate the correct ground states. To solve this problem, one can, in principle, use a new initial Hamiltonian with a new initial ground-state wave function $|\psi_0\rangle$.

A number of methods, including the time-dependent DMRG and TEBD introduced in §19.2, can be used to evaluate the time evolution of $|\Psi(t)\rangle$.

13

Continuous Matrix Product States

13.1 Lattice Discretization of Continuous Quantum Field Theory

Tensor networks allow encoding many-body quantum states whose dynamics are governed by local Hamiltonians with modest computational resources. In light of their successful applications for lattice models, it is natural to find a scheme to extend them to solve problems of quantum field theories in the continuum. In principle, this can be done in two ways. The first is to discretize the spatial coordinates and represent a quantum state as a lattice tensor network state. This approach is simple to implement. However, extrapolation to the zero lattice spacing limit has to be taken to obtain the results in the continuous space. The second way is to directly take the continuum limit of tensor network states so they can be applied to a continuum model without extrapolation. This approach preserves the spatial symmetries and avoids the perilous continuum extrapolations of the results.

The continuous MPS was first introduced by Verstraete and Cirac in 2010 [52]. It leads to a continuous representation of MPS whose entanglement structure is tailored to describe the low-energy states of quantum field theories.

Following Ref. [52], we take the Lieb–Liniger model [386, 387] in a periodic system of length L as an example to illustrate continuous MPS. The discussion given in this chapter can be readily extended to an interacting fermion system or systems with more than one species of bosons or fermions.

The Lieb–Liniger model is a nonrelativistic Hamiltonian of bosonic fields:

$$H = \int dx H(x) \tag{13.1}$$

with

$$H(x) = \frac{1}{2m} \partial_x \psi^\dagger(x) \partial_x \psi(x) - \mu \psi^\dagger(x) \psi(x) + g \psi^\dagger(x) \psi^\dagger(x) \psi(x) \psi(x). \tag{13.2}$$

μ is the chemical potential. The bosonic operator $\psi(x)$ fulfills the canonical commutation relation:

$$\left[\psi(x), \psi^\dagger(y)\right] = \delta(x-y). \tag{13.3}$$

This Hamiltonian can be discretized by dividing the continuous coordinate x from 0 to L into a N-site lattice of equidistant points $x_i = \varepsilon i$, with $\varepsilon = L/(N-1)$ the lattice spacing, and by replacing the integral with

$$\int dx \to \varepsilon \sum_i. \tag{13.4}$$

The discretization should not change the total number of particles. This imposes a constraint on the discretized boson operator:

$$\int dx \psi^\dagger(x)\psi(x) = \sum_i \int_{x_i}^{x_i+\varepsilon} dx \psi^\dagger(x)\psi(x)$$

$$\approx \varepsilon \sum_i \psi^\dagger(x_i)\psi(x_i) = \sum_i b_i^\dagger b_i. \tag{13.5}$$

b_i is the bosonic operator at site i. Thus, the bosonic field operator $\psi(x)$ in the continuum is related to b_i by the formula

$$\psi(x) \to \psi(x_i) = \frac{b_i}{\sqrt{\varepsilon}}. \tag{13.6}$$

b_i fulfills the lattice canonical commutation relation:

$$\left[b_i, b_j^\dagger\right] = \delta_{i,j}. \tag{13.7}$$

Using these discrete operators, we may write the kinetic energy term as

$$\int dx \partial_x \psi^\dagger(x)\partial_x \psi(x) \approx \varepsilon \sum_i \frac{\psi^\dagger(x_{i+1}) - \psi^\dagger(x_i)}{\varepsilon} \frac{\psi(x_{i+1}) - \psi(x_i)}{\varepsilon}$$

$$= 2\varepsilon^{-2} \sum_i b_i^\dagger b_i - \varepsilon^{-2} \sum_i \left(b_{i+1}^\dagger b_i + h.c.\right). \tag{13.8}$$

Similarly, the interaction term is

$$\int dx \psi^\dagger(x)\psi^\dagger(x)\psi(x)\psi(x) \approx \varepsilon \sum_i \psi^\dagger(x_i)\psi^\dagger(x_i)\psi(x_i)\psi(x_i)$$

$$= \varepsilon^{-1} \sum_i b_i^\dagger b_i^\dagger b_i b_i. \tag{13.9}$$

Thus in the lattice representation, the Lieb–Liniger model becomes

$$H = -\frac{1}{2m\varepsilon^2} \sum_i \left(b_{i+1}^\dagger b_i + h.c.\right) + \sum_i \left(\frac{1}{m\varepsilon^2} - \mu\right) b_i^\dagger b_i$$

$$+ \frac{g}{\varepsilon} \sum_i b_i^\dagger b_i^\dagger b_i b_i. \tag{13.10}$$

This discretized Hamiltonian acts on a lattice where each site i is described by the Fock space of the boson operators, characterized by the basis states of the boson number:

$$b_i^\dagger b_i |n_i\rangle = n_i |n_i\rangle, \tag{13.11}$$

where n_i takes any nonnegative integer number and $|n_i\rangle$ is the corresponding eigenstate.

13.2 Continuum Limit of MPS

We now consider how to regulate and parameterize an MPS in the discretized lattice space:

$$|\Psi\rangle = \sum_{n_1 \dots n_N} \text{Tr} A_1[n_1] \cdots A_N[n_N] |n_1 \cdots n_N\rangle, \tag{13.12}$$

so that it provides a faithful representation of the quantum state in the continuum limit $\varepsilon \to 0$, independent of the lattice spacing parameter ε. For doing this, let us first examine the scaling behavior of the density of bosons and other physical quantities with ε.

For a physical system, the local density of particles $n(x) = \langle \psi^\dagger(x)\psi(x)\rangle$, the kinetic energy $\langle \partial_x \psi^\dagger(x)\partial_x \psi(x)\rangle$, and the interaction $\langle \psi^\dagger(x)\psi^\dagger(x)\psi(x)\psi(x)\rangle$ should all be finite and not diverging. This suggests that in the lattice system, the corresponding terms in the lattice space should scale as

$$\langle b_i^\dagger b_i\rangle \approx \varepsilon \langle \psi^\dagger(x)\psi(x)\rangle = O(\varepsilon), \tag{13.13}$$

$$\langle b_i^\dagger b_i^\dagger b_i b_i\rangle \approx \varepsilon^2 \langle \psi^\dagger(x)\psi^\dagger(x)\psi(x)\psi(x)\rangle = O(\varepsilon^2). \tag{13.14}$$

Equation (13.13) implies that the probability of finding at least one boson at one site is of order ε on average, provided that the bosons are evenly distributed in space. So the probability of finding a boson at one site drops linearly with ε. Similarly, $\langle b_i^\dagger b_i^\dagger b_i b_i\rangle$ is finite if there are at least two bosons at site i, which implies the probability of finding two or more bosons at one site is of order ε^2. In general, the probability of finding at least n bosons at a site is of order ε^n.

Since a finite number of particles has to be distributed over a diverging number of lattice sites in the limit $\varepsilon \to 0$, most of the sites in the lattice are unoccupied on average. So the local tensor $A_i[n = 0]$ has to be the dominant matrix of order 1 up to a physically irrelevant normalization constant. Furthermore, to the leading order approximation, the probability of finding $n > 0$ particles at site i is determined predominantly by the tensor elements of $A_i[n]$. As the probability of finding n bosons at a site is of the order of ε^n, it suggests that $A_i[n]$ should scale as $\varepsilon^{n/2}$ in the small ε limit. Thus in the $\varepsilon \to 0$ limit, $A_i[n]$ can be parameterized as

$$A_i[0] = I + \varepsilon A(x_i) + O(\varepsilon^2), \tag{13.15}$$

$$A_i[1] = \sqrt{\varepsilon} B(x_i) + O(\varepsilon^{3/2}), \tag{13.16}$$

$$A_i[n] = \frac{1}{n}\varepsilon^{n/2} C^n(x_i) + O(\varepsilon^{1+n/2}), \quad (n \ge 2), \tag{13.17}$$

where $A(x)$, $B(x)$, and $C(x)$ are ε-independent. They are $D \times D$ matrices defined in the virtual space.

This suggests that the matrix-vector that generates the local tensor at site i:

$$\hat{A}_i \equiv \sum_n A_i[n]|n\rangle = \left(A_i[0] + A_i[1]b_i^\dagger + \frac{1}{\sqrt{2}}A_i[2]b_i^\dagger b_i^\dagger + \cdots \right)|0\rangle \tag{13.18}$$

can be expressed as

$$\hat{A}_i = \left[I + \varepsilon A(x_i) + \varepsilon B(x_i)\psi^\dagger(x_i) + O(\varepsilon^2) \right]|0\rangle. \tag{13.19}$$

$|0\rangle$ is the vacuum of the Fock space. In the limit $\varepsilon \to 0$, the local tensor becomes

$$\hat{A}_i \approx \exp\left[\varepsilon A(x_i) + \varepsilon B(x_i)\psi^\dagger(x_i) \right]|0\rangle$$

$$\approx \exp\left[\int_{x_i-\varepsilon/2}^{x_i+\varepsilon/2} dx A(x) + \varepsilon B(x)\psi^\dagger(x) \right]|0\rangle. \tag{13.20}$$

Thus in the continuum limit, the quantum state becomes

$$|\Psi\rangle = \mathrm{Tr}\hat{A}_1 \cdots \hat{A}_N$$

$$= \mathrm{Tr}\mathcal{P}\exp\left\{ \int_0^L dx \left[A(x) + B(x)\psi^\dagger(x) \right] \right\}|0\rangle, \tag{13.21}$$

where $\mathcal{P}\exp$ is the path ordered exponential, which orders its argument from left to right with increasing x. This is the continuous MPS representation of the ground state first proposed by Verstraete and Cirac [52]. In obtaining this expression, the inverse Trotter–Suzuki decomposition is used.

The above expression holds for a system with periodic boundary conditions, in which $A(L + x)$ and $B(L + x)$ are identical to $A(x)$ and $B(x)$, respectively.

If open boundary conditions are imposed, the left bond dimension of $\hat{A}_1$ and the right bond dimension of $\hat{A}_N$ become 1. Hence, $\hat{A}_1$ is a $1 \times D$ matrix or a row vector, and $\hat{A}_N$ is a $D \times 1$ matrix or a column vector. To construct the continuous MPS representation of $|\Psi\rangle$, we introduce two D-dimensional edge vectors, v_L and v_R, defined in the virtual basis space, so that both $\hat{A}_1$ and $\hat{A}_N$ are still $D \times D$ matrices and the ground state can be represented as

$$|\Psi\rangle = v_L^\dagger \hat{A}_1 \cdots \hat{A}_N v_R. \tag{13.22}$$

The corresponding continuous MPS is

$$|\Psi\rangle = \mathcal{P}v_L^\dagger \exp\left\{\int_0^L dx\left[A(x) + B(x)\psi^\dagger(x)\right]\right\} v_R|0\rangle. \tag{13.23}$$

Both v_L and v_R are variational parameters.

13.3 Expectation Values

The expectation value of any normal-ordered operator, in which all annihilation operators of bosons are moved to the right-hand side of all creation operators, is determined by acting with all annihilation operators on the ket $|\Psi\rangle$ and all creation operators on the bra $\langle\Psi|$. To evaluate these quantities, we first define a continuous MPS at an arbitrary interval from x to y:

$$R(x,y) = \mathcal{P}\exp\left\{\int_x^y dz\left[A(z) + B(z)\psi^\dagger(z)\right]\right\}. \tag{13.24}$$

$R(0, L)$ is just the continuous MPS that determines the ground state:

$$|\Psi\rangle = \mathrm{Tr}R(0, L)|0\rangle. \tag{13.25}$$

The path-ordered exponential function can be expanded using the variables in the exponent. From the definition, it is simple to show that

$$
\begin{aligned}
M_A(x,y) &\equiv \mathcal{P}\exp\left[\int_x^y dzA(z)\right] \\
&= \sum_{n=0}^\infty \int_{x\leq x_1\leq\cdots\leq x_n\leq y} dx_1\cdots dx_n A(x_1)A(x_2)\cdots A(x_n).
\end{aligned} \tag{13.26}
$$

More generally, if there are two variables on the exponent, we can expand the exponent using one of the variables. The result is

$$
\begin{aligned}
M_{A+B}(x,y) &= \sum_{n=0}^\infty \int_{x\leq x_1\leq\cdots\leq x_n\leq y} dx_1\cdots dx_n M_A(x,x_1)B(x_1) \\
&\quad M_A(x_1,x_2)B(x_2)\cdots B(x_n)M_A(x_n,y).
\end{aligned} \tag{13.27}
$$

Using this formula, we can expand $R(x,y)$ in terms of boson operators:

$$R(x,y) = \sum_{n=0}^\infty \int_{x\leq x_1\leq\cdots\leq x_n\leq y} dx_1\cdots dx_n R(x_1\cdots x_n)\psi^\dagger(x_1)\cdots\psi^\dagger(x_n), \tag{13.28}$$

where $R(x_1,\ldots,x_n)$ is a continuous MPS that determines the wave function of n bosons:

$$R(x_1,\ldots,x_n) = M_A(x,x_1)B(x_1)M_A(x_1,x_2)\cdots B(x_n)M_A(x_n,y). \tag{13.29}$$

$R(x, y)$ is, therefore, a superposition over all possible configurations at different particle number sectors in the Fock space.

Using the direct product identities of matrices

$$(A_1 B_1) \otimes (A_2 B_2) = (A_1 \otimes A_2)(B_1 \otimes B_2), \tag{13.30}$$

$$\exp(A) \otimes \exp(B) = \exp(A \otimes I + I \otimes B), \tag{13.31}$$

it is straightforward to show that

$$\begin{aligned} &R^*(x_1, \ldots, x_n) \otimes R(x_1, \ldots, x_n) \\ &= \overline{M}_A(x, x_1) \overline{B}(x_1) \overline{M}_A(x_1, x_2) \cdots \overline{B}(x_n) \overline{M}_A(x_n, y), \end{aligned} \tag{13.32}$$

where $\overline{O}$ is the direct product of O^* and O:

$$\overline{O}(x) = O^*(x) \otimes O(x). \tag{13.33}$$

$\overline{M}_A$ and $\overline{B}$ are given by

$$\overline{M}_A(x, y) = M_A^*(x, y) \otimes M_A(x, y) = M_{A^* \otimes I + I \otimes A}(x, y), \tag{13.34}$$

$$\overline{B}(x) = B^*(x) \otimes B(x). \tag{13.35}$$

It is clear that the basis states $\psi^\dagger(x_1) \cdots \psi^\dagger(x_n)|0\rangle$ are automatically orthogonal for different n, and

$$\langle 0 | \psi(y_n) \cdots \psi(y_1) \psi^\dagger(x_1) \cdots \psi^\dagger(x_n) |0\rangle = \delta(y_1 - x_1) \cdots \delta(y_n - x_n), \tag{13.36}$$

for any two sets of ordered arguments $x_1 \leq x_2 \leq \cdots \leq x_n$ and $y_1 \leq y_2 \leq \cdots \leq y_n$. The expectation value of $R^\dagger(x, y) \otimes R(x, y)$ in the vacuum can therefore be simplified as

$$\begin{aligned} &\langle 0 | R^\dagger(x, y) \otimes R(x, y) |0\rangle \\ &= \sum_{n=0}^\infty \int_{x \leq x_1 \leq \cdots \leq x_n \leq y} dx_1 \cdots dx_n R^*(x_1 \cdots x_n) \otimes R(x_1 \cdots x_n) \\ &= \sum_{n=0}^\infty \int_{x \leq x_1 \leq \cdots \leq x_n \leq y} dx_1 \cdots dx_n \overline{M}_A(x, x_1) \overline{B}(x_1) \overline{M}_A(x_1, x_2) \\ &\qquad\qquad\qquad \cdots \overline{B}(x_n) \overline{M}_A(x_n, y) \\ &= M_T(x, y), \end{aligned} \tag{13.37}$$

where $T(x)$ is the transfer matrix in the continuous limit

$$T(x) = A^*(x) \otimes I + I \otimes A(x) + B^*(x) \otimes B(x). \tag{13.38}$$

Graphically, it can be represented as

$$
\boxed{T(x)} \;=\; \begin{matrix} -\!\!\boxed{A^*}\!\!- \\ \rule{2.5em}{0.4pt} \end{matrix} \;+\; \begin{matrix} \rule{2.5em}{0.4pt} \\ -\!\!\boxed{A}\!\!- \end{matrix} \;+\; \begin{matrix} -\!\!\boxed{B^*}\!\!- \\ -\!\!\boxed{B}\!\!- \end{matrix}\;. \tag{13.39}
$$

It corresponds to the transfer matrix $T[x_i] = \sum_n A_i[n] \otimes A_i^*[n]$ in the evaluation of expectation values for an MPS on the lattice. Equation (13.37) is obtained by reverting the expansion of the path-ordered exponential given in Eq. (13.27).

Equation (13.37) is a useful equation for evaluating expectation values. For example, it can be used, in combined with the matrix identity

$$
(\mathrm{Tr}A) \cdot (\mathrm{Tr}B) = \mathrm{Tr}\,(A \otimes B)\,, \tag{13.40}
$$

to calculate the norm of the wave function:

$$
\begin{aligned}
\langle \Psi | \Psi \rangle &= \langle 0 | \left[\mathrm{Tr}R^\dagger(0, L) \right] \cdot \left[\mathrm{Tr}R(0, L) \right] | 0 \rangle \\
&= \mathrm{Tr}\langle 0 | R^\dagger(0, L) \otimes R(0, L) | 0 \rangle \\
&= \mathrm{Tr}M_T(0, L).
\end{aligned} \tag{13.41}
$$

Furthermore, $R(x, y)$ is factorizable:

$$
R(x, y) = R(x, z)R(z, y), \qquad x < z < y. \tag{13.42}
$$

This is another useful property for evaluating expectation values of continuous MPS. For example, the action of the annihilation operator $\psi(x)$ on the state $|\Psi\rangle$ can be derived by dividing $R(0, L)$ into a product of three terms according to their positions relative to $\psi(x)$:

$$
\begin{aligned}
\psi(x)|\Psi\rangle &= \mathrm{Tr}R(0, x^-)\mathcal{P}\left[\psi(x)R(x^-, x^+)\right]R(x^+, L)|0\rangle \\
&= \mathrm{Tr}R(0, x^-)B(x)R(x^+, L)|0\rangle \\
&= \mathrm{Tr}R(0, x)B(x)R(x, L)|0\rangle,
\end{aligned} \tag{13.43}
$$

where $x^\pm = x \pm \delta$ and $\delta \to 0$. The expectation value of the local density operator is

$$
\begin{aligned}
&\langle \Psi | \psi^\dagger(x)\psi(x) | \Psi \rangle \\
&= \langle 0 | \left[\mathrm{Tr}R^\dagger(0, x)B^*(x)R^\dagger(x, L) \right] \left[\mathrm{Tr}R(0, x)B(x)R(x, L) \right] | 0 \rangle \\
&= \mathrm{Tr}\langle 0 | \left[R^\dagger(0, x) \otimes R(0, x) \right] \overline{B}(x) \left[R^\dagger(x, L) \otimes R(x, L) \right] | 0 \rangle \\
&= \mathrm{Tr}M_T(0, x)\overline{B}(x)M_T(x, L).
\end{aligned} \tag{13.44}
$$

Similarly, the expectation value of the contact interaction is

$$
\langle \Psi | \psi^\dagger(x)\psi^\dagger(x)\psi(x)\psi(x) | \Psi \rangle = \mathrm{Tr}R(0, x)\overline{B^2}(x)R(x, L). \tag{13.45}
$$

The action of $\partial_x \psi(x)$ on $|\Psi\rangle$ is more convenient to be determined from the defining formula of the derivative:

$$\partial_x \psi(x)|\Psi\rangle = \lim_{\delta \to 0} \frac{\psi(x+\delta) - \psi(x)}{\delta}|\Psi\rangle. \qquad (13.46)$$

Substituting Eq. (13.43) into this equation, we find that

$$\partial_x \psi(x)|\Psi\rangle$$
$$= \lim_{\delta \to 0} \mathrm{Tr} R(0,x) \frac{R(x,x+\delta)B(x+\delta) - B(x)R(x,x+\delta)}{\delta} R(x+\delta, L)|0\rangle$$
$$= \mathrm{Tr} R(0,x) \{\partial_x B(x) + [A(x), B(x)]\} R(x, L)|0\rangle, \qquad (13.47)$$

and

$$\langle\Psi|\left[\partial_x \psi^\dagger(x)\right]\left[\partial_x \psi(x)\right]|\Psi\rangle = \mathrm{Tr} M_T(0,x)\overline{\partial_x B + [A, B]}(x) M_T(x, L). \qquad (13.48)$$

From these results, we find that the expectation value of the Lieb–Liniger Hamiltonian can be expressed as

$$\langle\Psi|H|\Psi\rangle = \int dx \mathrm{Tr} M_T(0,x)h(x)M_T(x, L), \qquad (13.49)$$

where

$$h(x) = \frac{1}{2m}\overline{\partial_x B + [A, B]} - \mu \overline{B}(x) + g\overline{B^2}(x). \qquad (13.50)$$

The correlation functions can be similarly calculated. For example, the single-particle correlation function is

$$\langle\Psi|\psi^\dagger(y)\psi(x)|\Psi\rangle$$
$$= \langle 0|\left[\mathrm{Tr} R^\dagger(0,y)B^*(y)R^\dagger(y, L)\right]\left[\mathrm{Tr} R(0,x)B(x)R(x, L)\right]|0\rangle$$
$$= \mathrm{Tr} M_T(0,y)\left[B^*(y) \otimes I\right] M_T(y, x)\left[I \otimes B(x)\right] M_T(x, L) \qquad (13.51)$$

if $y \leq x$, and

$$\langle\Psi|\psi^\dagger(y)\psi(x)|\Psi\rangle$$
$$= \mathrm{Tr} M_T(0,x)\left[I \otimes B(x)\right] M_T(x, y)\left[B^*(y) \otimes I\right] M_T(y, L) \qquad (13.52)$$

if $y > x$.

13.4 Canonicalization

In a lattice system, an MPS is invariant under the gauge transformation of virtual bond variables, meaning that if $A_i[n]$ is a tensor representation of a quantum state, then

$$A_i^g[n] = g^{-1}(x_i)A_i[n]g(x_i + \varepsilon) \qquad (13.53)$$

is another tensor representation that represents the same state. Here $g(x_i)$ is the gauge field defined on the bond linking $x_i - \varepsilon$ and x_i. In the limit $\varepsilon \to 0$, the gauge transformation becomes

$$A_i^g[n] = g^{-1}(x_i)A_i[n]\left[g(x_i) + \varepsilon\frac{dg(x_i)}{dx_i} + O(\varepsilon^2)\right].$$ (13.54)

More specifically, for the $n = 0$ term,

$$A_i^g[0] = g^{-1}(x_i)\left[I + \varepsilon A(x_i)\right]\left[I + \varepsilon\frac{d}{dx_i}\right]g(x_i) + O(\varepsilon^2)$$

$$= I + \varepsilon g^{-1}(x_i)\left[\frac{d}{dx_i} + A(x_i)g(x_i)\right]g(x_i) + O(\varepsilon^2).$$ (13.55)

Similarly, for the $n = 1$ term,

$$A_i^g[1] = \sqrt{\varepsilon}g^{-1}(x_i)B(x_i)g(x_i) + O(\varepsilon^{3/2}).$$ (13.56)

Thus in the continuum limit, the MPS is invariant under the following gauge transformation:

$$A^g(x) = g^{-1}(x)A(x)g(x) + g^{-1}(x)\frac{d}{dx}g(x),$$ (13.57)

$$B^g(x) = g^{-1}(x)B(x)g(x).$$ (13.58)

Here $g(x)$ should be second-order differentiable if A and B matrices are required first-order differentiable.

Like an MPS, we can use the gauge fixing condition to impose certain constraints on matrices $A(x)$ and $B(x)$. Again, $A(x)$ and $B(x)$ can be either left or right canonicalized.

13.4.1 Left Canonicalization

The left canonicalization of MPS boils down to imposing the constraint

$$\sum_n A_i^\dagger[n]A_i[n] = I.$$ (13.59)

In the small lattice spacing limit, this equation becomes

$$I + \varepsilon\left[A(x_i) + A^\dagger(x_i) + B^\dagger(x_i)B(x_i)\right] + O(\varepsilon^2) = I.$$ (13.60)

Thus in the continuum limit, the left canonical condition reduces to

$$A(x) + A^\dagger(x) + B^\dagger(x)B(x) = 0.$$ (13.61)

Graphically, it can be represented as

$$+ \quad \quad + \quad \quad = 0, \tag{13.62}$$

or simply

$$= 0, \tag{13.63}$$

which implies that the identity matrix is the null left eigenvector of the transfer matrix $T(x)$. A simple solution to this equation is

$$A(x) = iC(x) - \frac{1}{2}B^\dagger(x)B(x), \tag{13.64}$$

where $C(x)$ is a Hermitian matrix.

Furthermore, if the reduced density matrix at each bond is diagonalized, then A^i should also satisfy the second left canonical equation:

$$\sum_n A_i[n]\Lambda_i A_i^\dagger[n] = \Lambda_{i-1}, \tag{13.65}$$

where Λ_i is the diagonal eigenmatrix of the reduced density matrix on the bond linking i and $i+1$. This equation can also be expanded in the order of ε. Keeping the leading order terms in ε, we find that

$$A(x_i)\Lambda(x_i) + \Lambda(x_i)A^\dagger(x_i) + B(x_i)\Lambda(x_i)B^\dagger(x_i) = -\frac{d\Lambda(x_i)}{dx_i}. \tag{13.66}$$

Thus in the continuum limit, the second left canonical equation becomes

$$A(x)\Lambda(x) + \Lambda(x)A^\dagger(x) + B(x)\Lambda(x)B^\dagger(x) = -\frac{d\Lambda(x)}{dx}, \tag{13.67}$$

or graphically

$$= - \quad \frac{d\Lambda}{dx} \quad . \tag{13.68}$$

13.4.2 Right Canonicalization

Similarly, one can also impose the right canonicalization condition. In this case, the first canonical equation is

$$\sum_n A_i[n]A_i^\dagger[n] = I. \tag{13.69}$$

In the continuum limit, it reduces to

$$A(x) + A^\dagger(x) + B(x)B^\dagger(x) = 0,   \tag{13.70}$$

which implies that the identity matrix is the null right eigenvector of the transfer matrix

$$\boxed{T(x)} \quad = 0.   \tag{13.71}$$

A solution to this equation is

$$A(x) = iC(x) - \frac{1}{2}B(x)B^\dagger(x).   \tag{13.72}$$

Again, $C(x)$ is a Hermitian matrix.

In the canonical representation where the reduced density matrix is fully diagonalized, A_i should also satisfy the second right canonical equation:

$$\sum_n A_i^\dagger[n]\Lambda_{i-1}A_i[n] = \Lambda_i.   \tag{13.73}$$

In the continuum limit, it becomes

$$A^\dagger(x)\Lambda(x) + \Lambda(x)A(x) + B^\dagger(x)\Lambda(x)B(x) = \frac{d\Lambda(x)}{dx},   \tag{13.74}$$

or simply

$$\boxed{\Lambda \; T(x)} \quad = \quad \frac{d\Lambda}{dx} .   \tag{13.75}$$

13.5 Determination of Continuous MPS

There are several ways to determine the local tensors of a continuous MPS [220, 388, 389]. Among them, the energy minimization approach is relatively simple to implement. The goal of this approach is to optimize iteratively the local tensors $A(x)$ and $B(x)$ to lower the energy density:

$$e = \frac{1}{L}\frac{\langle \Psi|H|\Psi\rangle}{\langle \Psi|\Psi\rangle},   \tag{13.76}$$

according to its gradient. At each iteration, the most crucial step is to calculate the gradient of e. Below we take a translation-invariant system, in which both $A(x) = A$ and $B(x) = B$ do not depend on coordinate x, as an example to show how to determine this gradient.

In a translation-invariant system, the overlap of the wave function is still determined by Eq. (13.41), but the transfer matrix becomes position independent:

$$T = A^* \otimes I + I \otimes A + B^* \otimes B. \tag{13.77}$$

Based on this property, we find that the gradient of the wave function overlaps to be

$$
\begin{aligned}
\frac{\delta \langle \Psi | \Psi \rangle}{\delta(A, B)} &= \int_0^L dx \mathrm{Tr} M_T(0, x) \frac{\delta T}{\delta(A, B)} M_T(x, L) \\
&= \int_0^L dx \mathrm{Tr} \left[\frac{\delta T}{\delta(A, B)} M_T(0, L) \right] \\
&= L \mathrm{Tr} \left[\frac{\delta T}{\delta(A, B)} M_T(0, L) \right].
\end{aligned}
\tag{13.78}
$$

The periodic boundary condition is used in obtaining this expression. If the wave function is normalized, then the maximal eigenvalue of T equals zero. In this case, matrix M_T can be expanded using the eigenvectors of the transfer matrix as

$$M_T(x, y) = |l\rangle \langle r| + M_{\tilde{T}}(x, y), \tag{13.79}$$

where l and r are the left and right leading eigenvectors of T, respectively. $\tilde{T}$ is the transfer matrix excluding the contribution from the maximal eigenvectors:

$$\tilde{T} = T(I - |l\rangle \langle r|). \tag{13.80}$$

The eigenvalues of $\tilde{T}$ are less than 0.

For an infinite system, $L \to \infty$, $M_T(0, L)$ is determined purely by the maximal eigenvectors of T, and the derivative of $\langle \Psi | \Psi \rangle$ diverges linearly with the system size:

$$\frac{\delta \langle \Psi | \Psi \rangle}{\delta(A, B)} = L \left\langle l \left| \frac{\delta T}{\delta(A, B)} \right| r \right\rangle. \tag{13.81}$$

In this system, the expectation value of the Hamiltonian density is also translation-invariant:

$$\langle \Psi | H(x) | \Psi \rangle = \mathrm{Tr} \left[M_T(0, x) h(x) M_T(x, L) \right] = \mathrm{Tr} \left[h(0) M_T(0, L) \right]. \tag{13.82}$$

Its derivative is given by

$$
\begin{aligned}
\frac{\delta \langle \Psi | H(x) | \Psi \rangle}{\delta(A, B)} &= \int_0^L dx \mathrm{Tr} \left[h(0) M_T(0, x) \frac{\delta T}{\delta(A, B)} M_T(x, L) \right] \\
&+ \mathrm{Tr} \left[\frac{\delta h(0)}{\delta(A, B)} M_T(0, L) \right].
\end{aligned}
\tag{13.83}
$$

In the thermodynamic limit, $L \to \infty$, it can be simplified as

$$\frac{\delta \langle \Psi | H(x) | \Psi \rangle}{\delta(A, B)} = \left\langle l \left| \frac{\delta h(0)}{\delta(A, B)} - h(0) \frac{1}{\tilde{T}} \frac{\delta T}{\delta(A, B)} \right| r \right\rangle$$

$$+ eL \left\langle l \left| \frac{\delta T}{\delta(A, B)} \right| r \right\rangle, \tag{13.84}$$

where e is just the energy density:

$$e = \frac{\langle \Psi | H(x) | \Psi \rangle}{\langle \Psi | \Psi \rangle} = \langle l | h(0) | r \rangle. \tag{13.85}$$

The last term in Eq. (13.84) also diverges linearly with L. However, the derivative of the energy density e remains finite in the thermodynamic limit:

$$\frac{\delta e}{\delta(A, B)} = \frac{\delta \langle \Psi | H(x) | \Psi \rangle}{\delta(A, B)} - e \frac{\delta \langle \Psi | \Psi \rangle}{\delta(A, B)}$$

$$= \left\langle l \left| \frac{\delta h(0)}{\delta(A, B)} - h(0) \frac{1}{\tilde{T}} \frac{\delta T}{\delta(A, B)} \right| r \right\rangle. \tag{13.86}$$

Once the gradient of the energy density is determined, the local tensors can be updated by employing the steepest descent or the conjugate gradient method. In the method of steepest descent, for example, the local tensors are updated by minimizing the energy along the line extending from (A, B) in the direction of the negative gradient (i.e. the local downhill gradient).

14

Classical Transfer Matrix Renormalization

14.1 Classical Transfer Matrix

As discussed in Chapter 3, all classical statistical models with short-range interactions can be represented as tensor network models. These tensor network models can also be represented as a product of transfer matrices. Here we take the ferromagnetic Ising model on an $N_x \times N_y$ square lattice as an example to show how the transfer matrix is constructed.

Let us first construct the tensor network representation of the Ising model. For this model, the Boltzmann weight at each link can be decoupled as

$$e^{-\beta H_{ij}} = e^{\beta \sigma_i \sigma_j} = \quad \sigma_i \!-\!\boxed{u}\!-\!\boxed{u}\!-\!\sigma_j \tag{14.1}$$

where u is a 2×2 real symmetric matrix

$$u = \frac{\sqrt{\cosh \beta}}{\sqrt{2}} \begin{pmatrix} 1 & 1 \\ 1 & 1 \end{pmatrix} + \frac{\sqrt{\sinh \beta}}{\sqrt{2}} \begin{pmatrix} 1 & -1 \\ -1 & 1 \end{pmatrix}. \tag{14.2}$$

Following the derivation given in §3.3, we know that the partition function can be written as a tensor network:

$$Z = \qquad , \tag{14.3}$$

where the local tensor is defined by

$$
i \;\rule[0.5ex]{0.8em}{0.4pt}\;\boxed{\tau}\;\rule[0.5ex]{0.8em}{0.4pt}\; k \;=\; \sum_{\sigma} u_{\sigma i} u_{\sigma j} u_{\sigma k} u_{\sigma l}. \tag{14.4}
$$

In obtaining this expression, the property that u is real and symmetric is used.

In a system with open boundary conditions, the local tensor at the edge, for example, at the first row, is defined by

$$
\tau_{ijkl} = \sum_{\sigma} u_{\sigma j} u_{\sigma k} u_{\sigma l} \delta_{i,1}. \tag{14.5}
$$

Other edge tensors are similarly defined.

The transfer matrix T is defined by grouping all N_x local tensors at one row:

$$
T \;=\; \boxed{\tau}\;\rule[0.5ex]{0.8em}{0.4pt}\;\boxed{\tau}\;\rule[0.5ex]{0.8em}{0.4pt}\;\boxed{\tau}\;\rule[0.5ex]{0.8em}{0.4pt}\;\cdots\;\rule[0.5ex]{0.8em}{0.4pt}\;\boxed{\tau} \atop \quad 1 \quad\; 2 \quad\; 3 \quad \cdots \quad N_x \;. \tag{14.6}
$$

The partition function is determined by the product of all the transfer matrices along the y-direction:

$$
Z = \mathrm{Tr}\,T^{N_y}. \tag{14.7}
$$

In the limit $N_y \to \infty$, the partition function is governed by the largest eigenvalue of the transfer matrix. If we denote λ_1 as the largest eigenvalue of T, the partition function, and the free energy density f are then given by

$$
Z = \lim_{L_y \to \infty} \mathrm{Tr}\,T^{N_y} = \lambda_1^{N_y}, \tag{14.8}
$$

$$
f = -\frac{1}{\beta N_x N_y} \ln Z = -\frac{1}{\beta N_x} \ln \lambda_1. \tag{14.9}
$$

Thus by diagonalizing the transfer matrix, we can evaluate all thermodynamic quantities.

Transfer matrix renormalization group (TMRG) is a natural extension of DMRG to the classical two-dimensional systems. It is based on the RG transformation for the transfer matrix rather than the Hamiltonian. Like DMRG, TMRG is a variational method. It is to optimize thermodynamic calculations using a limited number of basis states. This method is superior to the Monte Carlo method for two-dimensional classical systems in accuracy, speed, and the possibility of treating much larger systems.

14.2 TMRG

The largest eigenvalue and the corresponding eigenvector of the transfer matrix T can be evaluated using DMRG introduced in Chapter 7, provided that T is Hermitian. Both the infinite and finite lattice algorithms of DMRG can be used. In the discussion here, free boundary conditions are assumed.

TMRG also divides a superblock into four parts, the left and right blocks and two added sites between them, and writes the transfer matrix as a product of four terms:

$$T = \boxed{T_L} - \boxed{\tau} - \boxed{\tau} - \boxed{T_R} , \tag{14.10}$$

where T_L and T_R are the transfer matrices of the left and right blocks, respectively. Initially, the superblock contains four sites, and T_L and T_R each contain one site. By diagonalizing the transfer matrix, one can obtain the largest eigenvalue λ_1 and the corresponding eigenvector $|\Psi\rangle$. The normalized thermal density matrix ρ is then determined by

$$\rho = \lim_{L_y \to \infty} \frac{e^{-\beta H}}{Z} = |\Psi\rangle\langle\Psi|. \tag{14.11}$$

H is the Hamiltonian of the whole system. The corresponding reduced density matrix for the system block, which contains the left block and the left added site, is then evaluated by integrating out all degrees of freedom in the environment block:

$$\rho^{\text{sys}} = \text{Tr}_{\text{env}}|\Psi\rangle\langle\Psi|. \tag{14.12}$$

ρ^{sys} is a positive semi-definite matrix. The eigenvalues of ρ^{sys} measure the probabilities of the corresponding Schmidt vectors in $|\Psi\rangle$. By diagonalizing ρ^{sys} and keeping the largest D eigenvectors, we obtain the isometric basis transformation matrix U, which is used to update the left transfer matrix:

$$\boxed{T'_L} - \quad = \quad \begin{array}{c} \boxed{U} \\ \boxed{T_L} - \boxed{\tau} - \\ \boxed{U^\dagger} \end{array} . \tag{14.13}$$

The augmented right transfer matrix and other physical quantities are similarly updated.

Repeating this TMRG step, one can evaluate the free energy at each given N_x using Eq. (14.9). Other thermodynamic quantities can be obtained from the derivatives of the free energy.

The magnetization, internal energy, and other quantities, which are the sum of local spin variables, can also be evaluated from the expectation value of the corresponding quantity in the leading eigenvector $|\Psi\rangle$. This avoids the calculation of numerical derivatives and allows these quantities to be more accurately calculated.

The magnetization density is determined by the equation

$$M = \frac{1}{N_x N_y} \sum_{x,y} \langle \sigma_{x,y} \rangle = \frac{1}{N_x} \sum_{i}^{N_x} \frac{\langle \Psi | T_i[\tau_\sigma]|\Psi\rangle}{\lambda_1 \langle \Psi|\Psi\rangle}, \tag{14.14}$$

where $T_i[\tau_\sigma]$ is a transfer matrix in which the ith local tensor is replaced by

$$(\tau_\sigma)_{ijkl} = \sum_{\sigma} \sigma u_{\sigma i} u_{\sigma j} u_{\sigma k} u_{\sigma l} \tag{14.15}$$

and

$$T_i[\tau_\sigma] = \boxed{\tau}\!\!-\!\cdots\!-\!\boxed{\tau}\!\!-\!\boxed{\tau_\sigma}\!\!-\!\boxed{\tau}\!\!-\!\cdots\!-\!\boxed{\tau} \atop \quad 1 \qquad\quad i-1\ \ i\ \ i+1 \qquad L_x \tag{14.16}$$

In obtaining Eq. (14.14), the translation invariance along the y-direction is assumed. The internal energy and other local quantities can be similarly calculated.

Along the y-direction, the correlation length ξ is determined by the two leading eigenvalues of the transfer matrix:

$$\xi = -\ln^{-1} \frac{|\lambda_2|}{\lambda_1}, \tag{14.17}$$

where λ_2 is the second-largest eigenvalue of T. For the anisotropic Ising model, the correlation function needs to be evaluated in order to determine the correlation length of spins along the x-axis.

The eigenvector $|\Psi\rangle$ obtained by TMRG is a variational wave function for the transfer matrix. The corresponding eigenvalue λ_1 is a variational lower bound for the true largest eigenvalue of the transfer matrix, and the difference is determined by the truncation error, which decreases with the increase D (i.e. the number of states retained). The accuracy can be improved by taking a finite lattice sweeping, similar to DMRG.

In this discussion, the transfer matrix is Hermitian. If the transfer matrix is non-Hermitian, the left eigenvector differs from the right one. This is the case encountered in the calculation of quantum transfer matrices. In this case, the definition of the density matrix needs to be modified. A detailed discussion is given in Chapter 16.

14.3 Fixed-Point MPS: One-Site Approach

One can also diagonalize the transfer matrix using the MPS wave function, represented by Eq. (11.2). The local tensors are determined by the eigenequation

$$T|\Psi\rangle = \lambda_1|\Psi\rangle,\qquad(14.18)$$

or graphically,

on an infinite lattice. This equation can be solved by iteratively applying the transfer matrix T to the wave function. At each iteration, the bond dimension of the wave function is increased, and an RG decimation is used to reduce the bond dimension back to the original one. However, fixing MPS does not mean all the local tensors are fixed because there are gauge ambiguities on all the virtual bonds. This means that the local tensors on the left and right sides of the above equation may not be the same. They can differ by a gauge transformation.

The local tensor can also be determined by maximizing the function

$$f(\Psi) = \frac{\langle\Psi|T|\Psi\rangle}{\langle\Psi|\Psi\rangle}.\qquad(14.20)$$

One way to fix the gauge of A is to left-right canonicalize the local tensors so that

where C is a bond matrix. Normalization of the wave function corresponds to the condition $\mathrm{Tr}C^\dagger C = 1$.

Now $\langle\Psi|T|\Psi\rangle$ can be written as

$$\langle\Psi|T|\Psi\rangle = \mathrm{Tr}\left(T_{A_L}^{N_L}T_C T_{A_R}^{N_R}\right)\qquad(14.22)$$

where N_L (or N_R) is the total number of sites on the left (or right) hand side of the bond on which the C-matrix is located. In the thermodynamic limit, both N_L and N_R equal infinity. If we absorb C into A_L or equivalently A_R, and define a new tensor

$$-\!\!\overset{|}{(A_C)}\!\!- \;=\; -\!\!\overset{|}{(A_L)}\!\!-\!\!(C)\!\!- \;=\; -\!(C)\!\!-\!\!\overset{|}{(A_R)}\!- \;, \tag{14.23}$$

the above expression can also be expressed as

$$\langle \Psi \, | T | \, \Psi \rangle = \mathrm{Tr}\left(T_L^{N_L-1} T_A T_R^{N_R} \right) \tag{14.24}$$

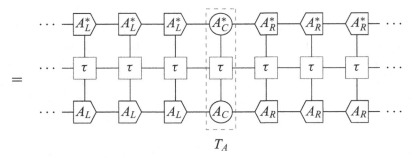

In the above equations, T_L, T_R, and T_A are transfer matrices, defined by

$$(T_S)_{\alpha\alpha\beta,\alpha'\alpha'\beta'} = \sum_{\sigma\sigma'} S_{\alpha\alpha'}[\sigma] \, \tau_{aa'}[\sigma\sigma'] S_{\beta\beta'}^*[\sigma'], \tag{14.25}$$

where $S = A_L$, A_R, or A_C. T_C is defined by

$$(T_C)_{\alpha\alpha\beta,\alpha'\alpha'\beta'} = C_{\alpha\alpha'} C_{\beta\beta'}^* \delta_{aa'}. \tag{14.26}$$

14.3.1 Fixed-Point Equations of Local Tensors

We now derive the fixed-point equations of local tensors, first obtained by Fishman et al. [390].

By taking the derivative of $f(\Psi)$ with respect to C^*, we can find the fixed-point equation for C. In the thermodynamic limit, $N_L = N_R \to \infty$, it is given by

$$\sum_{\beta\beta'a} E_{L,\alpha a\beta} C_{\alpha\alpha'} E_{R,\alpha'a\beta'} = \eta C_{\beta\beta'}, \tag{14.27}$$

where η is a normalization constant. Graphically, this equation can also be represented as

$$\overset{\displaystyle\frown}{(E_L)\!\!-\!\!-\!\!(E_R)} \atop {(C)} \;=\; \eta -\!(C)\!- \;. \tag{14.28}$$

E_R is obtained by applying T_R successively onto an arbitrary initial vector E_0, i.e.

$$E_R = \lim_{N_R \to \infty} T_R^{N_R} E_0,$$ (14.29)

which is nothing but the maximal right eigenvector of T_R, i.e.

$$= \lambda_R \quad \text{(14.30)}$$

Similarly, E_L is the maximal left eigenvector of T_L. It is obtained by applying successively all T_L onto an arbitrary initial left vector.

The fixed-point equation of A_C is obtained by taking the derivative of $f(\Psi)$ with respect to A_C^*. In the thermodynamic limit, it is given by

$$= \lambda \qquad \text{(14.31)}$$

where λ is the largest eigenvalue of the transfer matrix. Compared with DMRG, it is not difficult to understand that A_C is nothing but the maximal eigenvector of the full transfer matrix obtained from the one-site DMRG scheme. This kind of one-site renormalization scheme was first used in the momentum-space DMRG [56] for saving computational resources. In standard DMRG language, the superblock consists of a left block, a right block, and an added spin, whose basis states are represented by α, β, and σ, respectively. The relationship between A_C and the eigenfunction is

$$\Psi_{\alpha\sigma\beta} = (A_C[\sigma])_{\alpha,\beta}.$$ (14.32)

This one-site fixed-point tensor scheme is simple. However, it underestimates the renormalization effect of the environment, since the dimension of the environment equals D, which may cause the loss of accuracy in comparison with the two-site DMRG scheme.

We also need to use an iterative approach to find the fixed-point tensors. However, in comparison with the conventional optimization approaches introduced in Chapter 12, it has two advantages. First, the maximal eigenvector converges very quickly in the fixed-point tensor approach. It needs fewer iteration steps to converge to the final result than conventional optimization approaches. This saves the computational cost greatly, even though its cost at each iteration is even higher.

Second, in the conventional optimization approaches, the virtual bond dimension of the eigenvector is augmented at each step of the projection. To maintain the iteration, one has to truncate the bond dimension back to the original one. However, with the fixed-point tensor, there is no need to do this truncation explicitly.

14.3.2 Steps for Solving the Fixed-Point Equations

The fixed-point tensors can be found iteratively from the following steps:

(i) Find the left and right canonicalized tensors, A_L and A_R, starting from an arbitrary MPS.
(ii) Find the maximal right eigenvector E_R of T_R, and similarly, the maximal left eigenvector E_L of T_L.
(iii) Solve the fixed-point equation (14.31) to find A_C.
(iv) Determine C using equation

$$
\begin{array}{ccc}
\boxed{\begin{matrix} A^*_C \\ A_C \end{matrix}} & = & \boxed{\begin{matrix} C^* \\ C \end{matrix}} & = & C^\dagger - C
\end{array} \tag{14.33}
$$

This is done by diagonalizing the left part of the above equation:

$$
\boxed{\begin{matrix} A^*_C \\ A_C \end{matrix}} = V - \lambda^2 - V^\dagger \tag{14.34}
$$

where V is a unitary matrix and λ^2 are the eigenvalues. C is then found to be

$$
C = \lambda - V^\dagger . \tag{14.35}
$$

C can also be found by solving the fixed-point equation (14.27). However, it is more efficient to determine C by solving (14.33).

(v) Update A_L so that A_L is automatically left canonical. In principle, one can determine A_L using the equation

$$
A_L[\sigma] = A_C[\sigma]C^{-1} = A_C[\sigma]V\lambda^{-1}V^\dagger. \tag{14.36}
$$

However, this solution involves the inverse of λ, which may become numerically unstable when some of the eigenvalues λ become small.

Instead, we determine $A_L[\sigma]$ by minimizing $\|A_C[\sigma] - A_L[\sigma]C\|^2$ or equivalently the function

$$
g(A_L) = \min \sum_\sigma \|A_C[\sigma]C^\dagger - A_L[\sigma]CC^\dagger\|^2, \tag{14.37}
$$

subject to the constraint that $A_L [\sigma]$ is left canonical. The solution to this minimization problem is known. A_L is the isometry in the polar decomposition of $A_C C^\dagger$, which can be obtained from the following SVD:

$$ \text{---}(A_C)\text{---}(C^\dagger)\text{---} = \text{---}(U)\text{---}\langle \lambda^2 \rangle\text{---}(V^\dagger)\text{---} , \qquad (14.38) $$

where U is an isometric matrix. We then have

$$ \text{---}[A_L]\text{---} = \text{---}(U)\text{---}(V^\dagger)\text{---} . \qquad (14.39) $$

It is straightforward to show that λ and V are just the eigenvalue and eigenvector matrices of the density matrix found in Eq. (14.34).

(vi) Similarly, we can find A_R and a further updated C.

(vii) Repeat Steps (ii–vi) until A_L, A_R and C (or A_C) are converged.

This approach can be extended to determine the local tensors by minimizing the expectation value of the Hamiltonian. As discussed in §4.1, the Hamiltonian H of a one-dimensional quantum lattice model with local interactions can also be written as an MPO. In this case, the local Hamiltonian operator is also site-independent except at the endpoints of the lattice. In the thermodynamic limit, the effect of boundary tensors can be ignored, and the MPO becomes translation invariant. Finding the ground-state wave function of H is equivalent to finding the maximal eigenstate of $-H$ after a proper shift of the ground-state energy. In this sense, we can also regard $-H$ as a transfer matrix.

14.4 Fixed-Point MPS: Two-Site Approach

The one-site fixed-point tensor approach previously introduced is simple to implement. However, as the bond dimensions of local tensors are completely fixed in the whole process of iterations, no truncation is involved in solving the fixed-point equations, like in the one-site DMRG calculation [56]. This undermines the renormalization effect induced by the environment. Furthermore, if a symmetry, for example, an SU(2) symmetry, is used to block diagonalize the local tensors, the size of each block is fixed and cannot be renormalized. It implies that we have to set up the correct dimensions for all the blocks from the beginning, which is clearly not feasible.

Nevertheless, solving the fixed-point equations for the local tensors on two sites can remedy the problems. This two-site scheme is precisely a generalization of the standard DMRG scheme to a translation-invariant MPS.

The two-site approach works similarly to the one-site approach. Again, the initial MPS is first left-right canonicalized. We then solve Eq. (14.30) to determine the right eigenvector E_R, and similarly the left eigenvector E_L. But the fixed-point tensor A_C is now replaced by a tensor defined on two lattice sites, Φ, whose fixed-point equation is given by

$$ \tag{14.40} $$

η is the maximal eigenvalue of the normalized transfer matrix, and Φ is the corresponding eigenfunction. It is just the wave function determined in the TMRG calculation. We now decompose it by SVD:

$$ \tag{14.41} $$

The diagonal matrix λ has a higher dimension than the bond dimension of MPS. By truncating the basis space to retain the largest D singular values of λ, we approximate the above wave function by the formula

$$ \tag{14.42} $$

Both U and V are truncated. They are used to update the left and right canonicalized tensors in Eq. (14.22), respectively:

$$ \tag{14.43} $$

C is now updated by λ, which is already diagonalized.

The fixed-point MPS is obtained by repeating the calculations until A_L, A_R, and C are fully converged.

14.5 Corner Transfer Matrix Renormalization

The TMRG works when the lattice length along one of the spatial directions is infinite. The reduced density matrix is defined by tracing out all bond indices of tensors on the right-half network of the partition function. If the lattice size is finite, especially if the evolution operator is symmetric along both the x- and y-axis directions, it is better to perform the contraction along both directions simultaneously. This can be achieved by utilizing corner transfer matrices.

The corner transfer matrix formalism was originally derived by Baxter [24, 391]. It is based on the observation that the partition function can be obtained by contracting four corner transfer matrices. For the isotropic classical statistical model, for example, the Ising model discussed in §14.1, the partition function on a square lattice can be expressed as

$$Z = \mathrm{Tr} C^4 = \qquad , \qquad (14.44)$$

where

$$(14.45)$$

is the corner transfer matrix. Each corner contains one-quarter of all the building squares. C is defined by contracting out all internal bonds in the lower left corner (equivalently in other corners). In an isotropic system, the corner transfer matrix is symmetric

$$(14.46)$$

If the system is not isotropic, for example, if the coupling constants are different along x- and y-directions, the corner transfer matrices are different at the four corners. In this case, the partition function is just a product of these four corner transfer matrices. For simplicity, we only consider the isotropic case in the discussion that follows.

Nishino and Okunishi adapted the corner transfer matrix formalism to perform DMRG calculations in 1996 [70, 303]. The corner transfer matrix renormalization group (CTMRG) they proposed is an iterative scheme to contract the internal bond variables of the four corner transfer matrices. At each step, the corner transfer matrix is updated by a corner transfer matrix obtained from the previous step, a half-row transfer matrix E, a half-column transfer matrix E, and a local tensor τ:

$$\text{(C')}\!-\;=\;\begin{array}{c}\text{(E)}\!-\!\boxed{\tau}\!-\\[4pt]\text{(C)}\!-\!\text{(E)}\!-\end{array}.\tag{14.47}$$

In the original lattice representation, it is

$$\tag{14.48}$$

The initial corner transfer matrix and the half-row/column transfer matrix each contains just one local tensor. Hence, the partition function contains four corner transfer matrices, four half-row and four half-column transfer matrices, and four local tensors:

$$\tag{14.49}$$

14.5.1 Reduced Density Matrix

The reduced density matrix is determined by cutting one of the edges that connect two corner transfer matrices:

$$\rho = \left(C'\right)^4 = \begin{array}{c}\text{(C')}\!-\!\text{(C')}\\[2pt]\text{(C')}\!-\!\text{(C')}\end{array}.\tag{14.50}$$

Suppose U is the unitary matrix that diagonalizes the corner transfer matrix

$$C' = U\lambda U^\dagger.\tag{14.51}$$

The reduced density matrix is then found to be

$$\rho = U\lambda^4 U^\dagger.\tag{14.52}$$

λ is the eigenvalue matrix of the corner transfer matrix. The eigenvalues of ρ are given by λ^4. The partition function equals the trace of these eigenvalues:

$$Z = \text{Tr}\lambda^4.\tag{14.53}$$

After truncating the basis space by retaining the D largest eigenvectors of ρ, U becomes isometric and is used to update the corner transfer matrix:

$$\overset{|}{\textcircled{C}}- \quad = \quad \textcircled{C}-\boxed{U}- \quad . \tag{14.54}$$

Similarly, the half-row (or half-column) tensor is updated by the formula

$$-\overset{|}{\textcircled{E'}}- \quad = \quad -\boxed{U^\dagger}\,\overset{\boxed{\tau}}{\underset{\textcircled{E}}{}}\,\boxed{U}- \quad . \tag{14.55}$$

E' is then set up as the half-row tensor for the next iteration.

The main computational cost of CTMRG occurs in the step of diagonalizing the corner transfer matrix. It scales as $O(d^3 D^3)$, the same as in DMRG.

14.5.2 Simplified Scheme of CTMRG

In this discussion of CTMRG, the lattice is divided into 4×4 parts, and the partition function, as revealed by Eq. (14.49), is a product of four corner transfer matrices, eight edge tensors, and four local tensors. There are two edge tensors along each side of the lattice. However, other configurations can be used to carry out the CTMRG calculation. For example, one could use a 3×3 configuration of tensors to perform the calculation [71]. In this case, the partition function is represented as

$$Z = \;\; \begin{array}{ccc} \textcircled{C}-\textcircled{E}-\textcircled{C} \\ \textcircled{E}-\boxed{\tau}-\textcircled{E} \\ \textcircled{C}-\textcircled{E}-\textcircled{C} \end{array} \;\; . \tag{14.56}$$

The corresponding configuration in the original lattice is

$$Z = \qquad\qquad\qquad\qquad \begin{array}{c} C \\ E \\ C \end{array} \qquad . \tag{14.57}$$

In (14.56), the bottom row or right column is simply a three-site MPS. By comparison with Eq. (14.57), we know that the middle row or middle column is nothing but the transfer matrix:

$$t = \quad \text{(E)}-\boxed{\tau}-\text{(E)} \quad . \tag{14.58}$$

Instead of diagonalizing the reduced density matrix, both the corner and edge tensors, C and E, are determined by applying the transfer matrix to the bottom MPS. This generates a new MPS with thicker horizontal bonds:

$$\tag{14.59}$$

In this case, C' is at most a rank-D matrix, and the transformation matrix U is determined from C' by taking an LQ decomposition:

$$\tag{14.60}$$

After the decomposition, we use L to update the corner transfer matrix, $C = L$ up to a normalization constant, and U to update the edge tensor:

$$\tag{14.61}$$

We will eventually obtain the converged corner and edge tensors by repeating these steps.

14.5.3 Fixed-Point Tensor Approach

Now let us consider the property of fixed-point tensors. If C and E are the converged tensors obtained by solving Eqs. (14.59)–(14.61), they should satisfy the following eigenequation:

$$\tag{14.62}$$

where the bottom MPS is the maximal eigenvector of the transfer matrix and η_1 is the corresponding eigenvalue. Furthermore, as matrix L in Eq. (14.60) is proportional to C at the fixed point, the product of two corner transfer matrices should be the maximal eigenvector of the transfer matrix formed by two edge tensors,

$$\text{(diagram)} = \eta_2 \; \text{(C)—(U—U†)—(C)} = \eta_2 \; \text{(C)—(C)} , \qquad (14.63)$$

with η_2 the corresponding eigenvalue.

Equations (14.62) and (14.63) constitute a closed set of fixed-point equations for determining the corner and edge tensors. One could determine C and E by solving these two equations without explicitly using the transformation matrix U. It provides a simple and efficient approach for carrying out the CTMRG calculation.

These two fixed-point equations could be solved iteratively with an arbitrarily initialized but symmetrized edge tensor E. At each iterative step, we first solve the eigenequation (14.63) to find the corner transfer matrix C, and then use it to update the edge tensor E by solving Eq. (14.62). Repeating this iteration many times, we should be able to find the fixed-point tensors.

14.5.4 Expectation Values

In Eq. (14.56), if we use the vertical transfer matrix to act on the right vertical MPS formed by the corner and edge tensors at the fixed point for infinite many times, the partition function then becomes

$$Z = \begin{array}{c} \cdots -E-E-E-E-E- \cdots \\ \cdots -\tau-\tau-\tau-\tau-\tau- \cdots \quad T \\ \cdots -E-E-E-E-E- \cdots \quad |\Psi\rangle \end{array} \qquad (14.64)$$

up to a normalization constant. The middle row is nothing but the transfer matrix T. This expression implies that the translation-invariant MPS formed by the edge tensor E at the fixed point is just the leading eigenvector $|\Psi\rangle$ of the transfer matrix. Therefore, the maximal eigenvalue of the transfer matrix is given by

$$\langle T \rangle = \frac{\langle \Psi | T | \Psi \rangle}{\langle \Psi | \Psi \rangle}. \qquad (14.65)$$

Using Eq. (14.62), it is simple to show that

$$\langle \Psi | T | \Psi \rangle = \eta_1^{N_x} \; \begin{array}{c} \text{(C)—(C)} \\ \text{(E)—(E)} \\ \text{(C)—(C)} \end{array} , \qquad (14.66)$$

where N_x is the number of lattice sites along the x-axis. Similarly, from Eq. (14.63), the normalization constant is found to be

$$\langle \Psi | \Psi \rangle = \eta_2^{N_x} \quad \begin{array}{c} \text{©—©} \\ \text{©—©} \end{array} . \tag{14.67}$$

Thus, in the thermodynamic limit, the free energy per site is

$$f = -\lim_{N_x \to \infty} \frac{1}{\beta N_x} \ln \frac{\langle \Psi | T | \Psi \rangle}{\langle \Psi | \Psi \rangle} = -\frac{1}{\beta} \ln \frac{\eta_1}{\eta_2}, \tag{14.68}$$

determined by the ratio between η_1 and η_2.

We can determine η_1 and η_2 by solving the fixed-point equations (14.62) and (14.63). On the other hand, if C and E are determined from other methods, we could also determine η_1 and η_2 directly using these two tensors. Using Eq. (14.62), it is simple to show that

$$\eta_1 = \begin{array}{c} \text{©—(E)—©} \\ \text{(E)—[τ]—(E)} \\ \text{©—(E)—©} \end{array} \cdot \left(\begin{array}{c} \text{©—©} \\ \text{(E)—(E)} \\ \text{©—©} \end{array} \right)^{-1} . \tag{14.69}$$

Similarly, η_2 is given by

$$\eta_2 = \begin{array}{c} \text{©—(E)—©} \\ \text{©—(E)—©} \end{array} \cdot \left(\begin{array}{c} \text{©—©} \\ \text{©—©} \end{array} \right)^{-1} . \tag{14.70}$$

Other physical quantities can be similarly calculated. For example, the uniform magnetization per site could be represented using the transfer matrix defined in Eq. (14.16) as

$$M = \frac{\langle \Psi | T_0[\tau_\sigma] | \Psi \rangle}{\langle \Psi | T | \Psi \rangle}. \tag{14.71}$$

It can be further expressed using the fixed-point tensors as

$$M = \begin{array}{c} \text{©—(E)—©} \\ \text{(E)—[τ_σ]—(E)} \\ \text{©—(E)—©} \end{array} \cdot \left(\begin{array}{c} \text{©—(E)—©} \\ \text{(E)—[τ]—(E)} \\ \text{©—(E)—©} \end{array} \right)^{-1} . \tag{14.72}$$

In this discussion, we have assumed that the system is isotropic and that the corner transfer matrix is symmetric (or Hermitian). However, this assumption can be released. The algorithm can be extended to an arbitrary two-dimensional lattice, with or without translation invariance. In the case where the corner transfer matrix is not symmetric, U is no longer the matrix that diagonalizes the reduced density matrix. Instead, one should use the method introduced in §15.2 to determine the transformation matrix.

CTMRG could, in principle, be generalized to three dimensions. In that case, the system should be divided into eight subcubes. Then, one can introduce a corner transfer tensor to describe the Boltzmann weight at each subcube. However, the computational cost of this method is generally very high [392], and it has not been used to solve a problem that is difficult to solve with other numerical methods.

One can also sweep the system on a finite-size lattice to improve accuracy. The sweeping can be done similarly as in the finite lattice DMRG scheme. The sweep now needs to be done in both horizontal and vertical directions.

15

Criterion of Truncation: Nonsymmetric Systems

15.1 Nonsymmetric Density Matrix

In the study of many-body physical problems, we often have to diagonalize a non-Hermitian (or nonsymmetric) matrix to find its largest left and right eigenvectors. We encountered this problem in the study of TMRG. In evaluating expectation values from a PEPS wave function, an approach often adopted is to convert the inner product of the bra and ket of the PEPS as a product of a transfer matrix. In this case, we reencounter the non-Hermitian problem since this transfer matrix is generally nonsymmetric. In many cases, it may not even be semi-positive due to the accumulation of truncation errors.

Below we use the quantum transfer matrix, introduced in §4.3, as an example to show how the reduced density matrix is defined. The method introduced in this chapter, nevertheless, holds generally. It is useful not just for studying the thermodynamic properties of one-dimensional quantum lattice models but also for determining the optimized basis truncation scheme for almost all non-symmetric tensor network models. For example, it could be used to contract a scalar product of two PEPS using TEBD or CTMRG.

For a nonsymmetric quantum transfer matrix defined in the Trotter space, the density matrix is defined by demanding its trace to equal the partition function given by Eq. (4.40). In the framework of quantum transfer matrix, a natural extension of the thermal density matrix is

$$\rho = T_M^{N/2}. \tag{15.1}$$

Similarly, by integrating out all degrees of freedom of the environment, we define the reduced density matrix as

$$\rho^{\text{sys}} = \text{Tr}_{\text{env}} T_M^{N/2}. \tag{15.2}$$

In the thermodynamic limit $N \to \infty$, it becomes

$$\rho^{\text{sys}} = \text{Tr}_{\text{env}} |\Psi_R\rangle \langle \Psi_L|, \tag{15.3}$$

up to a normalization factor. $|\Psi_R\rangle$ and $\langle \Psi_L|$ are the largest right and left eigenvectors of $\mathcal{T}_M$. Like $\mathcal{T}_M$, ρ^{sys} is generally not a symmetric matrix. Physically, ρ^{sys} should be semi-positive definite. However, general proof for the semi-positivity of this matrix is not available.

Supposing $\Phi_{s,e}$ and $\Psi_{s,e}$ to be the wave functions of $|\Psi_R\rangle$ and $\langle \Psi_L|$ in the basis spaces of the system and environment blocks:

$$|\Psi_R\rangle = \sum_{s,e} \Psi_{s,e} |s, e\rangle, \tag{15.4}$$

$$\langle \Psi_L| = \sum_{s,e} \langle s, e| \Phi^*_{s,e}. \tag{15.5}$$

The right wave function is generally not equal to the left one, $\Phi \neq \Psi$ since $\mathcal{T}_M$ is nonsymmetric. $|s\rangle$ and $\langle s|$ are the biorthonormal basis states of the system

$$\langle s'|s\rangle = \delta_{s',s}. \tag{15.6}$$

Similarly, $|e\rangle$ and $\langle e|$ are the biorthonormal basis states of the environment

$$\langle e'|e\rangle = \delta_{e',e}. \tag{15.7}$$

Using these wave functions, we can represent ρ^{sys} as

$$\rho^{\text{sys}} = \Psi \Phi^{\dagger} = \qquad . \tag{15.8}$$

Similarly, the reduced density matrix of the environment is

$$\rho^{\text{env}} = \Psi^T \Phi^* = \qquad . \tag{15.9}$$

If Φ or Ψ, represented as a matrix, is invertible, we have

$$\left(\rho^{\text{env}}\right)^T = \Phi^{\dagger}\Psi = (\Psi)^{-1} \rho^{\text{sys}} \Psi = \Phi^{\dagger} \rho^{\text{sys}} \left(\Phi^{\dagger}\right)^{-1}. \tag{15.10}$$

Hence, ρ^{sys} and ρ^{env} have the same eigenspectra.

We can also define the bond density matrix by cutting the bond connecting the system and environment blocks of the right wave function $|\Psi_R\rangle$. For doing this, we first decompose Ψ into a product of two matrices, S and E, by taking SVD:

$$\Psi_{s,e} = \quad s - \!\!\text{S}\!-\!\text{E}\!- e. \tag{15.11}$$

The bond density matrix is then defined by

$$\rho^B = E\Phi^\dagger S = \boxed{} . \tag{15.12}$$

From this definition, it can be also shown that ρ^B has the same eigenspectra as ρ^{sys}:

$$\rho^B = S^{-1} S E \Phi^\dagger S = S^{-1} \rho^{\text{sys}} S. \tag{15.13}$$

Similarly, it can be shown that the bond density matrix defined on the bond of $\langle \Phi_L |$ also has the same eigenspectra as ρ^{sys}.

For a single impurity system whose partition function is given by Eq. (4.85) in the Trotter space, the corresponding thermal density matrix is defined by

$$\rho = \mathcal{T}_M^{N/2-1} \mathcal{T}^{\text{imp}}. \tag{15.14}$$

The reduced density matrix is simply defined by integrating out all basis states in the environment:

$$\rho^{\text{sys}} = \text{Tr}_{\text{env}} \left(\mathcal{T}_M^{N/2-1} \mathcal{T}^{\text{imp}} \right). \tag{15.15}$$

This is a generalization of Eq. (15.2). In the thermodynamic limit, $N \to \infty$, it becomes

$$\rho^{\text{sys}} = \text{Tr}_{\text{env}} \left(|\Psi_R\rangle \langle\Psi_L| \mathcal{T}^{\text{imp}} \right), \tag{15.16}$$

up to a normalization constant.

15.2 Transformation Matrices

In a system where the reduced density matrix is symmetric and semi-positive defined, for example, in the zero-temperature DMRG calculation, all the eigenvalues of the reduced density matrix are real and nonnegative. The basis truncation is done by retaining the largest D eigenvectors.

However, in a system where the reduced density matrix is not symmetric, the left and right eigenvectors of this reduced density matrix are no longer conjugate pairs. In this case, to find an optimal scheme to truncate the basis space of the system, we should find a set of biorthonormal basis pairs, $\{|\alpha\rangle, \langle\alpha|\}$, in the system to optimize the truncation scheme. The left basis vectors, $\{\langle\alpha|\}$, are orthonormal to the right ones, $\{|\alpha\rangle\}$:

$$\langle \alpha | \beta \rangle = \delta_{\alpha,\beta}. \tag{15.17}$$

However, the left basis vector is generally not the Hermitian conjugate of the right one:

$$\langle \alpha | \neq (|\alpha\rangle)^\dagger. \tag{15.18}$$

The transformation matrices between the basis states before and after truncation,

$$P_{s,\alpha} = \quad s \,-\!\boxed{P}\!-\, \alpha \quad = \langle s | \alpha \rangle, \tag{15.19}$$

$$Q_{\alpha,s} = \quad \alpha \,-\!\boxed{Q}\!-\, s \quad = \langle \alpha | s \rangle, \tag{15.20}$$

are a pair of oblique matrices. Using the completeness of the basis states before the truncation, we have

$$\langle \alpha | \beta \rangle = \sum_s \langle \alpha | s \rangle \langle s | \beta \rangle = \quad \alpha \,-\!\boxed{Q}\!-\!\boxed{P}\!-\, \beta \quad = \delta_{\alpha,\beta}. \tag{15.21}$$

Hence, P and Q are column orthonormalized:

$$-\!\boxed{Q}\!-\!\boxed{P}\!-\, = I_D \quad . \tag{15.22}$$

Truncation is to project a state into the space spanned by $\{|\alpha\rangle\}$. The projection operator is defined by

$$X = \sum_\alpha |\alpha\rangle \langle \alpha| = X^2. \tag{15.23}$$

It can be also expressed as

$$(PQ)^2 = PQ. \tag{15.24}$$

After the truncation, the left and right eigenvectors are reduced to

$$|\bar{\Psi}_R\rangle = X|\Psi_R\rangle = \sum_{\alpha,e} (Q\Psi)_{\alpha,e} |\alpha, e\rangle, \tag{15.25}$$

$$\langle \bar{\Psi}_L| = \langle \Psi_L|X = \sum_{\alpha,e} \langle \alpha, e| \left(\Phi^\dagger P\right)_{e,\alpha}. \tag{15.26}$$

The reduced density operator becomes

$$\bar{\rho}^{\text{sys}} = \text{Tr}_e |\bar{\Psi}_R\rangle\langle\bar{\Psi}_L| = \sum_{\alpha\beta} \left(Q\Psi\Phi^\dagger P\right)_{\alpha,\beta} |\alpha\rangle\langle\beta|. \tag{15.27}$$

The corresponding reduced density matrix is

$$\bar{\rho}^{\text{sys}} = Q\Psi\Phi^\dagger P = Q\rho^{\text{sys}}P. \tag{15.28}$$

It should be noted that the reduced density matrix ρ^{sys} is not gauge invariant. If we rotate the basis states of the system by a similar transformation matrix S, the reduced density matrix is changed to $S\rho^{\text{sys}}S^{-1}$. However, the eigenspectrum of ρ^{sys} is gauge invariant. Thus, the optimal transformation matrices P and Q should be determined by keeping the eigenspectra of ρ^{sys} and $\bar{\rho}^{\text{sys}}$ as close as possible.

15.3 Canonicalization of the Transformation Matrices

If the reduced density matrix is semi-positive definite, the transformation matrices P and Q for the system can be found by diagonalizing either ρ^{sys} or ρ^{env}.

If ρ^{sys} is symmetric and semi-positive definite, it is simple to show that $Q = P^\dagger$ is just the unitary matrix that diagonalizes ρ^{sys} after truncation. In other words, P is just the matrix that is formed by the first D largest eigenvectors of ρ^{sys}.

If ρ^{sys} is not symmetric but still semi-positive definite, to find the basis transformation matrices, we first diagonalize ρ^{sys} using a canonical matrix C:

$$\rho^{\text{sys}} = C \Lambda C^{-1}, \tag{15.29}$$

where Λ is the semi-positive eigenvalue matrix whose diagonal elements are descending ordered. In this case, it is optimal to set $\bar{\rho}^{\text{sys}}$ equal to the diagonal eigenmatrix by keeping only the first D eigenvalues with the largest absolute values; that is,

$$\bar{\rho}^{\text{sys}} = Q \rho^{\text{sys}} P = Q C \Lambda C^{-1} P = \Lambda_D, \tag{15.30}$$

where $\Lambda_D = \text{diag}(\Lambda_1, \ldots, \Lambda_D)$. This yields [75]:

$$P = (C)_{D_0 \times D}, \qquad Q = \left(C^{-1}\right)_{D \times D_0}. \tag{15.31}$$

D_0 is the dimension of the system before truncation. The truncation error is

$$\varepsilon = \sum_{m > D} \Lambda_m. \tag{15.32}$$

Alternatively, one can also find the canonical transformation matrix P and Q by diagonalizing the environment density matrix:

$$\left(\rho^{\text{env}}\right)^T = \Phi^\dagger \Psi = \begin{array}{c} \boxed{\Phi^*} \\ \boxed{} \\ \boxed{\Psi} \end{array} = X \Lambda X^{-1}. \tag{15.33}$$

ρ^{env} should have the same eigenvalue spectrum as ρ^{sys}. Again, we assume ρ^{env} is semi-positive defined and its eigenvalues are descending ordered.

To truncate the bond states that link Φ and Ψ from the left side, we insert a pair of projection operators, P and Q, between $\Phi^\dagger$ and Ψ so that

$$\Phi^\dagger P Q \Psi = \begin{array}{c} \boxed{\Phi^*} \\ \boxed{P} \\ \boxed{Q} \\ \boxed{\Psi} \end{array} = X_D \Lambda_D X_D^{-1}, \tag{15.34}$$

where Λ_D is the eigenvalue matrix after truncation. The truncated canonical matrix X and its inverse X^{-1} become X_D and X_D^{-1}, respectively. The dimension of X_D is $D_0 \times D$, and that of X_D^{-1} is $D \times D_0$.

We now set $\lambda_D = \sqrt{\Lambda_D}$ and

$$\Phi^\dagger P = X_D \lambda_D, \qquad Q\Psi = \lambda_D X_D^{-1}. \tag{15.35}$$

This yields the solution for the projection operators:

$$P = \left(\Phi^\dagger\right)^{-1} X_D \lambda_D, \qquad Q = \lambda_D X_D^{-1} \Psi^{-1}. \tag{15.36}$$

Using Eq. (15.33), we can also write these projection operators as

$$P = \Psi X_D \lambda_D^{-1}, \qquad Q = \lambda_D^{-1} X_D^{-1} \Phi^\dagger. \tag{15.37}$$

This avoids the calculation of the matrix inversion, Ψ^{-1} and $(\Phi^\dagger)^{-1}$. In the above derivation, we implicitly assume that both Φ and Ψ are square matrices.

After the truncation, the environment density matrix becomes

$$\left(\rho^{\text{env}}\right)^T = X_D \Lambda_D X_D^{-1}, \tag{15.38}$$

and its trace is

$$\text{Tr}\left(\rho^{\text{env}}\right)^T = \text{Tr}\Lambda_D. \tag{15.39}$$

However, it should be emphasized that the semi-positivity of a nonsymmetric density matrix ρ^{sys} or ρ^{env} (if exists) is not protected in numerical calculations. This nonpositivity may cause instability in the diagonalization of the transfer matrix due to the accumulation of truncation errors. To avoid this trouble, two approaches that do not involve the canonical diagonalization of ρ^{sys} or ρ^{env} were introduced [77, 129]. The first is based on the biorthogonalization of the truncated basis states obtained from the SVD of wave functions [77]. The second is based on the low-rank approach of the environment reduced density matrix ρ^{env}, even though it is the system basis space that is truncated [129]. These two approaches are not gauge invariant and are generally less accurate than the reduced density matrix approach once the latter is numerically stable.

15.4 Biorthonormalization

In a nonsymmetric system, there is no symmetry to ensure that the reduced density matrix is semi-positive definite. This makes it difficult to find an optimal scheme to determine the basis transformation matrices. A solution to this problem is the biorthonormalization first introduced by Huang in Ref. [77]. This is an approach based on a reorthogonalization of the truncated left and right basis states. To understand this more clearly, let us take SVD for Φ and Ψ:

$$\Psi = \;-\!\boxed{U_r}\!-\!\diamond\!-\!\boxed{V_r^\dagger}\!-\;, \tag{15.40}$$

$$\Phi = \;-\!\boxed{U_l}\!-\!\diamond\!-\!\boxed{V_l^\dagger}\!-\;, \tag{15.41}$$

where the singular spectra of λ_r and λ_l are descending ordered. To truncate the basis space by keeping the D largest singular values for both λ_r and λ_l, we obtain two isometric matrices $\bar{U}_r$ and $\bar{U}_l$, formed by the first D column vectors of U_r and U_l, respectively. These matrices are the optimal low-rank approximation for the basis space of Ψ and Φ, respectively.

However, $\bar{U}_r$ and $\bar{U}_l$ are not biorthonormalized. In order to find the basis transformation matrices, we need to biorthonormalize these two matrices. To do so, we first take SVD for $\bar{U}_l^\dagger \bar{U}_r$:

$$-\!\langle\bar{U}_l^\dagger|\!-\!|\bar{U}_r\rangle\!- \;=\; -\!\boxed{U}\!-\!\diamond\!-\!\boxed{V^\dagger}\!-\;, \tag{15.42}$$

where $U = (u_1, u_2, \ldots, u_D)$ and $V = (v_1, v_2, \ldots, v_D)$ are unitary matrices, and λ^2 is the diagonal singular matrix. This leads to

$$-\!\diamond\!-\!\boxed{U^\dagger}\!-\!\langle\bar{U}_l^\dagger|\!-\!|\bar{U}_r\rangle\!-\!\boxed{V}\!-\!\diamond\!- \;=\; I_D. \tag{15.43}$$

(with λ^{-1} on left and right ends)

We thus obtain the basis transformation matrices:

$$P = \;-\!|\bar{U}_r\rangle\!-\!\boxed{V}\!-\!\diamond\!-\;, \tag{15.44}$$

(with λ^{-1})

$$Q = \;-\!\diamond\!-\!\boxed{U^\dagger}\!-\!\langle\bar{U}_l^\dagger|\!-\;. \tag{15.45}$$

(with λ^{-1})

Eq. (15.42) can be also represented as

$$U^\dagger \bar{U}_l^\dagger \bar{U}_r V = \lambda^2. \tag{15.46}$$

Thus the singular value λ_i^2 measures the overlap between $\bar{U}_l u_i$ and $\bar{U}_r v_i$. If one of the singular values, say λ_D, is very small, then the corresponding left and right basis vectors, $\bar{U}_l u_D$ and $\bar{U}_r v_D$, are nearly orthogonal to each other. This may cause numerical instability if the above biorthonormalized basis states are used. To solve this instability, Huang *et al.* [393] suggested enlarging the dimension of basis states from D to $D + 1$ by appending a left column vector $(\bar{u}_l)_{D+1}$ to $\bar{U}_l$ and a right column vector $(\bar{u}_r)_{D+1}$ to $\bar{U}_r$, so that they have largest overlap with $\bar{U}_l u_D$ and $\bar{U}_r v_D$, respectively. A simple and natural choice is to take

$$(\bar{u}_l)_{D+1} = \bar{U}_r v_D, \qquad (\bar{u}_r)_{D+1} = \bar{U}_l u_D. \tag{15.47}$$

These augmented $\bar{U}_l$ and $\bar{U}_r$ will no longer have the small singular value problem after biorthonormalization.

This regularization scheme can be straightforwardly generalized to the case with more than one small singular value. If there are q small singular values, say λ_{D-k} $(k = 0, \ldots, q - 1)$, then one needs to add q vector pairs:

$$(\bar{u}_l)_{D+k+1} = \bar{U}_r v_{D-k}, \qquad (\bar{u}_r)_{D+k+1} = \bar{U}_l u_{D-k} \tag{15.48}$$

to $\bar{U}_l$ and $\bar{U}_r$, respectively. The augmented matrices now have a sufficiently large overlap with each other. This may remove the small singular value problem. However, this scheme deteriorates dramatically with the number of artificially introduced basis states q. Thus it works only in the limit $q \ll D$. If q is large, other regularization or even basis truncation schemes should be used.

15.5 Low-Rank Approximation to the Environment Density Matrix

In case the reduced density matrix is not semi-positive definite, one can also find the transformation matrices for the system by taking a low-rank approximation for the reduced density matrix of the environment. This approach was first introduced in Ref. [129]. It starts with a SVD decoupling of ρ^{env}:

$$\left(\rho^{\text{env}}\right)^T = -\!\boxed{\Phi^\dagger}\!-\!\boxed{\Psi}\!- \;=\; -\!\boxed{U}\!-\!\langle\lambda^2\rangle\!-\!\boxed{V^\dagger}\!- \;. \tag{15.49}$$

Since the bond variables that link Φ and Ψ are the system basis states, to truncate the basis space of the system block is to cut the dimension of the middle bond in the above diagram. To minimize the truncation error, one can take a low-rank approximation by keeping the D largest singular values in the above expression. This approximation can also be achieved by inserting the projection operator PQ between $\Phi^\dagger$ and Ψ so that

$$\Phi^\dagger PQ\Psi = U_D \lambda_D^2 V_D^\dagger, \tag{15.50}$$

where λ_D^2 is the singular value matrix after truncation. The truncated U and V matrices become U_D and V_D, respectively.

P and Q are now determined by the equations

$$\Phi^\dagger P = U_D \lambda_D, \qquad Q\Psi = \lambda_D V_D^\dagger. \tag{15.51}$$

The solution is

$$P = \left(\Phi^\dagger\right)^{-1} U_D \lambda_D = \Psi V_D \lambda_D^{-1}, \tag{15.52}$$

$$Q = \lambda_D V_D^\dagger \Psi^{-1} = \lambda_D^{-1} U_D^\dagger \Phi^\dagger. \tag{15.53}$$

In our derivation, we implicitly assume Φ and Ψ to be square matrices. If not, one should first take LQ decompositions for these two matrices and repeat the calculations using the squared L-matrices.

This approach is computationally stable. However, this scheme is not gauge invariant since there is a gauge ambiguity in the definition of ρ^{env}. It depends on the choice of the environment basis states. This approach is simple to implement, but it is generally less accurate than the approach introduced in §15.4.

16

Renormalization of Quantum Transfer Matrices

16.1 Quantum Transfer Matrix and Thermodynamics

As discussed in §4.3, a one-dimensional quantum lattice model can be mapped onto a two-dimensional classical model via the Trotter–Suzuki decomposition in the framework of path integral. In particular, the partition function of a one-dimensional quantum lattice model with short-range interactions can be represented as a product of transfer matrices defined in the imaginary time space. In the thermodynamic limit, the free energy is determined purely by the maximal eigenvalue of this transfer matrix. The quantum transfer matrix renormalization group (QTMRG) method was developed based on this representation [72, 75]. It extends DMRG from zero to finite temperature for quantum lattice systems.

QTMRG is a quantum generalization of the classical TMRG. A crucial difference between TMRG and QTMRG is that in the former case, the transfer matrix is often symmetric (or Hermitian), whereas in the latter case, the transfer matrix is often nonsymmetric (or non-Hermitian). Treating a nonsymmetric transfer matrix is much more challenging than a symmetric one in two respects. First, the concept of density matrix should be broadened and defined as a nonsymmetric matrix [75, 76], in accordance with the nonsymmetric transfer matrix. Second, since the reduced density matrix is not symmetric, it is significantly more difficult to determine the optimized basis transformation matrix with the associated basis truncation scheme, as discussed in §15.3.

Below we derive the formula for evaluating thermodynamic functions with the largest eigenvalue and eigenvectors of the transfer matrix. We only consider the case the lattice size N is even. It is simple to generalize to the case N is odd.

The free energy can be obtained directly from the partition function. Using Eq. (4.48), we find that the free energy density in the thermodynamic limit is determined solely by the largest eigenvalue, λ_1, of the quantum transfer matrix $\mathcal{T}_M$

$$F = - \lim_{N \to \infty} \frac{1}{\beta N} \ln Z = - \lim_{\varepsilon \to 0} \frac{\ln \lambda_1}{2\beta}. \tag{16.1}$$

In obtaining this formula, we assume that the order of $N \to \infty$ and $\varepsilon \to 0$ is interchangeable.

Once the free energy is known, other thermodynamic functions can be determined from its derivatives. However, there are errors in taking numerical derivatives. The higher the order of derivative, the larger the error. The internal energy, the magnetization, and other first derivatives of the free energy can be evaluated directly from the largest eigenvalue and eigenvectors of the quantum transfer matrix. This can eliminate the error resulting from the numerical derivatives. One can also use automatic differentiation to evaluate the derivatives. It can reduce the error in calculating derivatives, but the computational cost is generally high.

It is simple to show that for any physical quantity that is a sum of local operators, namely,

$$A = \sum_i A_{i,i+1}, \tag{16.2}$$

its thermal average can be evaluated through the maximal eigenvalue and eigenvectors of the transfer matrix. For the internal energy, $A_{i,i+1}$ is just the Hamiltonian density:

$$A_{i,i+1} = H_{i,i+1}. \tag{16.3}$$

For the magnetization, A is the total spin operator of the system and

$$A_{i,i+1} = \frac{S_i^z + S_{i+1}^z}{2}. \tag{16.4}$$

The thermal average of A is equal to the sum of the thermal average of $A_{i,i+1}$. In a translation-invariant system,

$$\langle A \rangle = \sum_i \langle A_{i,i+1} \rangle = N \langle A_{1,2} \rangle, \tag{16.5}$$

where

$$\langle A_{1,2} \rangle = \frac{1}{Z} \text{Tr} \left(e^{-\beta H} A_{1,2} \right). \tag{16.6}$$

The trace on the right-hand side of the equation can also be represented as a product of transfer matrices using the Trotter–Suzuki decomposition. Following the steps leading to Eq. (4.40), it can be shown that

$$\text{Tr} \left(A_{1,2} e^{-\beta H} \right) = \text{Tr} \left[\mathcal{T}_M(A_{1,2}) \mathcal{T}_M^{N/2-1} \right], \tag{16.7}$$

where $\mathcal{T}_M(A_{1,2})$ is defined by

$$\mathcal{T}_M(A_{1,2}) = T_1(A_{1,2}) T_2, \tag{16.8}$$

and

$$T_1(A_{1,2}) = \tilde{\tau}^{1,2}(A_{1,2})\tau^{3,4} \cdots \tau^{2M-1,2M}. \tag{16.9}$$

$T_1(A_{1,2})$ has a similar form as T_1, defined by Eq. (4.35), only τ^{12} needs replaced by $\tilde{\tau}^{1,2}(A_{1,2})$. The local transfer matrix $\tilde{\tau}_{1,2}(A_{1,2})$ is defined by

$$\langle \sigma_1^1 \sigma_2^1 | \tilde{\tau}^{1,2}(A_{1,2}) | \sigma_1^2 \sigma_2^2 \rangle = \langle s_1^1, s_1^2 | A_{1,2} v_{1,2} | s_2^1, s_2^2 \rangle, \tag{16.10}$$

and σ_i^m is related to s_i^m by Eq. (4.33).

In the thermodynamic limit, the thermal average of $A_{1,2}$ is given by

$$\langle A_{1,2} \rangle = \lim_{N \to \infty} \lim_{\varepsilon \to 0} \frac{1}{Z} \text{Tr}\left[\mathcal{T}_M(A_{1,2}) \mathcal{T}_M^{N/2-1} \right]$$
$$= \lim_{\varepsilon \to 0} \frac{\langle \Psi_L | \mathcal{T}_M(A_{1,2}) | \Psi_R \rangle}{\lambda_1}, \tag{16.11}$$

where Ψ_R are Ψ_L are the largest right and left eigenvectors of $\mathcal{T}_M$, respectively.

The internal energy is the thermal average of the Hamiltonian H. Using the translation symmetry, the internal energy per site is equal to the thermal average of $H_{1,2}$. Thus

$$U = \langle H_{1,2} \rangle = \lim_{\varepsilon \to 0} \frac{\langle \Psi_L | \mathcal{T}_M(H_{1,2}) | \Psi_R \rangle}{\lambda_1}. \tag{16.12}$$

Similarly, the uniform magnetization is

$$M_z = \lim_{N \to \infty} \frac{1}{N} \left\langle \sum_i S_i^z \right\rangle = \lim_{\varepsilon \to 0} \frac{\langle \Psi_L | \mathcal{T}_M(S_1^z + S_2^z) | \Psi_R \rangle}{2\lambda_1}. \tag{16.13}$$

From the derivatives of U and M_z, we can determine the specific heat

$$C_v = \frac{\partial U}{\partial T}, \tag{16.14}$$

as well as the magnetic susceptibility

$$\chi = \frac{\partial M_z}{\partial B}. \tag{16.15}$$

The zero-field magnetization is zero because there is no spontaneous symmetry breaking for the antiferromagnetic Heisenberg model. In this case, the zero field susceptibility is given by

$$\chi = \lim_{B \to 0} \frac{M_z}{B}. \tag{16.16}$$

Here we take the one-dimensional Heisenberg model, whose local Hamiltonian is defined by

$$H_{i,i+1} = J\mathbf{S}_i \cdot \mathbf{S}_{i+1} - \frac{B}{2}\left(S_i^z + S_{i+1}^z \right), \tag{16.17}$$

as an example to derive the local transfer matrices for the internal energy and the magnetization.

From the definition of the local evolution operator v, it is simple to verify that

$$
v_{i,i+1} = \begin{pmatrix}
e^{-(\frac{J}{4}-\frac{B}{2})\varepsilon} & 0 & 0 & 0 \\
0 & e^{\frac{J\varepsilon}{4}}\cosh\frac{J\varepsilon}{2} & -e^{\frac{J\varepsilon}{4}}\sinh\frac{J\varepsilon}{2} & 0 \\
0 & -e^{\frac{J\varepsilon}{4}}\sinh\frac{J\varepsilon}{2} & e^{\frac{J\varepsilon}{4}}\cosh\frac{J\varepsilon}{2} & 0 \\
0 & 0 & 0 & e^{-(\frac{J}{4}+\frac{B}{2})\varepsilon}
\end{pmatrix}
\tag{16.18}
$$

in the basis space spanned by $|s_i, s_{i+1}\rangle = (|\uparrow\uparrow\rangle, |\uparrow\downarrow\rangle, |\downarrow\uparrow\rangle, |\downarrow\downarrow\rangle)$. Here $v_{i,i+1}$ is block-diagonal and symmetric because $H_{i,i+1}$ conserves the total spin of the two sites. The corresponding local transfer matrix in the basis space $|\sigma^m, \sigma^{m+1}\rangle = (|\uparrow\uparrow\rangle, |\uparrow\downarrow\rangle, |\downarrow\uparrow\rangle, |\downarrow\downarrow\rangle)$, according to (4.33), is

$$
\tau^{m,m+1} = \begin{pmatrix}
-e^{\frac{J\varepsilon}{4}}\sinh\frac{J\varepsilon}{2} & 0 & 0 & 0 \\
0 & e^{\frac{J\varepsilon}{4}}\cosh\frac{J\varepsilon}{2} & e^{-(\frac{J}{4}-\frac{B}{2})\varepsilon} & 0 \\
0 & e^{-(\frac{J}{4}+\frac{B}{2})\varepsilon} & e^{\frac{J\varepsilon}{4}}\cosh\frac{J\varepsilon}{2} & 0 \\
0 & 0 & 0 & -e^{\frac{J\varepsilon}{4}}\sinh\frac{J\varepsilon}{2}
\end{pmatrix}.
\tag{16.19}
$$

τ is block-diagonal due to the spin conservation in the imaginary time space:

$$
\sigma_i^m + \sigma_i^{m+1} = \sigma_{i+1}^m + \sigma_{i+1}^{m+1}.
\tag{16.20}
$$

For the spin-1/2 Heisenberg model, it was shown that the largest eigenvector is nondegenerate and lies in the total σ spin $Q = \sum_m \sigma_i^m = 0$ subspace, irrespective of the sign of J and the value of B. For larger spins, it was found numerically that the largest eigenvectors of $\mathcal{T}_M$ are also in the $Q = 0$ subspace [157], but a rigorous proof is still not available.

To find $\tilde{\tau}^{1,2}(H_{1,2})$ and $\tilde{\tau}^{1,2}(S_1^z + S_2^z)$, we first evaluate the corresponding evolution matrices. By diagonalizing the local Hamiltonian in the basis states $(|\uparrow\uparrow\rangle, |\uparrow\downarrow\rangle, |\downarrow\uparrow\rangle, |\downarrow\downarrow\rangle)$, we find that

$$
H_{1,2}v_{1,2} = \begin{pmatrix}
J^- e^{-J^- \varepsilon} & 0 & 0 & 0 \\
0 & -\frac{J}{4}e^{\frac{J\varepsilon}{4}}a_J & \frac{J}{4}e^{\frac{J\varepsilon}{4}}b_J & 0 \\
0 & \frac{J}{4}e^{\frac{J\varepsilon}{4}}b_J & -\frac{J}{4}e^{\frac{J\varepsilon}{4}}a_J & 0 \\
0 & 0 & 0 & J^+ e^{-J^+ \varepsilon}
\end{pmatrix}.
\tag{16.21}
$$

Similarly, the evolution operator corresponding to the local spin operators is

$$
(S_1^z + S_2^z)v_{1,2} = \begin{pmatrix}
e^{-J^- \varepsilon} & 0 & 0 & 0 \\
0 & 0 & 0 & 0 \\
0 & 0 & 0 & 0 \\
0 & 0 & 0 & -e^{-J^+ \varepsilon}
\end{pmatrix}.
\tag{16.22}
$$

In these equations,

$$J^{\pm} = \frac{J}{4} \pm \frac{B}{2}, \tag{16.23}$$

$$a_J = 2\cosh\frac{J\varepsilon}{2} + \sinh\frac{J\varepsilon}{2}, \tag{16.24}$$

$$b_J = 2\sinh\frac{J\varepsilon}{2} + \cosh\frac{J\varepsilon}{2}. \tag{16.25}$$

Substituting these expressions into Eq. (16.10), we find that

$$\tilde{\tau}^{1,2}(H_{1,2}) = \begin{pmatrix} \frac{J}{4}e^{\frac{J\varepsilon}{4}}b_J & 0 & 0 & 0 \\ 0 & -\frac{J}{4}e^{\frac{J\varepsilon}{4}}a_J & J^- e^{-J^- \varepsilon} & 0 \\ 0 & J^+ e^{-J^+ \varepsilon} & -\frac{J}{4}e^{\frac{J\varepsilon}{4}}a_J & 0 \\ 0 & 0 & 0 & \frac{J}{4}e^{\frac{J\varepsilon}{4}}b_J \end{pmatrix}, \tag{16.26}$$

and

$$\tilde{\tau}^{1,2}\left(S_1^z + S_2^z\right) = \begin{pmatrix} 0 & 0 & 0 & 0 \\ 0 & 0 & e^{-J^- \varepsilon} & 0 \\ 0 & -e^{-J^+ \varepsilon} & 0 & 0 \\ 0 & 0 & 0 & 0 \end{pmatrix} \tag{16.27}$$

in the basis space spanned by $|\sigma^1, \sigma^2\rangle = (|\uparrow\uparrow\rangle, |\uparrow\downarrow\rangle, |\downarrow\uparrow\rangle, |\downarrow\downarrow\rangle)$. Both $\tilde{\tau}(H_{1,2})$ and $\tilde{\tau}^{1,2}(S_1^z + S_2^z)$ are block diagonal.

16.2 Correlation Functions

Besides thermodynamic quantities, the quantum transfer matrix can also be used to calculate the static and dynamical correlation functions at finite temperatures. For the static correlation function, what we are interested in is the spatial correlation of a physical quantity around its mean value,

$$\delta A_i = A_i - \langle A_i \rangle, \tag{16.28}$$

between sites i and $i + r$; that is, $\langle \delta A_i^\dagger \delta A_{i+r} \rangle$. In a translation-invariant system, this correlation function depends only on the distance between these two lattice sites, r,

$$G(r) = \langle \delta A_i^\dagger \delta A_{i+r} \rangle = \langle \delta A_0^\dagger \delta A_r \rangle. \tag{16.29}$$

This correlation function can also be represented using quantum transfer matrices. When r is even, both δA_0 and δA_r lie in the same sublattices. In the thermodynamic limit, it is simple to show that $G(r)$ is given by

$$G(r) = \frac{\langle \Psi_L | \mathcal{T}_M(\delta A_0^\dagger) \mathcal{T}_M^{r/2-1} \mathcal{T}_M(\delta A_r) | \Psi_R \rangle}{\lambda_1^{r/2+1}}. \tag{16.30}$$

Thus the correlation function $G(r)$ is determined purely by the expectation value of the product of transfer matrices over the leading eigenvectors of the transfer matrix. For odd r, the formula for evaluating $G(r)$ is similar, only the definition of $\mathcal{T}_M(A_r)$ needs to be slightly modified. For convenience in the discussion, we only consider the case r is even below.

For large r, the formula for evaluating $G(r)$; that is, Eq. (16.30), can be further simplified. To see this, let us substitute Eq. (4.47) into (16.30) and rewrite $G(r)$ as

$$G(r) = \sum_{n \neq 1} \frac{\langle \Psi_L | \mathcal{T}_M(\delta A_0^\dagger) | \psi_n^R \rangle \langle \psi_n^L | \mathcal{T}_M(\delta A_r) | \Psi_R \rangle}{\lambda_1 \lambda_n} \left(\frac{\lambda_n}{\lambda_1} \right)^{r/2}. \qquad (16.31)$$

Note here $\Psi_R = \psi_1^R$ and $\Psi_L = \psi_1^L$. The summation in this expression does not include the term $n = 1$ because the thermal average of δA_i is zero:

$$\langle \Psi_L | \mathcal{T}_M(\delta A_i) | \Psi_R \rangle = 0. \qquad (16.32)$$

In the limit $r \to \infty$, only the largest λ_α with

$$\langle \Psi_L | \mathcal{T}_M(\delta A_0^\dagger) | \psi_\alpha^R \rangle \langle \psi_\alpha^L | \mathcal{T}_M(\delta A_r) | \Psi_R \rangle \neq 0 \qquad (16.33)$$

has a contribution to the above summation. In this case, $G(r)$ is simplified to

$$G(r) = \frac{\langle \Psi_L | \mathcal{T}_M(\delta A_0^\dagger) | \psi_\alpha^R \rangle \langle \psi_\alpha^L | \mathcal{T}_M(\delta A_r) | \Psi_R \rangle}{\lambda_1 \lambda_\alpha} e^{-(\xi^{-1} - i\kappa)r}. \qquad (16.34)$$

If vector $\mathcal{T}_M(\delta A_0) | \Psi_R \rangle$ is in the same symmetry subspace as $| \Psi_R \rangle$ (for example, when $A_i = S_i^z$), then λ_α is usually the second-largest eigenvalue of $\mathcal{T}_M$ in that subspace. If, on the other hand, $\mathcal{T}_M(\delta A_0) | \Psi_R \rangle$ is in a different subspace (for example, when $A_i = S_i^x$), then λ_α is the largest eigenvalue of $\mathcal{T}_M$ in the subspace whose quantum numbers are the same as for $\mathcal{T}_M(\delta A_0) | \Psi_R \rangle$. In the former case, the two largest eigenvalues and eigenvectors in the maximal eigenvalue subspace of the transfer matrix need to be evaluated. In the latter case, we need to evaluate the largest eigenvalue and eigenvectors in two different subspaces. If the absolute value of λ_α is degenerate, then the contribution from all these degenerate states to $G(r)$ should be added up.

In Eq. (16.34), ξ is the thermal correlation length which is determined by λ_1 and λ_α

$$\xi^{-1} = \lim_{\varepsilon \to 0} \frac{1}{2} \ln \left| \frac{\lambda_1}{\lambda_\alpha} \right|. \qquad (16.35)$$

At finite temperature, ξ is always finite if there is no phase transition. κ is the characteristic wave vector of the most dominant fluctuations determined by the phase of λ_α

$$\kappa = \lim_{\varepsilon \to 0} \frac{1}{2} \arg \left(\frac{\lambda_\alpha}{\lambda_1} \right) + n\pi, \qquad (16.36)$$

where $n = 0$ or 1. The value of κ is not uniquely determined because $\mathcal{T}_M$ crosses over two sites and cannot distinguish between κ and $\kappa + \pi$. This unambiguity in determining the value of κ can be removed by requiring that it varies smoothly from low to high temperatures. At high temperatures, λ_α is generally real, and κ is either 0 or π. From physical considerations, one can readily determine whether κ is 0 or π. One can also remove this uncertainty by calculating the correlation function for the case r is odd.

Parameter κ determines the spatial variation of the phase factor of the correlation function $G(r)$. If A_i is a creation or annihilation operator of electrons, then κ is just the Fermi vector in the zero temperature limit. Thus, from the quantum transfer matrix calculation, we can determine the Fermi vector. It is useful for determining the Fermi vector of the periodic Kondo or Anderson lattice model [394]. In these models, there is a large Fermi surface formed by the interplay between itinerant electrons and localized spins, in addition to the conventional Fermi surface purely determined by the itinerant electrons. At high temperatures, the itinerant electrons and localized spins are decoupled, and only the small Fermi surface exists. At low temperatures, however, there are strong interactions between itinerant electrons and localized spins, and a large Fermi surface emerges at the Fermi level.

16.3 QTMRG

As mentioned, the thermodynamic properties of one-dimensional quantum lattice models are determined by the largest eigenvalue and eigenvectors of the quantum transfer matrix in the thermodynamic limit. Here we discuss how to extend the idea of DMRG to find the largest eigenvalue and the corresponding eigenvectors of the quantum transfer matrix.

The quantum transfer matrix is defined in the Trotter space. As the spatial dimension of the system is infinite, there is no finite lattice size effect. In the DMRG diagonalization, the system size equals $2M$ with M the number of time slices in cutting an inverse temperature β. At each iteration, the Trotter lattice size $2M$ increases by two. Given ε, increasing the Trotter lattice size is equivalent to lowering the temperature

$$T = \frac{1}{\varepsilon M}. \qquad (16.37)$$

In the diagonalization of quantum transfer matrix, a superblock contains four building units, the left and right blocks, and two added sites, in the Trotter space, same as in DMRG. As the periodic boundary conditions are imposed, the two added sites are located separately on the two sides of the left block. At each step of the iteration, the left block is augmented alternatively first by adding the site on its right-hand side and then the site on its left-hand side. The system block consists of the left block and one of the added sites. The other added site, together with the right block, forms the environment block. The quantum transfer matrix of the superblock is a product of the quantum transfer matrices of the system and environment blocks. Since the quantum transfer matrix is a product of local transfer matrices, unlike a Hamiltonian with local interactions, four kinds of system transfer matrices (similarly, four kinds of environment transfer matrices) are generated with the DMRG iterations. With these four kinds of system and environment transfer matrices, the following four superblock configurations are sequentially generated in each cycle of iterations:

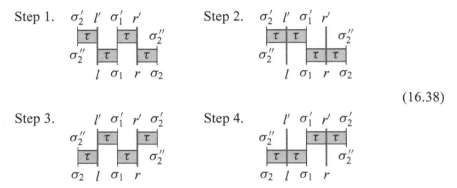

$$(16.38)$$

Initially, $M = 2$, the left and right blocks each contain one site, and the superblock contains four sites. The system transfer matrix is simply a product of two local transfer matrices:

$$S \left(\begin{array}{ccc} \sigma_2' & l' & \sigma_1'' \\ \sigma_2'' & l & \sigma_1 \end{array} \right) = \quad (16.39)$$

The matrix elements of the environment transfer matrix $\mathcal{E}$ are similarly defined. $\mathcal{T}^M$ is a product of S and $\mathcal{E}$.

In the infinite-lattice algorithm of DMRG, the sizes of the left and right blocks are increased at the same speed. As the superblock is always center reflection symmetric, for all the four kinds of configurations shown in Eq. (16.38), the matrix elements of S and $\mathcal{E}$ satisfy the equation:

$$\mathcal{E} \left(\begin{array}{ccc} \sigma_2' & l' & \sigma_1' \\ \sigma_2 & l & \sigma_1 \end{array} \right) = S \left(\begin{array}{ccc} \sigma_1 & l & \sigma_2 \\ \sigma_1' & l' & \sigma_2' \end{array} \right). \quad (16.40)$$

Thus only the matrix elements of S need to be updated and stored.

The whole transfer matrix $\mathcal{T}_M$ of the superblock is a product of $\mathcal{S}$ and $\mathcal{E}$:

$$\langle l'\sigma_1' r'\sigma_2' | \mathcal{T}_M | l\sigma_1 r\sigma_2 \rangle = \sum_{\sigma_1'' \sigma_2''} \mathcal{S}\left(\begin{matrix} \sigma_2' & l' & \sigma_1'' \\ \sigma_2'' & l & \sigma_1 \end{matrix} \right) \mathcal{E}\left(\begin{matrix} \sigma_1' & r' & \sigma_2'' \\ \sigma_1'' & r & \sigma_2 \end{matrix} \right). \quad (16.41)$$

The matrix elements of $\mathcal{T}_M$ for the other three configurations are similarly defined.

16.3.1 Eigenstates of the Transfer Matrix

Several methods can be applied to evaluate the largest eigenvalue and eigenvectors of $\mathcal{T}_M$. The simplest one is the power method introduced in §A.1. This method starts from an arbitrary initial vector that is not orthogonal to the largest eigenvector. It projects out the largest right eigenvector by successively applying $\mathcal{T}_M$ onto the vector generated in the previous iteration. The number of projections needed depends on the ratio between the second largest eigenvalue and the largest eigenvalue $\eta = |\lambda_2/\lambda_1|$. The logarithm of η is proportional to the correlation length. Therefore, the number of projections needed increases with the increase of the correlation length.

Other methods for determining the largest eigenvalue and eigenvectors of $\mathcal{T}_M$ include the implicitly restarted Arnoldi method and the look-ahead Lanczos method. They are more efficient than the power method and can be used to evaluate not only the largest eigenvalue but also the second largest or other eigenvalues of $\mathcal{T}_M$.

The left eigenvector $\langle \Psi_L |$ is similarly determined. If the transfer matrix is center reflection symmetric, the left eigenvector can be obtained directly from the right one using this symmetry.

In practical computations using the power or other sparse matrix diagonalization methods, one does not have to evaluate the matrix elements of $\mathcal{T}_M$ explicitly since $\mathcal{T}_M$ is a product of $\mathcal{S}$ and $\mathcal{E}$. Instead, one should successfully apply $\mathcal{S}$ and $\mathcal{E}$ onto a vector, which can significantly reduce the computational cost.

16.3.2 Reduced Density Matrix

The reduced density matrix of the system is determined by the maximal left and right eigenvectors of $\mathcal{T}_M$. In Step 1 and Step 3 of (16.38), the system contains the left block and the site on its right side, and the corresponding reduced density matrix is defined by

$$\rho^{\text{sys}}\left(l'\sigma_1', l\sigma_1\right) = \sum_{r\sigma_2} \Psi_L \left(l'\sigma_1' r\sigma_2\right) \Psi_R \left(l\sigma_1 r\sigma_2\right), \quad (16.42)$$

where Ψ_L and Ψ_R are the left and right maximal eigenvectors of $\mathcal{T}^M$. In Step 2 and Step 4, on the other hand, the system contains the left block and the site on its left side. The corresponding reduced density matrix is

$$\rho^{\text{sys}}\left(l'\sigma_2', l\sigma_2\right) = \sum_{r\sigma_1} \Psi_L\left(l'\sigma_1 r\sigma_2'\right) \Psi_R\left(l\sigma_1 r\sigma_2\right). \tag{16.43}$$

If the quantum transfer matrix is not symmetric (Hermitian), the reduced density matrix such defined is also not symmetric (Hermitian).

In Section 15, several approximation methods are introduced to find the transformation matrices. If the reduced density matrix is semi-positive definite, the canonical transformation is the most optimal one that should be used. However, this method may become unstable when the accumulated truncation errors become larger, particularly at low temperatures. The bi-orthogonalization method may not be optimal, but is more stable than the canonical transformation one.

16.3.3 Update of Transfer Matrices

Once the transformation matrices, P and Q, are determined, we now use them to augment the left and right blocks and update their matrix elements. At step 1, the left and right blocks are augmented to include the same site σ_1:

The system transfer matrix is then updated to

$$\mathcal{S}_u\left(\begin{array}{ccc} \sigma_2' & l_u' & \sigma_3' \\ \sigma_2'' & l_u & \sigma_3'' \end{array}\right) = \sum_{\sigma_1''} Q(l_u', l'\sigma_1')\mathcal{S}\left(\begin{array}{ccc} \sigma_2' & l' & \sigma_1'' \\ \sigma_2'' & l & \sigma_1 \end{array}\right)$$
$$\tau(\sigma_1', \sigma_3'|\sigma_1'', \sigma_3'')P\left(l\sigma_1, l_u\right). \tag{16.44}$$

The augmented system or environment transfer matrices in the other three steps are similarly defined.

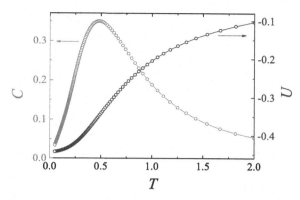

Figure 16.1 Temperature dependence of the internal energy (right) and the specific heat (left) for the spin-1/2 antiferromagnetic Heisenberg model obtained using QTMRG with $\varepsilon = 0.05$ and $D = 100$.

16.4 Thermodynamics of the Heisenberg Spin Chain

To gain a more quantitative idea of the accuracy of QTMRG, let us apply it to the spin-1/2 antiferromagnetic Heisenberg model (4.1) [157]. We set the antiferromagnetic coupling constant $J = 1$.

16.4.1 The Spin-1/2 Heisenberg Model

Figure 16.1 shows the temperature dependence of the internal energy $U(T)$ and the specific heat $C(T)$ for the S=1/2 antiferromagnetic Heisenberg model, obtained using QTMRG with $\varepsilon = 0.05$ and $D = 100$. As expected, the internal energy drops with decreasing temperature. The extrapolated zero-temperature internal energy is about -0.443, in agreement with the Bethe ansatz result of the ground-state energy $-\ln 2 + 0.25$, within numerical errors.

The specific heat varies nonmonotonically with temperature. It shows a broad peak at a temperature slightly lower than $T \sim 0.5$. The specific heat of the spin-1/2 Heisenberg model was first calculated numerically by Bonner and Fisher [395]. They found that the linear coefficient of the specific heat in the zero temperature limit is $\gamma(T) = C/T \sim 0.7$. On the other hand, Affleck [371] found that $\gamma(T \to 0) = 2/3$ from conformal field theory. The QTMRG of result of $\gamma(T \to 0)$ obtained with $D = 100$ is 0.66, close to the conformal field result [371].

Figure 16.2 shows the temperature dependence of the magnetic susceptibility $\chi(T)$ and the inverse correlation length $1/\xi$ for the S=1/2 Heisenberg model obtained by QTMRG. The exact results for the magnetic susceptibility, obtained from the Bethe ansatz, are also shown for comparison. The difference between the

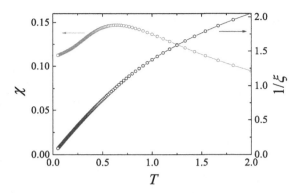

Figure 16.2 Uniform magnetic susceptibility (left) and inverse correlation length (right) versus temperature for the S=1/2 antiferromagnetic Heisenberg model. $\varepsilon = 0.05$ and $D = 100$ are used in the QTMRG calculation.

exact and QTMRG results for the susceptibility is less than 10^{-4} even at low temperatures. The magnetic susceptibility varies non-monotonically with temperature. The maximal susceptibility is approximately 0.147 at $T \approx 0.64$, consistent with the Bethe ansatz result [396]. The extrapolated zero-field and zero-temperature value of the spin susceptibility $\chi(T)$ is 0.109, in agreement with the Bethe ansatz result, $\chi = 1/\pi^2$, obtained by Griffiths [397].

At low temperatures, Eggert, Affleck, and Takahashi [396] showed that there is a logarithmic correction to the linear zero-field susceptibility when $T < 0.01$:

$$\chi = \frac{1}{2\pi v}\left[1 + \left(2\ln\frac{T_0}{T}\right)^{-1}\right], \tag{16.45}$$

where $v = \pi/2$ is the spin wave velocity and $T_0 \approx 7.7$. This logarithmic correction is also observed in the QTMRG calculation [157].

The correlation length at zero field grows monotonically with the decrease in temperature. It diverges in the zero-temperature limit. The slope of ξ^{-1} as a function of temperature is approximately 2, in agreement with the result obtained using the thermal Bethe ansatz and the SU(2)$_1$ Wess–Zumino–Witten model [398]

$$\xi^{-1} = T\left[2 - \left(\ln\frac{T_0}{T}\right)^{-1}\right], \qquad (T \to 0). \tag{16.46}$$

16.4.2 The Spin-1 Heisenberg Model

The spin-1 Heisenberg model has a finite energy gap Δ between the ground state and the first excited state at zero magnetic fields. Consequently, the thermodynamic

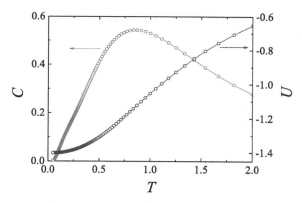

Figure 16.3 Temperature dependence of the internal energy (right) and the specific heat (left) for the spin-1 antiferromagnetic Heisenberg model obtained using QTMRG with $\varepsilon = 0.05$ and $D = 100$.

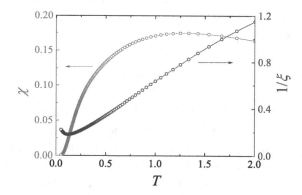

Figure 16.4 Uniform magnetic susceptibility (left) and inverse correlation length (right) versus temperature for the spin-1 antiferromagnetic Heisenberg model. $\varepsilon = 0.05$ and $D = 100$ are used in the QTMRG calculation.

quantities, including the internal energy $U(T)$, the specific heat $C(T)$, and the magnetic susceptibility $\chi(T)$, shown in Figs. 16.3 and 16.4, should drop exponentially with T at low temperatures. The numerical results are obtained by QTMRG with $\varepsilon = 0.05$ and $D = 100$.

The ground-state energy, extrapolated from the internal energy $U(T)$ (Fig. 16.3) in the $T \to 0$ limit, is about -1.4015, in agreement with the zero-temperature DMRG result [19]. The energy gap can be calculated using the zero-temperature DMRG by targeting separately the quantum states with total spin equal to 0 and 1, respectively. It can also be estimated from the low-temperature behaviors of the susceptibility and the specific heat.

Under the single-mode approximation, the low-lying excitation energy varies approximately with the wave vector k as [399]:

$$\varepsilon(k) = \Delta + \frac{v^2}{2\Delta}(k - \pi)^2 + O(|k - \pi|^3), \qquad (16.47)$$

where v is the spin wave velocity. Using this dispersion relation, it is simple to show that the susceptibility varies exponentially with temperature:

$$\chi(T) = \frac{1}{v}\sqrt{\frac{2\Delta}{\pi T}}e^{-\Delta/k_B T}, \qquad (16.48)$$

in the limit $k_B T \ll \Delta$. Similarly, the specific heat also drops exponentially:

$$C(T) = \frac{3\Delta}{v\sqrt{2\pi}}\left(\frac{\Delta}{T}\right)^{3/2} e^{-\Delta/k_B T}, \qquad (16.49)$$

at low temperatures. Taking the ratio between $C(T)$ and $\chi(T)$, the exponential functions in two expressions cancel out, leading to a simple equation that can be used to determine the energy gap:

$$\Delta = \lim_{T \to 0} \sqrt{\frac{2TC(T)}{3\chi(T)}}. \qquad (16.50)$$

Due to the existence of the excitation gap, the correlation length ξ becomes finite at zero temperature. We note, however, that ξ does not increase monotonically at low temperatures. It shows a maximum at a temperature T^*. This nonmonotonic behavior, as pointed out in Ref. [157], might result just from the truncation error since T^* drops with the increase in the number of states retained in the QTMRG calculation. If we do an extrapolation using the QTMRG data above T^*, we find that $\xi(T \to 0) \sim 6.0$, which is consistent with the zero-temperature DMRG result [19]. On the other hand, if the QTMRG data below T^* are used for the extrapolation, we find that $\xi(T \to 0) \sim 4.4$, which is less than the DMRG result.

16.4.3 Error Analysis

QTMRG treats an infinite lattice directly and is not bothered by the finite-size effect. The accuracy of QTMRG is determined by both the Trotter and truncation errors. To reduce the Trotter error, one can either reduce the Trotter interval ε or use a higher-order Trotter–Suzuki decomposition formula. To reduce the truncation error, one has to increase the number of retained states D in the basis truncation.

The left panel of Fig. 16.5 shows the error of the magnetic susceptibility as a function of ε. At high temperatures, such as $T = 0.5$, the truncation error is small, and the error comes mainly from the Trotter–Suzuki decomposition, which

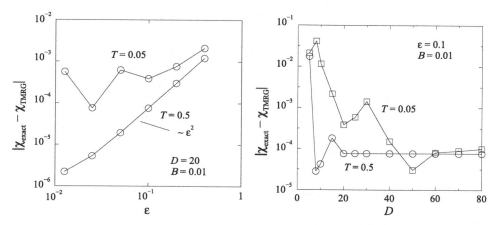

Figure 16.5 Errors of the magnetic susceptibility as a function of the Trotter interval ε (left panel) or the number of states retained D (right panel) for the magnetic susceptibility of the spin-1/2 antiferromagnetic Heisenberg model. The plots in the left and right panels are reproduced from figures 9 and 10 in Ref. [400], respectively.

varies approximately as ε^2 at a given D. At low temperatures, however, the error is dominated by the accumulated truncation error.

The right panel of Fig. 16.5 shows how the error varies with D. Given ε, we find that the systematical errors drop exponentially with increasing D initially and reach a value determined by the Trotter error when D becomes sufficiently large. This is because at finite temperatures, the correlation length is always finite, and there is a gap between the largest eigenvalue and the next-largest eigenvalues of the transfer matrix $\mathcal{T}_M$. Since the gap decreases with decreasing temperature, the systematical error should drop faster at high temperatures.

17

MPS Solution of QTMRG

17.1 Biorthonormal MPS

The standard DMRG (or TMRG discussed in Chapter 14) and MPS formulation is established based on a set of orthonormalized basis states obtained from the diagonalization of symmetric reduced density matrices. In the renormalization of a nonsymmetric transfer matrix, the corresponding MPS must be built on a set of reduced biorthonormal bases, referred to as biorthonormal MPS. Particularly, the left MPS eigenvector is not the Hermitian conjugate of the corresponding right eigenvector. To understand this more clearly, let us consider an infinite translation-invariant transfer matrix:

$$T = \quad \cdots \boxed{\tau} \boxed{\tau} \boxed{\tau} \boxed{\tau} \boxed{\tau} \cdots \quad . \tag{17.1}$$

We assume the following translation-invariant MPS,

$$\Psi_R = \quad \cdots -\!\!\left(A\right)\!-\!\!\left(A\right)\!-\!\!\left(A\right)\!-\!\!\left(A\right)\!-\!\!\left(A\right)\!-\cdots \ , \tag{17.2}$$

to be the leading right eigenvector of T, and

$$\Psi_L = \quad \cdots -\!\!\left(B^*\right)\!-\!\!\left(B^*\right)\!-\!\!\left(B^*\right)\!-\!\!\left(B^*\right)\!-\!\!\left(B^*\right)\!- \cdots \tag{17.3}$$

the corresponding left eigenvector. These two vectors can be canonicalized, following the ideas introduced in §15.4, such that

$$\Psi_R = \quad \cdots -\!\!\boxed{A_L}\!-\!\!\boxed{A_L}\!-\!\!\boxed{A_L}\!-\!\!\left(C\right)\!-\!\!\left\langle A_R\right|\!-\!\!\left\langle A_R\right|\!-\!\!\left\langle A_R\right| \cdots \tag{17.4}$$

$$= \quad \cdots -\!\!\boxed{A_L}\!-\!\!\boxed{A_L}\!-\!\!\boxed{A_L}\!-\!\!\left(A_C\right)\!-\!\!\left\langle A_R\right|\!-\!\!\left\langle A_R\right|\!-\!\!\left\langle A_R\right| \cdots \tag{17.5}$$

and

$$\Psi_L = \quad \cdots -\boxed{B_L^*}-\boxed{B_L^*}-\boxed{B_L^*}-D^*-\langle B_R^*|-\langle B_R^*|-\langle B_R^*|- \cdots \qquad (17.6)$$

$$= \quad \cdots -\boxed{B_L^*}-\boxed{B_L^*}-\boxed{B_L^*}-B_D^*-\langle B_R^*|-\langle B_R^*|-\langle B_R^*|- \cdots \qquad (17.7)$$

are left and right bicanonicalized.

In Eqs. (17.4)–(17.7), C and D are bond matrices, and

$$-A_C- \; = \; -\boxed{A_L}-C- \; = \; -C-\langle A_R|- \, , \qquad (17.8)$$

$$-B_D- \; = \; -\boxed{B_L}-D- \; = \; -D-\langle B_R|- \, . \qquad (17.9)$$

$A_L[\sigma]$ and $B_L[\sigma]$ form a pair of left canonical matrices:

$$\sum_\sigma B_L^\dagger[\sigma]A_L[\sigma] = I, \qquad (17.10)$$

and $B_R[\sigma]$ and $A_R[\sigma]$ form a pair of right canonical matrices:

$$\sum_\sigma A_R[\sigma]B_R^\dagger[\sigma] = I. \qquad (17.11)$$

The overlap between Ψ_R and Ψ_L is now given by

$$\langle \Psi_L | \Psi_R \rangle = \mathrm{Tr} D^\dagger C = \mathrm{Tr} B_D^\dagger A_C \qquad (17.12)$$

17.2 Biorthonormalization

Given A and B, we now consider how to left or right biorthonormalize them into the standard form represented by Eq. (17.4) or (17.6). The discussion below is for a pair of translation-invariant MPS. Nevertheless, the idea is applicable to a system without translation invariance.

17.2.1 Left Biorthonormalization

The left biorthogonalization can be done in three steps:

(i) Orthonormalize A and B separately so that they are independently left canonicalized, namely

$$\qquad \qquad = I. \qquad (17.13)$$

(ii) Define the left transfer matrix,

$$\mathcal{E}_{\alpha\alpha',\beta\beta'} = \qquad\qquad , \qquad (17.14)$$

and find its maximal left eigenvector X and the corresponding eigenvalue η:

$$\qquad\qquad = \eta \; (X) . \qquad (17.15)$$

X has a real solution if $\mathcal{E}$ is real and the absolute value of η is not degenerate.

(iii) Factorize X by SVD:

$$\qquad X \qquad = \qquad U^\dagger \qquad \lambda^2 \qquad V \qquad , \qquad (17.16)$$

and update the local tensors by the equations

$$A_L[\sigma] = \eta^{-1} P_L A[\sigma] P_L^{-1}, \qquad (17.17)$$

$$B_L[\sigma] = Q_L B[\sigma] Q_L^{-1}, \qquad (17.18)$$

where

$$P_L = \lambda V, \qquad Q_L = \lambda U. \qquad (17.19)$$

It is simple to show that A_L and B_L thus obtained satisfy approximately the left canonical equation (17.10). Eq. (17.10) is valid just approximately because the gauge transformation involves the inverse of λ whose small values may cause substantial errors in A_L and B_L. The accuracy is limited by the accuracy of λ^{-1}. If some singular values are small, the above gauge transformation is ill-conditioned.

If some of the singular values in λ become precisely zero, the bond dimensions of A_L and B_L are reduced by completely ignoring the basis states corresponding to these zero singular values.

(iv) Set $A = A_L$ and $B = B_L$ and repeat steps (ii)–(iii). It can improve the accuracy of left biorthogonalization because, in this case, the left eigenvector of the transfer matrix should be closer to the identity matrix; that is, $X \approx I$, and the inverse of the singular matrix λ becomes computationally stable. This is similar to the procedure of reorthogonalization of the Krylov basis states. This process can be repeated two to three times. To further repeat this process may not improve the accuracy significantly.

In this discussion, we assumed that the maximal eigenvalue η in its absolute value is real and nondegenerate. If it is a complex number, Eqs. (17.17) and (17.18) are still valid, but U and V generally become complex.

In step (ii), X can also be factorized by LU decomposition introduced in §2.3:

$$X = LU, \tag{17.20}$$

where L is an invertible lower triangle matrix and U is an invertible upper triangle matrix. In this case, P_L and Q_L change to

$$P_L = U, \qquad Q_L = L^\dagger. \tag{17.21}$$

To ensure that both L and U are invertible (assuming X is invertible), a partial pivoting to X should be performed before taking the LU decomposition.

17.2.2 Right Biorthonormalization

We can follow the same steps in the left biorthonormalization to right biorthonormalize A and B. Again we first orthonormalize A and B so that they are separately right canonicalized:

$$\tag{17.22}$$

We then determine the largest right eigenvector Y of the transfer matrix

$$\tag{17.23}$$

by solving the eigenequation

$$\tag{17.24}$$

By factorizing Y with SVD:

$$-\!\!\boxed{Y}\!\!- \;=\; -\!\!\boxed{U^\dagger}\!\!-\!\!\langle\lambda^2\rangle\!\!-\!\!\boxed{V}\!\!- \;, \tag{17.25}$$

we find the right orthonormalized tensors:

$$A_R[\sigma] = \eta^{-1} P_R A[\sigma] P_R^{-1}, \tag{17.26}$$

$$B_R[\sigma] = Q_R B[\sigma] Q_R^{-1}, \tag{17.27}$$

where

$$P_R = \lambda^{-1} U, \qquad Q_R = \lambda^{-1} V. \tag{17.28}$$

Generally, A_R and B_R thus obtained are just approximately right canonicalized. The accuracy of the right biorthonormal condition can be improved by setting $A = A_R$ and $B = B_R$, and repeating the above steps a few times.

17.2.3 Bicanonical Form

Our discussions indicate that we can find a pair of matrices P_L and Q_L to left bicanonicalize local tensor A and B:

$$A_L[\sigma] = P_L A[\sigma] P_L^{-1}, \tag{17.29}$$

$$B_L[\sigma] = Q_L B[\sigma] Q_L^{-1}. \tag{17.30}$$

Here the largest eigenvalue η is normalized to 1. Similarly, A and B can also be right bicanonicalized by a pair of matrices P_R and Q_R:

$$A_R[\sigma] = P_R A[\sigma] P_R^{-1}, \tag{17.31}$$

$$B_R[\sigma] = Q_R B[\sigma] Q_R^{-1}. \tag{17.32}$$

This leads to a left-right biorthonormalized MPS expressed by Eqs. (17.4) and (17.6) with

$$C = P_L P_R^{-1}, \qquad D = Q_L Q_R^{-1}. \tag{17.33}$$

However, by comparing (17.21) with (17.19), we see that there is a gauge ambiguity in defining the biorthogonalized local tensors: if (P_L, Q_L) is a pair of matrices that left biorthonormalize A and B, so is $(RP_L, (R^\dagger)^{-1} Q_L)$, where R is an arbitrary invertible matrix. In case the bond dimensions of local tensors A and B need to be truncated, this gauge ambiguity suggests that the biorthonormalization itself does not fix the condition for optimizing the biorthonormalized basis states retained. In a symmetric system, the left eigenvector Ψ_L is the Hermitian conjugate of the right eigenvector $\Psi_R^\dagger$ so that $A = B$. In this case, P_L must be equal to Q_L, and R is a unitary matrix, which does not alter the basis subspace on which the virtual bond states are represented.

To remove the gauge ambiguity in defining the virtual basis subspace, let us consider the reduced density matrix of the subblock that contains all the sites on the left of matrix C; that is,

$$\rho_L = \qquad\qquad (17.34)$$

From the biorthonormal condition of A_L and B_L, ρ_L can also be represented as

$$\rho_L = S\tilde{\rho}_L S^{-1}, \qquad\qquad (17.35)$$

where $\tilde{\rho}_L$ is a bond density matrix

$$\tilde{\rho}_L = CD^\dagger, \qquad\qquad (17.36)$$

and S^{-1} is the pseudo-inverse of S defined by

$$
\begin{aligned}
S_{\cdots\sigma_1\sigma_2\sigma_3,\alpha} &= \\
S^{-1}_{\alpha,\cdots\sigma_1\sigma_2\sigma_3} &=
\end{aligned}
\qquad (17.37)
$$

S and S^{-1} satisfy the equations

$$S^{-1}S = I, \qquad SS^{-1} \neq I. \qquad\qquad (17.38)$$

Apparently, $\tilde{\rho}_L$ has the same eigenspectrum as ρ_L. To diagonalize $\tilde{\rho}_L$ with a canonical matrix G_L:

$$\tilde{\rho}_L = G_L^{-1}\lambda G_L, \qquad\qquad (17.39)$$

λ is the eigenvalue matrix. It suggests that the matrices that left bicanonicalize A and B should be defined as

$$\tilde{P}_L = G_L P_L, \qquad \tilde{Q}_L = (G_L^{-1})^\dagger Q_L. \qquad\qquad (17.40)$$

Similarly, we can define a reduced bond density matrix that is isometric to the reduced density matrix:

$$\tilde{\rho}_R = D^\dagger C. \qquad\qquad (17.41)$$

It should have the same eigenspectrum as $\tilde{\rho}_L$. By diagonalizing this matrix with a canonical matrix G_R,

$$\tilde{\rho}_R = G_R^{-1}\lambda G_R, \tag{17.42}$$

we obtain the matrices that right bicanonicalize A and B:

$$\tilde{P}_R = G_R P_R, \qquad \tilde{Q}_R = (G_R^\dagger)^{-1}Q_R. \tag{17.43}$$

By utilizing these transformation matrices, the left biorthonormalized tensors become

$$\tilde{A}_L[\sigma] = \tilde{P}_L A[\sigma]\tilde{P}_L^{-1}, \tag{17.44}$$
$$\tilde{B}_L[\sigma] = \tilde{Q}_L B[\sigma]\tilde{Q}_L^{-1}, \tag{17.45}$$

and the right bicanonicalized tensors become

$$\tilde{A}_R[\sigma] = \tilde{P}_R A[\sigma]\tilde{P}_R^{-1}, \tag{17.46}$$
$$\tilde{B}_R[\sigma] = \tilde{Q}_R B[\sigma]\tilde{Q}_R^{-1}. \tag{17.47}$$

Correspondingly, the two middle tensors become

$$\tilde{C} = \tilde{P}_L \tilde{P}_R^{-1} = G_L C G_R^{-1}, \tag{17.48}$$
$$\tilde{D} = \tilde{Q}_L \tilde{Q}_R^{-1} = (G_L^\dagger)^{-1}D G_R^\dagger. \tag{17.49}$$

$\tilde{C}$ and $\tilde{D}^\dagger$ mutually commute and satisfy the equation

$$\tilde{C}\tilde{D}^\dagger = \tilde{D}^\dagger\tilde{C} = \lambda. \tag{17.50}$$

These local tensors remove the gauge ambiguity in the biorthonormalization.

17.3 Fixed-Point Equations

The fixed-point tensors A and B are determined by the eigen equations

$$T|\Psi_R\rangle = \lambda|\Psi_R\rangle, \tag{17.51}$$
$$\langle\Psi_L|T = \lambda\langle\Psi_L|, \tag{17.52}$$

where λ is the largest eigenvalue of T. Eq. (17.51) can also be represented as

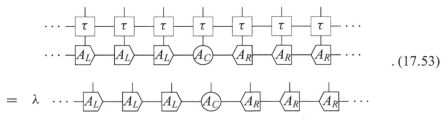

$$. \tag{17.53}$$

Multiplying both sides by $\langle \Psi_L |$ but without B_C, this equation becomes

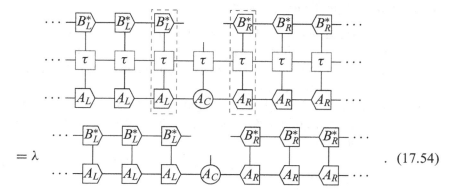

$$= \lambda \qquad\qquad\qquad\qquad . \quad (17.54)$$

In the thermodynamic limit, it is simplified to

$$(17.55)$$

λ_1 is an eigenvalue that is related to λ. This is just the fixed-point equation for determining A_C. In Eq. (17.55), E_R is the largest right eigenvector of the transfer matrix defined by the product of tensors enclosed by the right dashed rectangle in Eq. (17.54):

$$= \lambda_R \qquad\qquad , \quad (17.56)$$

and λ_R is the corresponding eigenvalue. Similarly, E_L is the maximal left eigenvector of the transfer matrix defined by the tensor product enclosed by the left dashed rectangle in Eq. (17.54).

Following these steps, we can also obtain the fixed-point equation for determining B_D:

$$= \lambda_1 \qquad\qquad . \quad (17.57)$$

Similar to a symmetric transfer matrix, the fixed-point tensor equations can be solved iteratively in the following steps:

(i) Biorthonormalize the local tensors so that A_L and B_L satisfy the left canonical condition, and A_R and B_R satisfy the right canonical condition.
(ii) Solve Eq. (17.56) to find the maximal right eigenvector E_R, and similarly the maximal left eigenvector E_L.
(iii) Solve Eqs. (17.55) and (17.57) to find the fixed-point tensors A_C and B_D
(iv) Given A_C and B_D, find updated A_L, B_L, C, and D. We first find updated C, and D using the formula

$$\sum_\sigma B_D^\dagger[\sigma]A_C[\sigma] = D^\dagger C. \tag{17.58}$$

After diagonalizing the left-hand side using a canonical matrix P,

$$\sum_\sigma B_D^\dagger[\sigma]A_C[\sigma] = P\Lambda P^{-1}, \tag{17.59}$$

we can update C and D as

$$C = P|\Lambda|^{1/2}P^{-1}, \tag{17.60}$$

$$D^\dagger = P\left(e^{-i\theta}|\Lambda|^{1/2}\right)P^{-1}, \tag{17.61}$$

where θ is the diagonal phase matrix of Λ:

$$\Lambda = |\Lambda|e^{i\theta}. \tag{17.62}$$

A_L and B_L are then updated by the formula

$$A_L[\sigma] = A_C[\sigma]C^{-1}, \tag{17.63}$$

$$B_L[\sigma] = B_C[\sigma]D^{-1}. \tag{17.64}$$

If A_L and B_L thus obtained are not precisely left canonicalized, one should take a few more steps to further left orthonormalize them using the biorthogonalization method introduced in §15.4. Of course, at each step of biorthogonalization, C and D should be updated.

(v) Repeat steps (ii)–(iv) until A_L, A_R, B_L, B_R, C, and D are all converged.

17.4 Translation Invariant System with a Finite Unit Cell

In this section, we extend the preceding discussion to a system with a unit cell that contains n and m sites along the vertical and horizontal directions, respectively.

The discussion holds for both two-dimensional classical and one-dimensional quantum lattice models. Now the transfer matrix is defined by

$$T = T^n T^{n-1} \cdots T^2 T^1, \tag{17.65}$$

where T^j is a transfer matrix at the jth row

$$T^j = \quad \cdots \boxed{\tau_1^j} \cdots \boxed{\tau_m^j} \boxed{\tau_1^j} \cdots \boxed{\tau_m^j} \cdots \quad . \tag{17.66}$$

One can define other $n - 1$ transfer matrices equivalent to T from all these row transfer matrices. These transfer matrices are defined by a cycle rotation of $T^1, \ldots, T^n$, one row at one time:

$$T^{[j]} = T^{j-1} \cdots T^1 T^n \cdots T^j. \tag{17.67}$$

$T^{[1]}$ is just the transfer matrix defined by Eq. (17.65). The largest right and left eigenvectors of $T^{[j]}$, $\left| \Psi_R^j \right\rangle$ and $\langle \Psi_L^j |$, are determined by the eigen equations

$$T^{[j]} | \Psi_R^j \rangle = \lambda_j | \Psi_R^j \rangle, \tag{17.68}$$

$$\langle \Psi_L^j | T^{[j]} = \lambda_j \langle \Psi_L^j |, \tag{17.69}$$

where λ_j is the maximal eigenvalue of $T^{[j]}$.

Multiplying T^j from the left to both sides of Eq. (17.68), we obtain the equation

$$T^{[j+1]} \left(T^j | \Psi_R^j \rangle \right) = \lambda_j \left(T^j | \Psi_R^j \rangle \right). \tag{17.70}$$

It implies that

$$T^j | \Psi_R^j \rangle = \eta_j | \Psi_R^{j+1} \rangle \tag{17.71}$$

and $\lambda_j = \lambda_{j+1}$ after all eigenvectors are converged. η_j is a normalization constant. Similarly, one can show that

$$\langle \Psi_L^j | T^{j-1} = \eta_j' \langle \Psi_L^{j-1} |. \tag{17.72}$$

η_j' is also a normalization constant.

In the framework of MPS, Ψ_R^j and Ψ_L^j can be represented as

$$\Psi_R^j = \cdots A_1^j[\sigma_1] \cdots A_m^j[\sigma_m] A_1^j[\sigma_{m+1}] \cdots A_m^j[\sigma_{2m}] \cdots$$

$$\begin{array}{ccccccc} & \sigma_1 & & \sigma_m & \sigma_{m+1} & & \sigma_{2m} \\ = & \cdots \!-\!\!\bigcirc\!\!\!A_1^j\!\!-\! & \cdots & -\!\!\bigcirc\!\!\!A_m^j\!\!-\! & -\!\!\bigcirc\!\!\!A_1^j\!\!-\! & \cdots & -\!\!\bigcirc\!\!\!A_m^j\!\!-\! \cdots \end{array} \tag{17.73}$$

$$\Psi^j_L = \mathrm{Tr}\left(\cdots B^{j*}_1[\sigma_1]\cdots B^{j*}_m[\sigma_m]B^{j*}_1[\sigma_{m+1}]\cdots B^{j*}_m[\sigma_{2m}]\cdots\right)$$

$$= \quad \cdots -\!\!\underset{\sigma_1}{\overset{}{\textcircled{B^{j*}_1}}}\!- \cdots -\!\!\underset{\sigma_m}{\overset{}{\textcircled{B^{j*}_m}}}\!\!-\!\!\underset{\sigma_{m+1}}{\overset{}{\textcircled{B^{j*}_1}}}\!- \cdots -\!\!\underset{\sigma_{2m}}{\overset{}{\textcircled{B^{j*}_m}}}\!- \cdots \qquad (17.74)$$

Each pair of these local tensors, A^j_i and B^j_i, can be left biorthonormalized so that they are left canonicalized:

$$\sum_\sigma B^{j\dagger}_{L,i}[\sigma]A^j_{L,i}[\sigma] = I. \qquad (17.75)$$

Similarly, A^j_i and B^j_i can also be right canonicalized:

$$\sum_\sigma A^j_{R,i}[\sigma]B^{j\dagger}_{R,i}[\sigma] = I. \qquad (17.76)$$

Using these left and right canonical tensors, Ψ^j_R can be generally represented as

$$\Psi^j_R = \mathrm{Tr}\left(\cdots A^j_{L,1}[\sigma_1]\cdots A^j_{L,i}[\sigma_i]C^j_i A^j_{R,i+1}[\sigma_{i+1}]\cdots A^j_{R,m}[\sigma_m]\cdots\right)$$

$$= \quad \cdots -\!\!\triangleright\!\!- \cdots -\!\!\triangleright\!\!-\!\!\bigcirc\!\!-\!\!\triangleleft\!\!- \cdots -\!\!\triangleleft\!\!- \cdots \quad , \qquad (17.77)$$

where C^j_i is a bond matrix. All the local tensors on the left- and right-hand sides of C are left and right orthonormalized, respectively. By absorbing C^j_i into its left or right local tensor, and defining

$$A^j_{C,i}[\sigma_i] = A^j_{L,i}[\sigma_i]C^j_i = C^j_{i-1}A^j_{R,i}[\sigma_i], \qquad (17.78)$$

we can also represent Ψ^j_R as

$$\Psi^j_R = \quad \cdots -\!\!\triangleright\!\!- \cdots -\!\!\triangleright\!\!-\!\!\bigcirc\!\!-\!\!\triangleleft\!\!- \cdots -\!\!\triangleleft\!\!- \cdots \quad . \qquad (17.79)$$

The corresponding left eigenvector, Ψ^j_L, is

$$\Psi_L^j = \quad \cdots \quad \overset{B_{L,1}^{j*}}{\underset{\sigma_1}{-\square-}} \quad \cdots \quad \overset{B_{L,i}^{j*}}{\underset{\sigma_i}{-\square-}} \overset{D_i^{j*}}{-\bigcirc-} \overset{B_{R,i+1}^{j*}}{\underset{\sigma_{i+1}}{-\square-}} \quad \cdots \quad \overset{B_{R,m}^{j*}}{\underset{\sigma_m}{-\square-}} \quad \cdots \tag{17.80}$$

or

$$\Psi_L^j = \quad \cdots \quad \overset{B_{L,1}^{j*}}{\underset{\sigma_1}{-\square-}} \quad \cdots \quad \overset{B_{L,i-1}^{j*}}{\underset{\sigma_{i-1}}{-\square-}} \overset{B_{D,i}^{j*}}{\underset{\sigma_i}{-\bigcirc-}} \overset{B_{R,i+1}^{j*}}{\underset{\sigma_{i+1}}{-\square-}} \quad \cdots \quad \overset{B_{R,m}^{j*}}{\underset{\sigma_m}{-\square-}} \quad \cdots \tag{17.81}$$

where D_i^j is a bond matrix, and

$$B_{D,i}^j[\sigma_i] = B_{L,i}^j[\sigma_i]D_i^j = D_{i-1}^j B_{R,i}^j[\sigma_i], \tag{17.82}$$

Fixed-Point Equations

To find the fixed-point equations, we first take a derivative of Ψ_L^{j+1} with respect to $B_{R,i+1}^{j+1,*}$, and then multiply the resulting MPS to Eq. (17.71). This leads to

$$\partial_{B_{R,i+1}^{j+1,*}} \left\langle \Psi_L^{j+1} | T^j | \Psi_R^j \right\rangle = \eta_j \partial_{B_{R,i+1}^{j+1,*}} \langle \Psi_L^{j+1} | \Psi_R^{j+1} \rangle. \tag{17.83}$$

Graphically, this equation can be represented as

After simplification, we obtain the fixed-point equation of $A^j_{C,i}$

$$E^j_{L,i} \; \boxed{\tau^j_i} \; E^j_{R,i} \;\; = \eta'_i \;\; A^j_{C,i} \;,$$

$$\text{(17.85)}$$

η'_i is a parameter related to η_i. Similarly, $B^j_{D,i}$ is determined by the equation

$$B^{j*}_{D,i} \; E^j_{L,i} \; \boxed{\tau^j_i} \; E^j_{R,i} \;\; = \eta'_i \;\; B^{j*}_{D,i} \;.$$

$$\text{(17.86)}$$

In the above equations, $E^j_{L,i}$ is the maximal left eigenvector of the left transfer matrix:

$$L^j_i = \;\; B^{j+1,*}_{L,i} \;\boxed{\tau^j_i}\; A^j_{L,i} \; \cdots \; B^{j+1,*}_{L,m} \,B^{j+1,*}_{L,1} \;\boxed{\tau^j_m}\,\boxed{\tau^j_1}\; A^j_{L,m}\,A^j_{L,1} \; \cdots \; B^{j+1,*}_{L,i-1} \;\boxed{\tau^j_{i-1}}\; A^j_{L,i-1} \;,$$

$$\text{(17.87)}$$

namely,

$$E^j_{L,i} \cdot L^j_i = E^j_{L,i} \xi^j_{L,i}$$

$$\text{(17.88)}$$

with $\xi^j_{L,i}$ the corresponding maximal eigenvalue. Since

$$E^j_{L,i} \;\boxed{\tau^j_i}\; \left[\begin{array}{c} B^{j+1,*}_{L,i} \\ A^j_{L,i} \end{array}\right] \cdot L^j_{i+1} = E^j_{L,i} \;\boxed{\tau^j_i}\; \left[\begin{array}{c} B^{j+1,*}_{L,i} \\ A^j_{L,i} \end{array}\right] \xi^j_{L,i},$$

$$\text{(17.89)}$$

we have

$$E^j_{L,i} \quad \boxed{\tau^j_i} \quad = \quad E^j_{L,i+1} \qquad , \qquad (17.90)$$

and $\xi^j_{L,i}$ should be i-independent (i.e. $\xi^j_{L,i} = \xi^j_L$). Eq. (17.90) indicates that $E^j_{L,i}$ can be iteratively obtained by applying one column transfer matrix one time.

Similarly, $E^j_{R,i}$ is the maximal right eigenvector of the right transfer matrix

$$R^j_i = \cdots \boxed{\tau^j_{i+1}} \cdots \boxed{\tau^j_m} \boxed{\tau^j_1} \cdots \boxed{\tau^j_i} , \qquad (17.91)$$

and

$$R^j_i E^j_{R,i} = \xi^j_R E^j_{R,i}. \qquad (17.92)$$

Here the eigenvalue ξ^j_R is also i independent. From the fact that

$$L^j_{i-1} \cdot \quad \boxed{\tau^j_i} \quad E^j_{R,i} \quad = \quad \xi^j_R \quad \boxed{\tau^j_i} \quad E^j_{R,i} \qquad (17.93)$$

we obtain the recursive equation for determining $E_{R,i}^j$:

$$\begin{array}{c} B_{R,i}^{j+1,*} \\ \hline \\ \tau_i^j \quad \bigcirc E_{R,i}^j \\ \hline \\ A_{R,i}^j \end{array} = \bigcirc E_{R,i-1}^j \quad . \tag{17.94}$$

18

Dynamical Correlation Functions

18.1 Spectral Functions

The dynamical response of a quantum system to a time-dependent perturbation is determined by the dynamical correlation function or the Green's function at zero temperature:

$$G(t) = -i\langle\psi_0|A^\dagger(t)A(0)|\psi_0\rangle,\tag{18.1}$$

where A is the physical operator whose physical property is analyzed and $|\psi_0\rangle$ is the ground state. In the frequency space, Green's function reads

$$\begin{aligned}G(\omega + i\eta) &= \left\langle\psi_0|A^\dagger\frac{1}{\omega - H + E_0 + i\eta}A|\psi_0\right\rangle\\ &= \left\langle A|\frac{1}{\omega - H + E_0 + i\eta}|A\right\rangle\end{aligned}\tag{18.2}$$

with $|A\rangle = A|\psi_0\rangle$. E_0 is the ground-state energy, and η is a small positive number introduced to smooth out the singularity of the Green's function. It broadens the discrete spectrum of finite systems to mimic a system in the thermodynamic limit.

The spectral function is determined by the imaginary part of Green's function:

$$I(\omega + i\eta) = -\frac{1}{\pi}\mathrm{Im}G(\omega + i\eta).\tag{18.3}$$

The physical spectral function is determined by $I(\omega + i\eta)$ in the limit $\eta \to 0$:

$$I(\omega) = \lim_{\eta\to 0} I(\omega + i\eta) = \langle A|\delta\left(\omega - H + E_0\right)|A\rangle.\tag{18.4}$$

Precise calculation of this quantity is necessary because it can be directly compared with experimental measurements.

In the spectral or Lehmann representation, η introduces a Lorentzian broadening to the spectral function:

$$I(\omega + i\eta) = \frac{1}{\pi} \int d\omega' I(\omega') \frac{\eta}{(\omega - \omega')^2 + \eta^2}. \qquad (18.5)$$

The DMRG is an excellent method for calculating ground states and selected excited states. It has also been used to evaluate dynamical correlation functions. Green's function involves the calculation of the inverse of $\omega - H + E_0$, which can be done, for example, using the continued-fraction method first introduced by Gagliano and Baliseiro in the Lanczos exact diagonalization [80]. Hallberg first exploited this method in the DMRG calculation of spectral function in 1995 [81]. It runs very efficiently but yields reliable results only at low frequencies. A significant improvement to this method was achieved by taking the MPS representation of the basis vectors in two different schemes. One is to generate the basis vectors still using the Lanczos method [90]. The other is to use the Chebyshev expansion to create the basis vectors [88]. Both work more accurately and efficiently than the Hallberg method. However, the basis vectors generated in either scheme are not orthogonal. The accuracy can be further improved by reorthogonalizing these vectors [89].

An alternative approach for generating dynamical spectra is the correction vector method introduced by Soos and Ramasesha in the framework of exact diagonalization [82]. This method was first applied to the DMRG calculation of dynamical correlation functions by Ramasesha *et al.* [83] in 1997 and refined by Kühner and White [84] in 1999. The idea of correction vectors is to take Green's function at a particular frequency as a target state and calculate it by solving a set of large but sparse linear equations. This method can generate very accurate results at a given frequency, but the computational cost is very high. Later on, two schemes were proposed to improve this method. One is the dynamical DMRG, introduced by Jeckelmann in 2002 [85]. Instead of solving sparse linear equations, this method determines the correction vector by variationally minimizing a cost function defined by Eq. (18.72). The other is the Krylov-space decomposition, introduced by Nocera and Alvarez in 2016 [87]. It is to generate a set of basis states using the Lanczos method and find the correction vector by directly diagonalizing the Hamiltonian within that basis space.

18.2 Continued-Fraction Expansion

From the ground state wave function $|\psi_0\rangle$ obtained by DMRG or other methods, one can further use the Lanczos tridiagonalization to calculate the dynamical correlation function at zero temperature [401]. Gagliano and Balseiro [80] first exploited this approach in the framework of exact diagonalization. It is one of the most appealing features of the Lanczos method and was first adopted by Hallberg to calculate dynamical correlation functions with DMRG [81].

This method starts with the initial basis vector:

$$|e_0\rangle = \frac{|A\rangle}{\langle A|A\rangle^{1/2}}. \tag{18.6}$$

It then follows the steps introduced in §A.2 to generate a set of orthogonal basis states $|E\rangle = (|e_0\rangle, |e_1\rangle, \cdots)$ by applying H iteratively to these states. This leads to a tridiagonal representation of the Hamiltonian:

$$H = \begin{pmatrix} \alpha_0 & \beta_1 & 0 & 0 & \cdots \\ \beta_1 & \alpha_1 & \beta_2 & 0 & \cdots \\ 0 & \beta_2 & \alpha_2 & \beta_3 & \cdots \\ 0 & 0 & \beta_3 & \alpha_3 & \cdots \\ \vdots & \vdots & \vdots & \vdots & \ddots \end{pmatrix}, \tag{18.7}$$

where all the matrix elements are real and $\beta_i \geq 0$.

Using this set of basis states, $I(\omega)$ can be expressed as

$$I(\omega) = -\frac{1}{\pi} \langle A|A\rangle \,\mathrm{Im}\langle e_0| \frac{1}{z - H} |e_0\rangle, \tag{18.8}$$

where $z = \omega + E_0 + i\eta$ and

$$z - H = \begin{pmatrix} z - \alpha_0 & -\beta_1 & 0 & 0 & \cdots \\ -\beta_1 & z - \alpha_1 & -\beta_2 & 0 & \cdots \\ 0 & -\beta_2 & z - \alpha_2 & -\beta_3 & \cdots \\ 0 & 0 & -\beta_3 & z - \alpha_3 & \cdots \\ \vdots & \vdots & \vdots & \vdots & \ddots \end{pmatrix}. \tag{18.9}$$

According to Cramer's rule, the inverse of $z - H$ is proportional to the adjoint of $z - H$. By applying this rule to $(z - H)_{11}^{-1}$, we obtain

$$(z - H)_{11}^{-1} = \langle e_0|(z - H)^{-1}|e_0\rangle = \frac{\det B_2}{\det B_1}, \tag{18.10}$$

where B_n is obtained from Eq. (18.9) by removing the first $(n - 1)$ rows and columns:

$$B_n = \begin{pmatrix} z - \alpha_{n-1} & -\beta_n & 0 & 0 & \cdots \\ -\beta_n & z - \alpha_n & -\beta_{n+1} & 0 & \cdots \\ 0 & -\beta_{n+1} & z - \alpha_{n+1} & -\beta_{n+2} & \cdots \\ 0 & 0 & -\beta_{n+2} & z - \alpha_{n+2} & \cdots \\ \vdots & \vdots & \vdots & \vdots & \ddots \end{pmatrix}. \tag{18.11}$$

From the algorithm of determinant, it is straightforward to show that B_n satisfies the following recursion relation:

$$\det B_n = (z - \alpha_{n-1})\det B_{n+1} - \beta_n^2 \det B_{n+2}. \tag{18.12}$$

By dividing both sides with $\det B_{n+1}$, Eq. (18.12) becomes

$$\frac{\det B_{n+1}}{\det B_n} = \frac{1}{z - \alpha_n - \beta_{n+1}^2 \dfrac{\det B_{n+2}}{\det B_{n+1}}}. \tag{18.13}$$

Substituting Eq. (18.13) into Eq. (18.10) allows us to express $I(\omega)$ as a continuous fraction:

$$I(\omega) = -\frac{1}{\pi}\text{Im}\cfrac{\langle A|A \rangle}{z - \alpha_0 - \cfrac{\beta_1^2}{z - \alpha_1 - \cfrac{\beta_2^2}{z - \alpha_2 - \cdots}}}. \tag{18.14}$$

The eigenvalues of the Hamiltonian determine the positions of the poles.

In real calculation, the above continuous fraction has to be terminated. If this termination is done at the Mth order, the ratio $\det B_{M+1}/\det B_M$ is undetermined. A simple approach is to set $\beta_M = 0$, in which

$$\frac{\det B_{M+1}}{\det B_M} = \frac{1}{z - \alpha_{M-1}}. \tag{18.15}$$

It is a good approximation if $I(\omega)$ varies slowly with ω. However, the approximation becomes poor if $I(\omega)$ or its derivative diverges or shows other singularities.

Another kind of approximation is to assume that the value of $\det B_{n+1}/\det B_n$ does not change when $n \geq M$ for sufficiently large M. This approximation can significantly change the singular behaviors of $I(\omega)$. Other more sophisticated methods to terminate the iteration can be found from Refs. [402, 403].

18.3 Dynamical Moments

From the tridiagonal Hamiltonian (18.7), one can also determine the spectral function by moment expansion. From the definition (18.6), the spectral function can be also written as

$$I(\omega) = \langle A|A \rangle \langle e_0|\delta(\omega - H + E_0)|e_0 \rangle. \tag{18.16}$$

In the space spanned by $|E\rangle = (|e_0\rangle, |e_1\rangle, \cdots)$, H is defined by Eq. (18.7) and its eigen equation is given by

$$H|n\rangle = \lambda_n|n\rangle. \tag{18.17}$$

By solving this equation, we obtain the energy eigenvalues E_n ($n = 0, 1, 2, \cdots$) and the corresponding eigenvectors $|n\rangle$:

$$|n\rangle = \sum_i U_{ni}|e_i\rangle, \tag{18.18}$$

where U_{ni} is the wave function of $|n\rangle$ in the basis representation of $|E\rangle = (|e_0\rangle, |e_1\rangle, \cdots)$.

After inserting the completeness condition

$$\sum_n |n\rangle\langle n| = 1 \tag{18.19}$$

into the right-hand side of Eq. (18.16), the spectral function becomes

$$I(\omega) = \langle A|A\rangle \sum_n \delta(\omega - \lambda_n + E_0)|U_{n,0}|^2. \tag{18.20}$$

The mth order of moment of the spectral function is defined by

$$I_m(\omega) = \int_{-\infty}^{\omega} d\varepsilon\, \varepsilon^m I(\varepsilon). \tag{18.21}$$

Therefore, the spectral function is the derivative of the zeroth order of moment $I_0(\omega)$. Substituting Eq. (18.20) into the expression yields

$$I_m(\omega) = \langle A|A\rangle \sum_n (\lambda_n - E_0)^m |U_{n,0}|^2 \theta(\omega - \lambda_n + E_0), \tag{18.22}$$

where $\theta(x)$ is the step function that equals 1 if $x > 0$ and 0 otherwise.

If only the zeroth order of moment is considered, from Eq. (18.22) we know that the total spectral weight of $I(\omega)$ between $\omega_1 = (\lambda_{n-1} + \lambda_n)/2$ and $\omega_2 = (\lambda_n + \lambda_{n+1})/2$ is approximately given by $I_0(\omega_2) - I_0(\omega_1)$. If $\lambda_{n+1} - \lambda_{n-1}$ is small, the spectral function at $\omega = \lambda_n$ is approximately given by

$$I(\lambda_n) \approx \frac{2\,[I_0(\omega_2) - I_0(\omega_1)]}{\lambda_{n+1} - \lambda_{n-1}} = \frac{2\langle A|A\rangle |U_{n,0}|^2}{\lambda_{n+1} - \lambda_{n-1}}. \tag{18.23}$$

If higher order of moment is included, the spectral function can be more accurately determined.

The precision of dynamical correlation functions can be checked by comparison with the equal-time correlation functions independently obtained using the sum rules:

$$I_0\,(+\infty) = \langle \psi_0|A^\dagger A|\psi_0\rangle, \tag{18.24}$$

$$I_1\,(+\infty) = \langle \psi_0|A^\dagger\,[H, A]\,|\psi_0\rangle, \tag{18.25}$$

$$I_2\,(+\infty) = \langle \psi_0|\,[A^\dagger, H]\,[H, A]\,|\psi_0\rangle. \tag{18.26}$$

The right-hand sides of the above equations are all equal-time correlation functions. They can be more accurately calculated using DMRG.

18.4 Lanczos-DMRG Method

To calculate Green's functions, one needs to evaluate the inversion of the matrix $X = \omega - H + E_0 + i\eta$. This matrix is generally a large sparse matrix. Its inversion in the Lanczos-DMRG method [81] is performed in two steps:

(i) A Lanczos iteration is taken to tridiagonalize the Hamiltonian H:

$$H = \begin{pmatrix} \alpha_0 & \beta_1 & 0 & \cdots & & 0 \\ \beta_1 & \alpha_1 & \beta_2 & \cdots & & 0 \\ 0 & \beta_2 & \alpha_2 & \cdots & & 0 \\ \vdots & \vdots & \vdots & \ddots & & \beta_{M-1} \\ 0 & 0 & 0 & \beta_{M-1} & & \alpha_{M-1} \end{pmatrix}, \qquad (18.27)$$

where α_i and β_i are the Lanczos coefficients. M is the total number of Lanczos vectors generated. X is also tridiagonal in this representation.

(ii) The tridiagonal X is inverted to calculate Green's function based on the continued-fraction expansion introduced in §18.2.

The Lanczos-DMRG is a simple method for calculating dynamical correlation functions. It relies on the accurate calculation of the Lanczos vectors. However, the Hamiltonian H itself is available only approximately due to the basis truncation in the DMRG calculation. It describes low energy states, especially the ground state, better than high energy excited states. Only the first few Lanczos vectors can be accurately generated using this approximate Hamiltonian. DMRG, if applied naively, will introduce increasingly severe and systematic errors for the higher Lanczos vectors.

An improvement to this method is to target several Lanczos vectors in addition to the ground state so that the reduced Hilbert space could more accurately describe the relevant excited states. Unfortunately, this changes the approximations made in the representation of the Hamiltonian, and the DMRG truncation errors increase rapidly with the number of target Lanczos vectors. Therefore, only the first few Lanczos vectors are targeted in most applications of the Lanczos DMRG. In that case, the Hamiltonian does not necessarily converge to an optimal representation for all excited states. Consequently, this method is suitable only for spectra consisting of a few discrete peaks [84], and the calculated spectrum does not resolve correlators in the whole frequency range very well. This difficulty is encountered in the Lanczos-DMRG calculation: if only a few Lanczos vectors are targeted, these vectors are represented accurately, but all others poorly; on the other hand, if all Lanczos vectors are targeted, then all of them are represented at similar but low accuracy.

The Lanczos-DMRG generally gives the best convergence to the extremal eigenvectors that are contained in $|A\rangle$. Other eigenvectors will only converge after a large number of Lanczos iterations. However, numerical errors may destroy the orthogonality of these Lanczos vectors, leading to the appearance of spurious eigenvalues. Loss of orthogonality occurs in almost all Lanczos calculations. It can be mended by reorthonormalizing these Lanczos vectors.

Dargel *et al.* [86] proposed an adaptive Lanczos vector method to calculate high energy states efficiently. At each step of Lanczos iterations, they proposed using a multitargeting approach but restricting the targeted Lanczos vectors only to the last three. Their approach involves continuously changing the truncated basis states as the iteration evolves. They found that this approach is more efficient than the original one and allows the position of the singularity in Green's function to be more accurately calculated. However, this approach is more vulnerable to the loss of orthonormality of Lanczos vectors. Furthermore, the adaptive changes of basis states make the reorthogonalization of the Lanczos vectors difficult.

18.5 Dynamical Calculations with MPS

18.5.1 Lanczos MPS Method

The MPS formulation can be also combined with the Lanczos method to calculate dynamical spectra [90]. In this scheme, the ground state $|\psi_0\rangle$ and all the Lanczos vectors $(|e_0\rangle, |e_1\rangle, \dots, |e_{M-1}\rangle)$ are represented by MPS. These vectors can be generally represented as

$$|e_n\rangle = \mathrm{Tr} A_1^{(n)}[\sigma_1] A_2^{(n)}[\sigma_2] \cdots A_N^{(n)}[\sigma_N] |\sigma_1 \sigma_2 \cdots \sigma_N\rangle, \qquad (18.28)$$

where N is the lattice size, $A_i^{(n)}$ and $|\sigma_i\rangle$ $(i = 1, \dots, N)$ are the local matrices and local basis states, respectively.

The ground-state wave function $|\psi_0\rangle$ can be obtained using either DMRG or an MPS-based algorithm. The first Lanczos vector is defined by $|A\rangle$. However, as $|A\rangle$ is generally a sum of several MPS, it cannot be determined simply using Eq. (18.6). Instead, it is determined by minimizing a cost function defined by

$$f(\phi_0) = \|\,|A\rangle - |\phi_0\rangle\,\|^2, \qquad (18.29)$$

where $|\phi_0\rangle$ is the MPS to be determined. The first Lanczos vector is the normalized $|\phi_0\rangle$ that minimizes the above cost function:

$$|e_0\rangle = \frac{|\phi_0\rangle}{\sqrt{\langle\phi_0|\phi_0\rangle}}. \tag{18.30}$$

For other Lanczos vectors, we cannot calculate them simply using the formula introduced in §A.2. Nevertheless, we can still find them recursively. Suppose we have already found all the Lanczos vectors from $|e_0\rangle$ to $|e_{n-1}\rangle$; we determine the $(n+1)$th Lanczos vector $|e_n\rangle$ by minimizing the cost function defined by

$$f(\phi_n) = \|H|e_{n-1}\rangle - \beta_{n-1}|e_{n-2}\rangle - \alpha_{n-1}|e_n\rangle - |\phi_n\rangle\|^2, \tag{18.31}$$

where $\beta_0 = 0$ and

$$\alpha_n = \langle e_n|H|e_n\rangle. \tag{18.32}$$

β_n is just the norm of $|\phi_n\rangle$:

$$\beta_n = \sqrt{\langle\phi_n|\phi_n\rangle}, \tag{18.33}$$

and $|e_n\rangle$ is the normalized $|\phi_n\rangle$:

$$|e_n\rangle = \frac{1}{\beta_n}|\phi_n\rangle. \tag{18.34}$$

Repeating this procedure, we generate all Lanczos vectors.

If the MPS-represented Lanczos vectors $(|e_0\rangle, |e_1\rangle, \cdots)$ are orthogonal to each other, α_n is simply the diagonal matrix element of the Hamiltonian, defined in Eq. (18.27). However, as the MPS vectors generated here may not be orthogonal to each other, H is not tridiagonal in this set of basis states and α_n is not truly the diagonal matrix element of H.

In our minimization, either the application of H to a state or the addition of two MPS generates a new MPS whose bond dimension is larger than the original one. Therefore, minimizing the cost function is equivalent to finding an optimal but low-rank approximation for this newly generated MPS. This approximation is taken at every step of the Lanczos iteration.

The Lanczos-MPS is more accurate and more efficient than the Lanczos-DMRG. It has the following two advantages:

(i) The Hamiltonian is exactly treated in the MPS representation. However, it is approximately represented in DMRG due to the truncation of the basis states. Clearly, this MPS representation eliminates the error in the calculation of Lanczos vectors introduced by the truncation of the Hamiltonian.

(ii) It eliminates the multiple-targeting problem, unlike in the Lanczos-DMRG calculation. This is because each state carries its own choice of basis states in the MPS formulation and is approximated more precisely by its own reduced density matrix.

However, the vectors $(|e_0\rangle, |e_1\rangle, \cdots)$ generated by the Lanczos-MPS are not orthogonal to each other. Therefore, the Hamiltonian is not tridiagonal in this basis space, and, rigorously speaking, $(|e_0\rangle, |e_1\rangle, \cdots)$ are not truly the Lanczos vectors. More importantly, the Lanczos coefficients α_n and β_n are only approximately calculated. However, in the Lanczos-DMRG, these coefficients are exactly calculated, at least in the truncated Hilbert space determined by DMRG. The errors in these coefficients will pass to all the Lanczos vectors that are determined afterward, eventually destabilizing the algorithm.

The problem of numerical loss of orthogonality between Lanczos states can be removed by an ex-post reorthogonalization using the Gram–Schmidt process. The orthonormalized basis states are represented by the linear superpositions of all the MPS Lanczos vectors. After that, one can tridiagonalize the Hamiltonian using these orthonormalized basis states and calculate the spectral function with the continued-fraction method.

Alternatively, one can fully diagonalize the Hamiltonian in the space spanned by these Lanczos vectors and evaluate the spectra just using the moment expansion method introduced in §18.3. The additional cost for this full diagonalization is minor because it is done just in the truncated Hilbert space. Nevertheless, it allows direct access to the spectral poles and weights without broadening. It also leads to a simple scheme to do finite-size extrapolations to the thermodynamic limit by directly defining the spectral function as the spectral weight per unit frequency interval. A detailed discussion on this is given in §18.5.3.

18.5.2 Chebyshev MPS Method

Like the Lanczos expansion, the Chebyshev expansion can also generate a sequence of basis states from which spectral functions can be evaluated in the framework of MPS [88]. It offers a numerically efficient representation with *uniform resolution* over the entire spectral width.

Chebyshev polynomials of the first kind, $T_n(x)$, henceforth simply called Chebyshev polynomials, form an orthonormal set of polynomials on the interval $x \in [-1, 1]$, in which any piecewise smooth and continuous function $f(x)$ can be expanded with a weight function $\left(\pi\sqrt{1-x^2}\right)^{-1}$. They are defined by the recurrence relation

$$T_{n+1}(x) = 2xT_n(x) - T_{n-1}(x), \tag{18.35}$$

$$T_0(x) = 1, \quad T_1(x) = x, \tag{18.36}$$

and satisfy the relation

$$T_{n+n'}(x) = 2T_n(x)T_{n'}(x) - T_{n-n'}(x). \tag{18.37}$$

Two explicit representations are

$$T_n(x) = \cos[n \cos^{-1}(x)] = \cosh[n \cosh^{-1}(x)]. \tag{18.38}$$

The Chebyshev expansion that is used for the present purpose has the form

$$f(x) = \frac{1}{\pi\sqrt{1-x^2}}\left[\mu_0 + 2\sum_{n=1}^{\infty}\mu_n T_n(x)\right], \tag{18.39}$$

where

$$\mu_n = \int_{-1}^{1} dx f(x) T_n(x) \tag{18.40}$$

are defined as the Chebyshev moments.

This series is truncated in practical calculations, leading to an approximate representation of order M if only the first M terms are retained. This truncation, however, introduces an artificial oscillation of period $1/M$ called Gibbs oscillations. The oscillation can be smoothed by employing a broadening kernel, which rearranges the series before truncation. It leads to a modified Chebyshev expansion given by

$$f(x) \approx \frac{1}{\pi\sqrt{1-x^2}}\left[g_0\mu_0 + 2\sum_{n=1}^{\infty}g_n\mu_n T_n(x)\right], \tag{18.41}$$

where g_n is a damping factor. In most cases, it is sufficient to take the Jackson damping:

$$g_n = \frac{M-n+1}{M+1}\cos\frac{\pi n}{M+1} + \frac{1}{M+1}\sin\frac{\pi n}{M+1}\cot\frac{\pi}{M+1}. \tag{18.42}$$

To calculate the spectral function defined by Eq. (18.4), we first rescale and shift the frequency and the Hamiltonian such that the nonzero parts of the spectral function concentrate on the frequency interval $(-1, 1)$. The spectral function then becomes

$$I(\omega') = \frac{1}{a}\langle A|\delta(\omega' - H')|A\rangle, \tag{18.43}$$

where $a = W/2$,

$$\omega' = \frac{\omega}{a} - 1, \quad H' = \frac{H - E_0}{a} - 1, \tag{18.44}$$

and W is the spectral width.

To expand the δ-function using Eq. (18.41), the spectral function becomes

$$I(\omega') = \frac{1}{a\pi\sqrt{1-\omega'^2}}\left[g_0\mu_0 + 2\sum_{n=1}^{M-1} g_n\mu_n T_n\left(\omega'\right)\right], \qquad (18.45)$$

where

$$\mu_n = \int_{-1}^{1} dx\langle A|\delta(x-H')T_n\left(x\right)|A\rangle = \langle A|T_n\left(H'\right)|A\rangle. \qquad (18.46)$$

The Chebyshev moments μ_n are determined by the inner product between $|A\rangle$ and the Chebyshev vectors:

$$|e_n\rangle = T_n(H')|A\rangle. \qquad (18.47)$$

Using Eq. (18.35), we find that

$$|e_0\rangle = |A\rangle, \qquad |e_1\rangle = H'|e_0\rangle, \qquad (18.48)$$

and more generally

$$|e_n\rangle = 2H'|e_{n-1}\rangle - |e_{n-2}\rangle. \qquad (18.49)$$

These Chebyshev vectors $|e_n\rangle$ span a Krylov basis space.

In the MPS representation, Eqs. (18.47)–(18.49) are valid because all the vectors, including the ground state $|\psi_0\rangle$ and all the Chebyshev vectors $|e_n\rangle$, are standard MPS. The first two MPS-represented Chebyshev vectors are determined by variationally minimizing the functions:

$$f_0\left(e_0\right) = \||e_0\rangle - |A\rangle\|^2, \qquad (18.50)$$
$$f_1\left(e_1\right) = \||e_1\rangle - H'|e_0\rangle\|^2. \qquad (18.51)$$

Similarly, other Chebyshev vectors are determined by minimizing the cost function:

$$f_n\left(e_n\right) = \||e_n\rangle - 2H'|e_{n-1}\rangle + |e_{n-2}\rangle\|^2. \qquad (18.52)$$

The Chebyshev moments are determined by

$$\mu_n = \langle A|e_n\rangle. \qquad (18.53)$$

For a given M, the spectrum obtained with the above formula has a resolution of order $O(W/M)$ in the original energy scale. When the spectrum varies drastically, or the spectral width is broadened, we have to increase N to raise the resolution.

In our calculation, two approximations are used. First, the infinite Chebyshev series is truncated by a finite polynomial. Second, more severely, the Chebyshev vectors are represented by MPS. This approximation breaks the orthonormal condition of the Chebyshev series, and the error is accumulated in the recursive calculation of the Chebyshev vectors. This nonorthonormalization problem is similar to that encountered in the Lanczos-MPS calculation. It can be remedied using the reorthogonalization scheme introduced in the next section [89].

18.5.3 Reorthogonalization

In the framework of MPS, the M basis vectors generated by the Lanczos iteration or the Chebyshev expansion, $(|e_0\rangle, \ldots, |e_{M-1}\rangle)$, are not orthogonal to each other. It leads to errors in the calculation using the standard Lanczos continued fraction or Chebyshev expansion formula. This problem can be resolved by reorthonormalizing these vectors [89] using the Gram–Schmidt orthogonalization process:

$$|\Psi_\alpha\rangle = c_\alpha \left(1 - \sum_{\beta < \alpha} |\Psi_\beta\rangle\langle\Psi_\beta| \right) |e_\alpha\rangle, \tag{18.54}$$

such that $|\Psi_\alpha\rangle$ are the orthonormalized basis states:

$$\langle\Psi_\alpha|\Psi_\beta\rangle = \delta_{\alpha\beta}, \tag{18.55}$$

and c_α are the normalization constants, starting from $|\Psi_0\rangle = |e_0\rangle$. The cost for this reorthonormalization is small because only the inner products between any pair of MPS need to be evaluated.

In the orthonormalized basis space, the matrix elements of the Hamiltonian are given by

$$H_{\alpha\beta} = \langle\Psi_\alpha|H|\Psi_\beta\rangle. \tag{18.56}$$

We can fully diagonalize this matrix to obtain its eigenspectrum:

$$H|n\rangle = \lambda_n|n\rangle \qquad (n = 1 \cdots M), \tag{18.57}$$

where λ_n and $|n\rangle$ are the eigenpairs of H, which are arranged in an ascending order; that is, $\lambda_n < \lambda_m$ if $n < m$. In this truncated Hilbert space, the spectral function is given by

$$I(\omega) = \sum_{n=1}^{M} W_n \delta(\omega - \lambda_n + E_0), \tag{18.58}$$

and

$$W_n = |\langle n|A\rangle|^2 \tag{18.59}$$

is the spectral weight corresponding to λ_n.

To reduce the finite-size effect, one can broaden the δ-function on the right-hand side of the equation by a Lorentzian function. However, a more convenient way to carry out the finite-size scaling analysis is to define the spectral weight from the derivative of the zeroth moment of the spectral function as defined in §18.3.

Let us denote a_n as the midpoint between λ_{n-1} and λ_n:

$$a_n = \frac{\lambda_{n-1} + \lambda_n}{2}, \qquad (1 < n \leq M). \qquad (18.60)$$

The average spectral weight within the interval between a_n and a_{n+m} ($m > 0$) is given by

$$I(\bar{\omega}_{n,m}) = \frac{1}{\Delta_{n,m}} \sum_{0 \leq j < m} W_{n+j}, \qquad (18.61)$$

where

$$\Delta_{n,m} = a_{n+m} - a_n \qquad (18.62)$$

is the energy difference between a_{n+m} and a_n, and

$$\bar{\omega}_{n,m} = \frac{a_n + a_{n+m}}{2} \qquad (18.63)$$

is the average energy of these two points. Here m should be taken such that the interval between a_n and a_{n+m} contains just one prominent peak.

In this calculation, some peaks are identified as prominent peaks, some not. To distinguish a peak, we should calculate the residual r_n of an eigenstate $|n\rangle$, defined by

$$r_n = \langle r_n | r_n \rangle = \langle n | (H - \lambda_n)^2 | n \rangle, \qquad (18.64)$$

where H is the full Hamiltonian, not the truncated one. A peak is regarded as a ghost rather than a prominent peak if its spectral weight W_n is smaller than an empirical cutoff W_{cut} and the residual r_n is larger than a threshold that is empirically determined.

18.6 Correction-Vector Method

Within the DMRG framework, the correction-vector method is an alternative approach for calculating spectral functions directly in frequency space. This method can achieve high precision but at a higher numerical cost than the Lanczos vector method.

The correction vector associated with the Green's function is defined by

$$|\psi\,(\omega + i\eta)\rangle = \frac{1}{\omega - H + E_0 + i\eta} \, |A\rangle, \qquad (18.65)$$

where $|A\rangle$ is assumed to be real. The correction vector determines Green's function, which is an inner product between $|\psi\rangle$ and $|A\rangle$:

$$G(\omega + i\eta) = \langle A|\psi \, (\omega + i\eta)\rangle. \tag{18.66}$$

$|\psi \, (\omega + i\eta)\rangle$ is a complex vector. The spectral function is determined by its imaginary part:

$$I(\omega + i\eta) = -\frac{1}{\pi}\langle A|\psi_{\mathrm{Im}} \, (\omega + i\eta)\rangle, \tag{18.67}$$

where

$$|\psi \, (\omega + i\eta)\rangle = |\psi_{\mathrm{Re}} \, (\omega + i\eta)\rangle + i|\psi_{\mathrm{Im}} \, (\omega + i\eta)\rangle, \tag{18.68}$$

and $|\psi_{\mathrm{Re}} \, (\omega + i\eta)\rangle$ and $|\psi_{\mathrm{Im}} \, (\omega + i\eta)\rangle$ are the real and imaginary parts of the correction vector, respectively.

In the DMRG calculation, four states (or vectors) – the ground state wave function $|\psi_0\rangle$, and vectors $|A\rangle$, $|\psi_{\mathrm{Re}}\rangle$, and $|\psi_{\mathrm{Im}}\rangle$ – at a given frequency are targeted. The finite-lattice DMRG algorithm must be used to sweep the lattice until the results converge. To obtain the dynamical correlation function over a finite range of frequencies, one has to repeat this calculation for a number of frequencies. If the DMRG calculation for each correction vector is performed independently, the computational effort is roughly proportional to the number of frequencies targeted.

It is possible to carry out the DMRG calculation for several frequencies and include all the corresponding correction vectors in the target states. A commonly used scheme is to simultaneously target two close frequencies ω_1 and ω_2 and calculate the dynamical correlation function at a frequency between these two by interpolation. This scheme could significantly reduce the computational cost for determining the spectrum over a frequency range, but the results obtained by interpolation are less accurate. The interpolation works if the spectral function varies smoothly with frequency. To ensure a good description of the spectra between these two frequencies, a broadening parameter of the order of $\omega_2 - \omega_1$ should be used.

Accurate determination of $|\psi \, (\omega + i\eta)\rangle$ is the most crucial step in the correction vector method. Several different methods have been proposed for calculating this correction vector.

18.6.1 Conjugate Gradient Method

A direct approach for calculating the correction vector is to solve the equation:

$$(\omega - H + E_0 + i\eta) \, |\psi \, (\omega + i\eta)\rangle = |A\rangle \, . \tag{18.69}$$

This equation could be solved, for example, using a generalized minimal residual method. However, as $\omega - H + E_0 + i\eta$ is not Hermitian, a converged result with the desired accuracy is not always efficiently obtained by directly solving this equation.

To avoid dealing with a non-Hermitian matrix, one can first calculate the imaginary part of the correction vector by solving the equation:

$$\left[(\omega - H + E_0)^2 + \eta^2\right] |\psi_{\text{Im}} (\omega + i\eta)\rangle = -\eta |A\rangle . \tag{18.70}$$

The real part of the correction vector, $|\psi_{\text{Re}} (\omega + i\eta)\rangle$, is then determined through the equation:

$$|\psi_{\text{Re}} (\omega + i\eta)\rangle = -\frac{\omega - H + E_0}{\eta} |\psi_{\text{Im}} (\omega + i\eta)\rangle . \tag{18.71}$$

Since the matrix $(\omega - H + E_0)^2 + \eta^2$ is Hermitian and positive definite, Eq. (18.70) can be solved reliably and efficiently with the conjugate-gradient method. This method was first adopted to calculate the correction vectors. However, its demand for computer resources is very high, which sets a high bar for applications.

18.6.2 Dynamical DMRG

Dynamical DMRG is more precise and efficient than the conjugate gradient method. Instead of solving the linear equation, Eq. (18.70), it recasts the calculation of the correction vector as a variational problem [85]. It is straightforward to show that solving Eq. (18.70) is equivalent to minimizing the following functional for given ω and η:

$$W[Y] = \langle Y| (\omega - H + E_0)^2 + \eta^2 |Y\rangle + \eta\langle A|Y\rangle + \eta\langle Y|A\rangle . \tag{18.72}$$

In other words, the minimum of this functional, Y_{min}, is just the solution of Eq. (18.70):

$$\left[(\omega - H + E_0)^2 + \eta^2\right] |Y_{\text{min}}\rangle = -\eta |A\rangle , \tag{18.73}$$

hence

$$|\psi_{\text{Im}} (\omega + i\eta)\rangle = |Y_{\text{min}}\rangle . \tag{18.74}$$

Furthermore, at the minimum,

$$\begin{aligned} W[Y_{\text{min}}] &= \langle Y_{\text{min}}| (\omega - H + E_0)^2 + \eta^2 |Y_{\text{min}}\rangle + \eta\langle A|Y_{\text{min}}\rangle + \eta\langle Y_{\text{min}}|A\rangle \\ &= \eta\langle A|Y_{\text{min}}\rangle = -\pi\eta I (\omega + i\eta) . \end{aligned} \tag{18.75}$$

Therefore, the spectral function is completely determined by the minimum of $W[Y]$. Thus from the minimization, we can directly calculate the spectral function without explicitly using the correct vector. An advantage of this approach is that the calculated spectral function has an error much smaller than that obtained with the correction-vector method. To understand this, let us assume the error in the solution $|Y\rangle$ of the minimized functional W to be ϵ, namely

$$|Y\rangle = |Y_{\min}\rangle + \epsilon\,|\phi\rangle\,, \tag{18.76}$$

where $|Y_{\min}\rangle$ is the rigorous solution and $|\phi\rangle$ is a normalized error vector which is orthogonal to $|Y_{\min}\rangle$:

$$\langle Y_{\min}|\phi\rangle = 0, \qquad \langle\phi|\phi\rangle = 1. \tag{18.77}$$

The error in the functional or the spectral function is then given by

$$\delta W = W\,(Y,\omega,\eta) - W\,(Y_{\min},\omega,\eta) \tag{18.78}$$

$$= \epsilon^2\,\langle\phi|\,(\omega - H + E_0)^2 + \eta^2\,|\phi\rangle\,. \tag{18.79}$$

It is of the order of ϵ^2, one order of ϵ less than the error in the spectral function obtained with Eq. (18.67).

In DMRG, the ground-state wave function can be variationally evaluated. We can also use this variational approach to minimize the functional $W[Y]$. Thus the minimization of $W[Y]$ can be readily integrated into the standard DMRG algorithm. This is another advantage of this method.

18.6.3 Krylov-Space Decomposition Method

In the calculation of the correction vector, either by solving the linear equation (18.70) using the conjugate gradient method or by minimizing the functional $W[Y]$, the matrix

$$\Delta = (\omega - H + E_0)^2 + \eta^2 \tag{18.80}$$

is used. In either case, the convergence error of the correction vector is controlled by the condition number [404] of this matrix, $\kappa\,(\Delta)$. This condition number is determined by the ratio between the largest and the smallest singular values of Δ. If Δ is singular, the condition number is infinite. In general, the larger the condition number, the greater the number of iterations needed to get a converged result, and the smaller the improvement of the solution at each iteration step.

On the other hand, if we calculate the correction vector directly using Eq. (18.69), we only need to treat the matrix:

$$X = \omega - H + E_0 + i\eta, \tag{18.81}$$

whose condition number $\kappa\,(X)$ is roughly equal to the square root of $\kappa\,(\Delta)$, hence $\kappa\,(X) \ll \kappa\,(\Delta)$. This advantage in using X instead of Δ has not been considered in either the conjugate gradient or the dynamical DMRG methods.

However, solving the linear equation (18.69), as already mentioned, is generally not stable because X is not Hermitian. To avoid this problem, Nocera and Alvarez proposed a Krylov-space decomposition approach to evaluate the correction vector [87]. Their method takes a Lanczos iteration to tridiagonalize the Hamiltonian, the same as in the Lanczos DMRG. After diagonalizing this tridiagonal Hamiltonian, the inverse of X can be rigorously calculated within this space spanned by the Lanczos vectors.

There are two steps for calculating the correction vector after obtaining the ground-state wave function $|\psi_0\rangle$:

(i) Generate all the Lanczos vectors, $(|e_0\rangle, |e_1\rangle, \ldots, |e_{N-1}\rangle)$, starting from the initial normalized basis state $|e_0\rangle$ defined by Eq. (18.6). The Hamiltonian is tridiagonalized in this set of basis states, as given by Eq. (18.27).

(ii) H is a real symmetric matrix. It can be diagonalized by an orthogonal matrix U:

$$H = U^\dagger \Lambda U, \tag{18.82}$$

where $\Lambda = \mathrm{diag}(\lambda_1, \lambda_2, \ldots, \lambda_N)$ is a diagonal matrix and λ_i is an eigenvalue of H. The Hamiltonian operator can be represented as

$$H = |E\rangle U^\dagger \Lambda U \langle E| = \sum_{ijl} |e_{i-1}\rangle U_{li} \lambda_l U_{lj} \langle e_{j-1}|. \tag{18.83}$$

The correction vector is then given by

$$
\begin{aligned}
|\psi(\omega + i\eta)\rangle &= \frac{\sqrt{\langle A|A\rangle}}{\omega - H + E_0 + i\eta} |e_0\rangle \\
&= |E\rangle U^\dagger \frac{\sqrt{\langle A|A\rangle}}{\omega - \Lambda + E_0 + i\eta} U \langle E|e_0\rangle \\
&= \sum_{il} |e_{i-1}\rangle U_{li} \frac{\sqrt{\langle A|A\rangle}}{\omega - \lambda_l + E_0 + i\eta} U_{l1}.
\end{aligned}
\tag{18.84}
$$

This method is more accurate and efficient than the conjugate gradient. It is similar to the Lanczos-DMRG in that it uses a tridiagonal decomposition of the Hamiltonian to compute the correction vector and integrates well with the ground-state DMRG. However, it only targets the correction vector at a given frequency, not the Lanczos vectors in constructing the reduced density matrix.

18.7 Spin Structure Factor of the Heisenberg Model

To demonstrate the accuracy the methods previously introduced can achieve, we apply them to the one-dimensional spin-1/2 Heisenberg model defined in Eq. (4.1).

Open boundary conditions are assumed. In practical calculations, any of the methods introduced in §18.5 and §18.6 could be used to perform this calculation. Here we show the results of spin structure factor obtained using the Chebyshev MPS method [89]. The results obtained with the Lanczos MPS [90] or correction-vector methods [87] are qualitatively similar.

The spin structure factor $S(k, \omega)$ is defined by the spin-spin correlation function:

$$S(k, \omega) = \langle \psi_0 | S^z_{-k} \delta(\omega - H + E_0) S^z_k | \psi_0 \rangle. \tag{18.85}$$

It can be measured by inelastic neutron scattering spectroscopy. In (18.85), E_0 is the ground-state energy, and S^z_k is the spin operator defined in the wave-vector space:

$$S^z_k = \sqrt{\frac{2}{N+1}} \sum_{r=1}^{N} S^z_r \sin(rk). \tag{18.86}$$

In an open boundary system, k is the quasi-momentum defined using the standing wave rather than the plane wave.

For the isotropic Heisenberg model, it is known that the low-lying excitations are the two-spinon states [407, 408]. Each spinon carries an $S = 1/2$ spin. In a finite spin chain, the spinons are confined. The lowest excited states are the des Cloizeaux–Pearsons triplet with total spin $S = 1$ [405] formed by two spinons. The singlet excitations formed by two spinons have slightly higher energies. The gap between the lowest spin singlet and the lowest spin triplet is finite in a finite spin chain. However, it approaches zero in the large size limit [409], implying that the lowest-lying excitations are fourfold degenerate and the spinons are deconfined in the thermodynamic limit. Furthermore, from the Bethe ansatz solution, it was shown that the two-spinon excitations form a continuum in the (k, ω) plane, which is bounded by the lower and upper branches [405, 406, 407]:

$$\omega_{\min} = \frac{\pi}{2} |\sin k|, \quad \omega_{\max} = \pi \left| \sin \frac{k}{2} \right|. \tag{18.87}$$

Figure 18.1 shows the intensity plot of $S(k, \omega)$ in the (k, ω) plane for the spin-half Heisenberg model of $N = 100$, obtained with 300 Chebyshev vectors. The numerical result correctly produces the behavior of the spinon continuum spectra. In particular, the lower and upper bounds of the continuum agree well with the exact results.

For the spin-1/2 Heisenberg model, the asymptotic behavior of the dynamical spin structure factor is known from the Bethe ansatz [410, 411]. It diverges at $k = \pi$ in the low-frequency ($\omega \to 0$) and thermodynamic limit:

$$S(\pi, \omega) \sim \frac{\sqrt{-\ln \omega}}{\omega}. \tag{18.88}$$

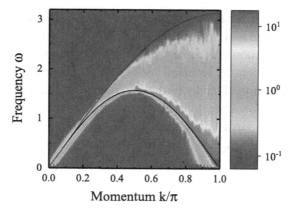

Figure 18.1 Intensity plot of the dynamical spin correlation function $S(k, \omega)$ obtained using the Chebyshev MPS expansion with the Gram–Schmidt reorthogonalization for the one-dimensional spin-1/2 Heisenberg model of $N = 100$ with open boundary conditions. In obtaining the result, 300 Chebyshev vectors are used. The bond dimension is $D = 64$. The two solid lines represent the lower and upper bounds, $\omega_{\min}$ and $\omega_{\max}$ defined in Eq. (18.87), stemmed from the two-spinon excitations [405, 406, 407]. Reprinted from fig. 6 of Ref. [89]. Copyright by the American Physical Society.

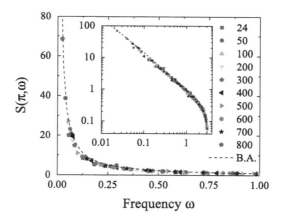

Figure 18.2 Energy dependence of $S(\pi, \omega)$ obtained using Eq. (18.61) for the one-dimensional spin-1/2 Heisenberg model with the system size, bond dimension, and order of Chebyshev polynomials (N, D, M) varying as $(24, 128, 200)$, $(50, 128, 200)$, $(100, 128, 100)$, $(200, 256, 100)$, $(300, 256, 80)$, $(400, 256, 60)$, $(500, 128, 80)$, $(600, 128, 60)$, $(700, 128, 50)$, and $(800, 64, 200)$. The inset is the log-log plot of the data. The dashed line denotes the Bethe ansatz results in the thermodynamic limit [410, 411]. Reprinted from fig. 8 of Ref. [89]. Copyright by the American Physical Society.

The finite-size effect suppresses the divergence of $S(\pi, \omega)$. However, the energy gap between the ground and first excited states drops with the lattice size. Thus the singularity of $S(\pi, \omega)$ should become more and more pronounced with the increase of the lattice size, which, as revealed by Fig. 18.2, is indeed what we see from the numerical calculation. Thanks to the divergence of the spectral weight in the zero-energy limit, the low-energy spectrum can be very accurately determined with relatively small D in a large lattice system. The size-dependent spectral weight is obtained using Eq. (18.61). It agrees with the exact scaling behavior of $S(\pi, \omega)$ obtained from the Bethe ansatz solution.

19

Time-Dependent Methods

19.1 Pace-Keeping DMRG

Cazalilla and Marston [91] first used DMRG to solve the time-dependent Schrödinger equation out-of-equilibrium. They applied DMRG directly to the time-dependent problem without considering the complication introduced by the time evolution. Their approach correctly produces the short-time result but fails to produce a reliable result on a long time scale. This is not that surprising because their approach, which involved the use of DMRG by targeting only the ground state, is essentially static with respect to the evolution of the Hilbert space.

To solve this problem, Luo, Xiang, and Wang [92] proposed a pace-keeping approach to track the time evolution of wave function. A particular scheme they suggested, which is later referred to as the pace-keeping DMRG, is to target not just the static ground state but also the time evolution of the wave function in the construction of the reduced density matrix [92]. This is similar to the case when DMRG is extended to calculate dynamical correlation functions where the correction vectors are targeted.

The pace-keeping DMRG is applicable to any quantum system with time-dependent interactions in any dimensions. It is particularly useful in treating a system with nonlocal interactions where TEBD introduced in §19.2 is not applicable. Here we take a quantum dot system in an applied bias voltage

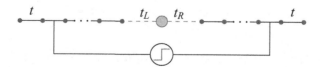

as an example to elucidate this method. The Hamiltonian of this system reads

$$H = H_0 + H_v(t), \tag{19.1}$$

where

$$H_0 = -\frac{1}{4}\sum_{i\neq0,-1}(c_i^\dagger c_{i+1} + h.c.) + \varepsilon n_0 - (t_L c_{-1}^\dagger c_0 + t_R c_0^\dagger c_1 + h.c.), \quad (19.2)$$

$$H_v = V\theta(t - t_0)(N_L - N_R). \quad (19.3)$$

The quantum dot is located at site $i = 0$, linked by two leads under a bias voltage. $n_i = c_i^\dagger c_i$ is the number operator at site i and ε is the energy of fermion at site 0. The band widths for both the left ($i < 0$) and right ($i > 0$) leads are set to 1. N_L and N_R are the number of electrons in the left and right leads, respectively. The bias between two leads is switched on at $t = t_0$, and V is the bias voltage applied. $\theta(t)$ is a step-raising function which changes from 0 to 1 in a characteristic time scale t_s.

The time-dependent Schrödinger equation reads:

$$i\frac{\partial}{\partial t}|\Psi(t)\rangle = [H_0 + H_v(t)]|\Psi(t)\rangle. \quad (19.4)$$

The system becomes out-of-equilibrium when a finite voltage is applied between the two leads. It implies that the state of the system cannot be simply expressed in terms of the eigenstates of the Hamiltonian. To eliminate this non-equilibrium effect, we perform a unitary transformation for the wave function

$$|\overline{\Psi}(t)\rangle = U(t)|\Psi(t)\rangle, \qquad U(t) = T_t \exp\left[-i\int_{t_0}^t d\tau H_v(\tau)\right], \quad (19.5)$$

where T_t is the time-ordering operator. The Schrödinger equation then becomes

$$i\frac{\partial}{\partial t}|\overline{\Psi}(t)\rangle = [H_0 + H_I(t)]|\overline{\Psi}(t)\rangle, \quad (19.6)$$

where

$$H_I(t) = \left[1 - e^{i\Phi(t)}\right](t_L c_{-1}^\dagger c_0 + t_R c_0^\dagger c_1) + h.c., \quad (19.7)$$

$$\Phi(t) = \int_{t_0}^t d\tau V\theta(\tau - t_0). \quad (19.8)$$

Schematically, the system becomes

After the transformation, this out-of-equilibrium problem is recast into an equilibrium one (without the bias voltage) but with a time-dependent hopping term. Moreover, the current operator,

$$J = \frac{1}{2}i(t_L c_{-1}^\dagger c_0 + t_R c_0^\dagger c_1) + h.c. \quad (19.9)$$

also becomes time-dependent:

$$J(t) = \frac{1}{2} i e^{i\Phi(t)} (t_L c_{-1}^\dagger c_0 + t_R c_0^\dagger c_1) + h.c.. \tag{19.10}$$

The measurement current is determined by the expectation value of $J(t)$:

$$I(t) = \langle \overline{\Psi}(t) | J(t) | \overline{\Psi}(t) \rangle. \tag{19.11}$$

The wave function can be solved by integrating over a small time interval starting from the ground state wave function at $t = t_0$:

$$\left| \overline{\Psi}(t + \Delta t) \right\rangle = e^{iA(t,\Delta t)} \left| \overline{\Psi}(t) \right\rangle, \tag{19.12}$$

where

$$A(t, \Delta t) = -\int_t^{t+\Delta t} d\tau \, [H_0 + H_I(\tau)]. \tag{19.13}$$

The phase factor in Eq. (19.12) needs to be calculated numerically. Up to the fourth order in Δt, the wave function is

$$\left| \overline{\Psi}(t + \Delta t) \right\rangle \approx \left(1 + iA - \frac{1}{2} A^2 - \frac{i}{6} A^3 + \frac{1}{24} A^4 \right) \left| \overline{\Psi}(t) \right\rangle. \tag{19.14}$$

The normalization condition, i.e. $|\exp(iA)| = 1$, is satisfied up to the fifth order in Δt. The error due to the fourth-order truncation in the expansion is negligible for small Δt.

At each step of time evolution, the pace-keeping DMRG is done in three steps:

(i) Evaluate the ground-state wave function $\left| \overline{\Psi}(t_0) \right\rangle = |\Psi(t_0)\rangle$ of H_0 for a finite lattice system using the standard DMRG. The time evolution of the wave function is then calculated by taking the expansion (19.14) iteratively from t_0 to a size-dependent time scale t_T at which the current reaches one of the open ends. After that, the current will be reflected back, yielding a dramatic reduction in the current. This reduction is purely a finite lattice size effect. Therefore, we should only consider $t_0 < t < t_L$ as a valid time scale within which transport properties of out-of-equilibrium systems can be reliably studied. For the noninteracting system considered here, the Fermi velocity is $v_F \sim 1$. As a result, the reflection time t_T is estimated to be $t_T = N/2v_F \sim N/2$, which is determined by the half chain length.

(ii) Define the density matrix by targeting both the ground state and N_t sampled points of $|\overline{\Psi}(t)\rangle$:

$$\rho = \alpha \, |\Psi(t_0)\rangle \langle \Psi(t_0)| + \sum_{l=1}^{N_t} \beta_l \left| \overline{\Psi}(t_l) \right\rangle \left\langle \overline{\Psi}(t_l) \right|, \tag{19.15}$$

where α and β_l ($l = 1, 2, \ldots, N_t$) are the weighting factors for the corresponding states. They are normalized to unity:

$$\alpha + \sum_{l=1}^{N_t} \beta_l = 1. \tag{19.16}$$

The sampling points can be determined according to the variance of the wave function. In general, more points should be taken in the time domain where the wave function or some relevant physical quantity, such as the current discussed here, varies fast. Empirically, one can set all β_l equal and take the sample points equally spaced in time; that is, at $t_l = t_0 + l \Delta \tau$, where $\Delta \tau = (t_F - t_0)/N_t \geq \Delta t$ and $t_F < t_T$ is the final time evaluated. N_t can vary with the system size and be smaller than the total time points evaluated.

(iii) Diagonalize the reduced density matrix and renormalize the Hamiltonian and other operators using its D dominant eigenvectors.

For the noninteracting model considered here, the Hamiltonian can be rigorously diagonalized for very large N. Figure 19.1 shows how the current evolves with time for several different N. After the bias is switched on, the current first rapidly increases and then begins to oscillate. After an initial decay, the currents tend to have a steady value. The oscillation period is lattice size-independent, but the amplitude of the oscillations depends strongly on the lattice size. The oscillations tend to disappear in the thermodynamic limit $N \to \infty$. When the lattice size is small, $N = 64$ and 128, a fast drop in the current occurs at $t_T \sim 70$ and 130, respectively. This is clearly due to the reflection of the current at the open end. For larger N, the fast drop of the current induced by the reflection is not observed in the time scale shown in Fig. 19.1.

Figure 19.2 shows the time dependence of the current for the quantum dot system obtained by keeping 256 states in the DMRG calculation. In the case $N_t = 0$, namely only the ground state is targeted, the current $I(t)$ begins to deviate from the exact result roughly at $t \sim 25$. However, if the time-evolving wave functions are included in the targeted state, the long-time behavior of the current is improved. For example, when $N_t = 5$, the pace-keeping result agrees with the exact one up to $t \sim 42$. By further increasing N_t to 30, the agreement of the pace-keeping result with the exact one extends to $t \sim 47$. However, there is no sizeable improvement by increasing the sampling points N_t. The deviation of the DMRG result from the exact one at $t > 47$ is caused by the truncation error. This error can be systematically eliminated by keeping more states in the DMRG truncation.

There is no particular rule to determine the weighting factors α and β_l. However, the weight of the ground state α cannot be too small. Otherwise, the ground state would not be accurately calculated with DMRG. Empirically, it was found that the

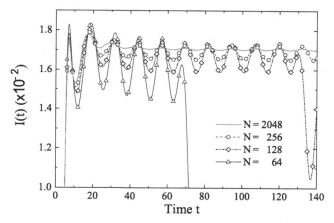

Figure 19.1 Time evolution of the exact current $I(t)$ for the noninteracting quantum dot system with different lattice size N. The current collapses due to the reflection of the current at the end of the chain. The parameters used are $t_L = t_R = 0.15$, $\varepsilon = -0.25$, $\Delta t \simeq 0.05$, $V = 0.25$, and the bias is switched on at $t_0 = 4$ with a rising time $t_s = 0.1$.

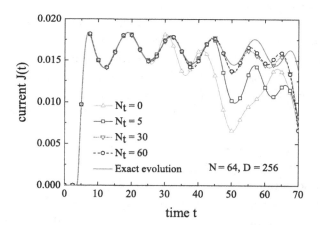

Figure 19.2 Evolution of the current $I(t)$ with time for the quantum dot system obtained by the pace-keeping DMRG with several different N_t. The exact result is shown for comparison. The chain length is fixed to $N = 64$, and $D = 256$ states are retained in the DMRG calculation. The weight factor of the ground state is set to $\alpha = 1$ for the case $N_t = 0$ and 0.5 otherwise. Other parameters used are the same as for Fig. 19.1.

pace-keeping DMRG result converges well once α is between 0.1 and 0.9. As an example, Fig. 19.3 compares the pace-keeping DMRG results obtained with two different α.

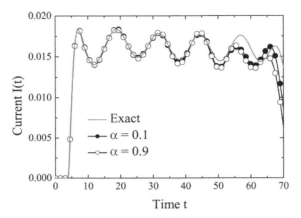

Figure 19.3 Comparison between the pace-keeping DMRG results of the current $I(t)$ as a function of time for the quantum dot system obtained with $\alpha = 0.1$ and 0.9. The exact result is also shown for comparison. $N = 64$, $N_t = 40$, and 256 states are retained in the DMRG truncation. Other parameters used are the same as for Fig. 19.1.

The above adaptive pace-keeping scheme is an infinite-lattice algorithm of DMRG at which the wave function is evaluated over the entire time period before determining the basis states by diagonalizing the density matrix. It can be extended to combine with the finite-lattice algorithm to perform the finite-lattice sweep. The reduced density matrix defined by Eq. (19.15) is evaluated at each sweep step, which can significantly improve the accuracy. However, performing a full-time evolution of the wave function at every step of finite-lattice sweep is time-consuming.

Feiguin and White suggested another adaptive time-dependent DMRG algorithm called the time-step targeting method [412]. Instead of evaluating the wave function in the entire time period in a single step, they suggested to construct the density matrix from the wave packets at four time points $(t, t + \tau/3, t + 2\tau/3, t + \tau)$ for each time step of evolution from t to $t + \Delta t$, here $\tau \sim 10\Delta t$ is the time interval used for targeting the basis states. To improve accuracy, one can take a finite-lattice sweep at every step of evolution.

This time-step targeting method, as pointed out by Dutta and Ramasesha [413], is less accurate than the pace-keeping DMRG because the short-time step used in targeting states is not sufficient to represent the Hilbert space of the time-evolving system—it does not contain all the information about the future states along the trajectory of time propagation. It is also not more efficient than the pace-keeping method because many steps of finite-lattice sweeps are taken in the time evolution.

19.2 Time-Evolving Block Decimation

Time-evolving block decimation (TEBD) is an efficient algorithm for evaluating the real-time dynamical evolution of a quantum state [38]. The time evolution involved is numerically stable since the only source of non-unitarity comes from the truncation error, but it is limited to systems with short-range interactions in one dimension and is difficult to apply to a system with long-range interactions. In the case of narrow ladders with nearest-neighbor interactions, one can group all sites in a rung into an effective site. The computational cost grows linearly with the system size. This algorithm is dubbed time-evolving block decimation because it dynamically identifies the relevant low-dimensional Hilbert subspaces in the dynamical evolution.

The algorithm is based on the MPS representation of quantum wave functions. It works efficiently because the real-time evolution operator is unitary, and a unitary evolution preserves the canonical form of MPS. From the orthogonality of the canonical matrices, the contribution of the environment can be read out just from the relevant bond vectors without explicitly contracting the entire MPS. The MPS can therefore be updated just by performing local matrix manipulations.

It is more convenient to use the $\lambda - \Gamma$ representation of MPS to carry out a TEBD calculation. We assume that the initial MPS

$$\Psi_0 = \langle\lambda_0\rangle\!-\!(\Gamma_1)\!-\!\langle\lambda_1\rangle\!-\!(\Gamma_2)\!-\!\langle\lambda_2\rangle\!-\;\cdots\;-\!(\Gamma_N)\!-\!\langle\lambda_N\rangle \qquad (19.17)$$

is canonicalized. Open boundary conditions are assumed.

The dynamical evolution of this state with time is determined by the equation

$$|\Psi(t)\rangle = e^{-iHt}|\Psi_0\rangle, \qquad (19.18)$$

The Hamiltonian H is assumed to be time-independent. By slightly modifying this formula, the algorithm can also be extended to the case H is time-dependent. To understand how the method works, let us first consider a quantum system with only on-site interactions,

$$H = \sum_i H_i, \qquad (19.19)$$

where H_i is the interaction at site i. For simplicity, we further assume that the local interactions on different lattices commute with each other, i.e. $[H_i, H_j] = 0$. In this case, the time evolution of the state can be done separately on each lattice site. Equation (19.18) then becomes

$$|\Psi(t)\rangle = \prod_i e^{-iH_i t}|\Psi_0\rangle. \qquad (19.20)$$

$\exp(-iH_i t)$ is a unitary operator. It generates a time evolution of the local tensor at site i, with Γ_i updated by

$$\underline{\Gamma}_i[\sigma] = \sum_{\sigma'} \langle \sigma | e^{-iH_i t} | \sigma' \rangle \Gamma_i[\sigma'] . \tag{19.21}$$

It is simple to show that $\underline{\Gamma}_i$ is canonical and satisfies the equations

$$\lambda_{i-1}^2 \qquad = I, \qquad \qquad \lambda_i^2 \quad = I. \tag{19.22}$$

The other Γ's and all λ's are unchanged. It shows that the local unitary transformation preserves the canonical form of MPS. One can then update the MPS by setting $\underline{\Gamma}_i$ as the new Γ_i and perform the evolution on other sites. Only a local unitary transformation for the local matrix at each site needs to be performed. There is no truncation to the basis space.

Now let us extend the above discussion to a quantum system with nearest-neighbor interactions. The Hamiltonian is now defined by

$$H = \sum_i H_{i,i+1}. \tag{19.23}$$

As the local Hamiltonians, $H_{i,i+1}$ generally do not commute with each other. The dynamical evolution at a given time t cannot be done just in one step. To carry out the dynamical evolution, we need to divide t into many small steps and perform the evolution step by step by utilizing the Trotter–Suzuki decomposition at each small step:

$$e^{-iH\tau} \approx e^{-iH_{\text{even}}\tau} e^{-iH_{\text{odd}}\tau} + o(\tau^2), \tag{19.24}$$

where τ is a small time step, and

$$H_{\text{odd}} = \sum_{i \in \text{odd}} H_{i,i+1}, \tag{19.25}$$

$$H_{\text{even}} = \sum_{i \in \text{even}} H_{i,i+1}. \tag{19.26}$$

After each step, by performing the time evolution with H_{odd} and H_{even} successively, the wave function evolves to

$$|\Psi(t+\tau)\rangle = e^{-iH\tau} |\Psi(t)\rangle. \tag{19.27}$$

The local interaction terms in H_{even} (similarly in H_{odd}) commute with each other. Thus the time evolution with H_{even} or H_{odd} can be done separately for each individual term. The error in the Trotter decomposition is due to the non-commutativity of bond Hamiltonians. By applying $\exp\left(-iH_{i,i+1}\tau\right)$ to the wave function, the local tensors at site i and $i+1$ evolve to

$$
-\!\!\boxed{M}\!\!- \;=\; \begin{array}{c} \boxed{\exp(-iH_{i,i+1}\tau)} \\[4pt] -\bigcirc-\diamond-\bigcirc- \\[2pt] \Gamma_i \quad \lambda_i \quad \Gamma_{i+1} \end{array}\;. \tag{19.28}
$$

Here $\exp\left(-iH_{i,i+1}\tau\right)$ is a unitary operator. If i and $i+1$ effectively merge into a big site in a dimerized system, $\exp\left(-iH_{i,i+1}\tau\right)$ serves as an on-site time evolution operator. From the preceding discussion on the one-site evolution, we know that the local operator $\exp\left(-iH_{i,i+1}\tau\right)$ affects only the matrices on the sites that the operator acts on. Thus only $\Gamma_i[\sigma_i]$, λ_i, and $\Gamma_{i+1}[\sigma_{i+1}]$ will be changed by the evolution operator. To update these matrices, we need to consider the renormalization effect of the environment to M.

The canonical representation of MPS shows that the bond vector λ measures the entanglement between the left and right blocks connected by this bond. This implies that the environment contribution to M is simply determined by the entanglement vectors λ_{i-1} on the left bond connecting to i, and λ_{i+1} on the right bond connecting $i+1$. It further suggests that the renormalization effect of the environment on M is well accounted by the matrix

$$
\alpha -\!\!\boxed{W}\!\!- \beta \;=\; \alpha -\diamond-\!\!\boxed{M}\!\!-\diamond- \beta\;. \tag{19.29}
$$

Taking SVD for this matrix, we get

$$
\alpha -\!\!\boxed{W}\!\!- \beta \;=\; \alpha -\boxed{U}-\diamond-\boxed{V}- \beta. \tag{19.30}
$$

where λ is a semi-positive definite diagonal matrix measuring the entanglement on the bond linking i and $i+1$. The dimension of λ is dD with d the dimension of σ_i. We then truncate the bond dimension of λ by keeping the D largest singular values. After the truncation, U and V become isometric.

Now we update the bond matrix with the truncated λ (i.e. $\lambda_i = \lambda$), and the Γ-matrices at sites i and $i+1$ by the formula

$$
\alpha -\bigcirc- \beta \;=\; \alpha -\diamond-\boxed{}- \beta\;, \tag{19.31}
$$

$$\sigma_{i+1} \qquad\qquad \sigma_{i+1}$$

$$\alpha \ \underset{\Gamma_{i+1}}{-\bigcirc-} \ \beta \ = \ \alpha \ \underset{V \ \lambda_{i+1}^{-1}}{-\square-\diamond-} \ \beta \ . \tag{19.32}$$

It is straightforward to show that Γ_i satisfies the left canonical condition

$$\sum_{\sigma_i} \Gamma_i^\dagger[\sigma_i]\Lambda_{i-1}\Gamma_i[\sigma_i] = U^\dagger U = I, \tag{19.33}$$

and the right canonical condition

$$\sum_{\sigma_i} \Gamma_i[\sigma_i]\Lambda_i\Gamma_i^\dagger[\sigma_i] = I, \tag{19.34}$$

where $\Lambda_i = \lambda_i^2$. Similarly, it can be shown that Γ_{i+1} are left and right canonicalized.

These updated tensors, together with other unrenormalized Γ's and λ's, form a canonical MPS for the next iteration. This dynamical evolution involves the manipulation of a few local matrices and is simple to implement in practical calculations.

TEBD is an accurate algorithm for carrying out a unitary evolution. It works efficiently because a unitary evolution preserves the canonical form of MPS and allows the global optimization to be implemented locally. As discussed in §12.3.3, this method is generalizable to treat a nonunitary evolution problem, provided that the evolution operator is not far from unitary. In that case, the local update becomes less accurate. Nevertheless, it is applicable in the ground-state calculation because the truncation error is not accumulated with the imaginary time evolution (see §12.3.3). In the simulation of imaginary time evolution at finite temperature, however, truncation errors accumulate rapidly in the absence of the canonical form. Therefore, a global canonicalization before the basis truncation is desired after each step of evolution.

19.3 Adaptive Time-Dependent DMRG

The algorithm of TEBD can be incorporated into the framework of DMRG by taking the finite lattice sweeping to simulate the dynamical evolution at each Trotter time step. The wave function is evolved by taking the unitary bond evolution first from left to right using H_{odd} and then from right to left using H_{even}. In literature, this algorithm is dubbed adaptive time-dependent DMRG [93] because the wave function is dynamically updated with the evolution. It differs from DMRG in two respects: First, there is no need to diagonalize the Hamiltonian, and the time evolution on each bond is to perform a unitary transformation. Second, in the standard

DMRG calculation, only one site is shifted, while in the dynamical evolution with the Trotter–Suzuki decomposition, two sites are shifted at each step of the iteration.

It is more convenient to use the left-right representation of MPS to implement the adaptive time-dependent DMRG. This representation could avoid using White's prediction method [228] in preparing the initial wave function. To understand this, let us consider the time evolution by applying the evolution operator $\exp\left(-iH_{2i-1,2i}\tau\right)$ successively to the wave function from left to right. Suppose we have already performed i iterations so that the wave function becomes

$$
\Psi \;=\; \underset{A_1}{\square}\!-\cdots-\underset{A_{2i}}{\square}\!-\underset{\lambda_{2i}}{\diamond}\!-\underset{B_{2i+1}}{\square}\!-\cdots-\underset{B_N}{\square}\;, \tag{19.35}
$$

where local tensors A's and B's are left and right canonicalized, respectively. The next step is to apply the evolution operator $\exp\left(-iH_{i+1,i+2}\tau\right)$ to the wave function. This changes the wave function to

$$
\Psi' \;=\; \underset{A_1}{\square}\!-\cdots-\underset{A_{2i}}{\square}\!-\underset{\lambda_{2i}}{\diamond}\!-\boxed{\;\underset{B_{2i+1}\;B_{2i+2}}{\overline{e^{-iH_{2i+1,2i+2}\tau}}}\;}\!-\cdots-\underset{B_N}{\square}\;. \tag{19.36}
$$

By introducing the matrix $\psi_{l\sigma_{2i+1},\sigma_{2i+2}k}$,

$$
\underset{l}{\overset{\sigma_{2i+1}\,\sigma_{2i+2}}{\,\psi\,}}\,r \;=\; \underset{l}{\diamond}\!-\underset{\lambda_i}{}\boxed{\;\overset{\sigma_{2i+1}\;\sigma_{2i+2}}{e^{-iH_{i+1,i+2}\tau}}\;}\!-\,r \;, \tag{19.37}
$$

we rewrite Ψ' as

$$
\Psi' \;=\; \underset{A_1}{\square}\!-\cdots-\underset{A_{2i}}{\square}\!-\underset{}{\psi}\!-\underset{B_{2i+3}}{\square}\!-\cdots-\underset{B_N}{\square}\;. \tag{19.38}
$$

Since all the local tensors on the left or right of ψ are left or right canonicalized, $\psi_{l\sigma_{2i+1},\sigma_{2i+2}r}$ is just the wave function used in the DMRG calculation. l and r represent the basis states for the left and right blocks, respectively.

The matrix dimension of ψ is dD if the dimensions of l and r are both equal to D. Taking the Schmidt decomposition for this wave function,

$$
\underset{l}{\overset{\sigma_{2i+1}\,\sigma_{2i+2}}{\,\psi\,}}\,r \;=\; \underset{l}{\overset{\sigma_{2i+1}}{U}}\!-\diamond\!-\overset{\sigma_{2i+2}}{V}\!-\,r \;, \tag{19.39}
$$

where U and V are two unitary matrices of dimension dD. One then needs to truncate the dimension for the bond between $(2i + 1)$ and $(2i + 2)$ from dD to D by keeping the largest D singular values. After that, we set the truncated λ as the new λ_{2i+1} and

$$
\underset{A_{2i+1}}{\diamond\!\!-} = \underset{}{-\!\!\boxed{U}\!\!-} \;,\qquad \underset{B_{2i+2}}{-\!\!\diamond} = \underset{}{-\!\!\boxed{V}\!\!-} \;. \tag{19.40}
$$

The wave function then becomes

$$
\Psi = \underset{A_1}{\diamond\!\!-}\cdots\underset{A_{2i+1}\,\lambda_{2i+1}\,B_{2i+2}}{-\!\!\diamond\!\!-\!\!\Diamond\!\!-\!\!\diamond}\cdots\underset{B_N}{-\!\!\diamond} \;. \tag{19.41}
$$

From the discussion in the previous section, it is not difficult to show that the two newly defined matrices (i.e. A_{2i+1} and B_{2i+2}), are all canonical, satisfying the left and right canonical conditions, respectively.

This derivation is to repeat the TEBD steps introduced in the previous section, using a language close to that used in DMRG. However, in this algorithm, there is no need to find the inverse of the bond matrices λ's. This avoids an instability that may encounter in the TEBD iterations since the inverse of λ may have a large error if some elements of λ become small.

To perform the next iteration using the evolution operator $\exp\left(-iH_{2i+3,2i+4}\tau\right)$, we should start with a wave function where the singular value vector λ is located to the left of site $(2i+2)$, rather than at site $(2i+1)$. It means we cannot do the evolution directly using Ψ'. To solve this problem, we need to take a basis transformation to shift the singular vector one site from left to right. To do this, let us take SVD for the following matrix:

$$
\phi_{l\sigma_{2i+2},\sigma_{2i+3}r} = \underset{\lambda_{2i+1}\quad B_{2i+2}\quad B_{2i+3}}{\overset{\sigma_{2i+2}\quad\sigma_{2i+3}}{l\!-\!\Diamond\!\!-\!\!\diamond\!\!-\!\!\diamond\!-r}} \;. \tag{19.42}
$$

Again, $\phi_{l\sigma_{2i+2},\sigma_{2i+3}r}$ represents a DMRG wave function when σ_{2i+2} and σ_{2i+3} are taken as the two added spins; l and r are the basis states for the left and right blocks, respectively. Taking SVD, ϕ becomes

$$
\phi_{l\sigma_{2i+2},\sigma_{2i+3}r} = \overset{\sigma_{2i+2}\qquad\qquad\sigma_{2i+3}}{l\!-\!\boxed{U'}\!\!-\!\!\langle\lambda'\rangle\!\!-\!\!\boxed{V'}\!-r} \tag{19.43}
$$

The dimension of λ' is dD. To construct a standard MPS wave function in the left-right representation, we need to truncate its dimension to D. This approximation is not used in TEBD. After that, we set λ' equal to λ_{2i+2}, and update the local tensors at sites $2i + 2$ and $2i + 3$ by truncated U' and V', respectively:

$$
\underset{A_{2i+2}}{\longrightarrow} = \underset{}{-\boxed{U'}} \;\; , \quad \underset{B_{2i+3}}{\longrightarrow} = \underset{}{-\boxed{V'}} \;\; . \tag{19.44}
$$

These two local tensors are left and right canonical, respectively. Finally, the wave function becomes

$$
\Psi = \underset{A_1}{\longrightarrow} \cdots \underset{A_{2i+2}\lambda_{2i+2}B_{2i+3}}{\longrightarrow} \cdots \underset{B_N}{\longrightarrow} . \tag{19.45}
$$

It is the initial wave function for the local time evolution at sites $(2i+3)$ and $(2i+4)$.

After the sweep from left to right, one can do a similar sweep using the evolution operator $\exp(-iH_{even}\tau)$ from right to left, with A replaced by B and vice versa. Repeating these steps, one can calculate the wave function evolution and physical observables up to the desired time scale.

Compared to TEBD, an advantage of the adaptive time-dependent DMRG is that it does not need to evaluate the inverse of singular vectors. However, in implementing this algorithm, one has to take an extra basis truncation at each step when the wave function is shifted by one lattice site.

19.4 Folded Transfer Matrix Method

In TEBD, the dynamical evolution of wave function is computed by applying the evolution operator successively to an initial MPS. In this framework, the evolution operator is decomposed into a two-dimensional tensor network by making use of the Trotter–Suzuki decomposition formula. Within each small time step, the evolution operator is represented as an MPO, a product of local evolution operators. The action of a sequence of MPO on the initial MPS is to contract the tensor network along the direction of time. Each contraction yields an MPS with a larger bond dimension, which has to be truncated. However, as the entanglement entropy of the wave function raises linearly with time [96], the accuracy of this method drops exponentially with the increase of time. It leads to an exponential increase of the computational resources and limits the maximal reachable evolution time, depending on the desired accuracy. To solve this problem, Bañuls *et al.* suggested defining a folded transfer matrix for this tensor network state and contracting it along the spatial direction [97]. Their approach is similar to the calculation of thermodynamic quantities where the free energy is determined by diagonalizing the quantum transfer matrix with DMRG [72, 75]. A quantum transfer matrix extension of this method to finite temperatures was proposed by Huang *et al.* [98].

To elucidate how the folded transfer matrix method [97] works, let us consider the calculation of the expectation value of a local operator A in a time-evolving state in one dimension:

$$|\Psi(t)\rangle = e^{-itH}|\Psi_0\rangle, \tag{19.46}$$

where $|\Psi_0\rangle$ is the initial state represented by an MPS. The Hamiltonian is assumed to contain only nearest-neighbor interactions. The time evolution of the operator O is determined by the formula:

$$O(t) = \langle\Psi(t)|O|\Psi(t)\rangle. \tag{19.47}$$

If $|\Psi_0\rangle$ is an eigenstate of H or O commutes with the Hamiltonian, $O(t)$ is time-independent:

$$O(t) = \langle\Psi_0|O|\Psi_0\rangle. \tag{19.48}$$

Following the steps presented in §4.2 and §4.3, we can express $\langle\Psi(t)|O|\Psi(t)\rangle$ as a two-dimensional tensor network using the Trotter–Suzuki decomposition. We first divide time t into many small steps and express $\exp(-itH)$ as

$$e^{-itH} = \prod_{l=1}^{M} e^{-i\varepsilon H}, \tag{19.49}$$

where $\varepsilon = t/M$ is the Trotter time scale, and M is the total number of steps used. At each time step, the evolution operator $\exp(-i\varepsilon H)$ can be approximately expressed as an MPO using the second-order Trotter–Suzuki formula:

$$e^{-i\varepsilon H} \approx \prod_{i \in \text{even}} e^{-i\varepsilon H_{i,i+1}} \prod_{i \in \text{odd}} e^{-i\varepsilon H_{i,i+1}} = \prod_{i \in \text{even}} v_{i,i+1} \prod_{i \in \text{odd}} v_{i,i+1}, \tag{19.50}$$

where

$$v_{i,i+1} = e^{-i\varepsilon H_{i,i+1}} \tag{19.51}$$

is a local evolution operator. $v_{i,i+1}$ acts on two neighboring sites. By decoupling this tensor using SVD,

$$\tag{19.52}$$

we obtain the diagonal singular matrix γ, and two unitary matrices U and V. By absorbing $\sqrt{\gamma}$ into U and V, we define two new matrices:

$$\boxed{x} \!- \;=\; \boxed{U}\!-\!\diamondsuit_{\sqrt{\lambda}} \;,\qquad -\!\boxed{y} \;=\; \diamondsuit_{\sqrt{\lambda}}\!-\!\boxed{V} \tag{19.53}$$

and represent v as

$$v = xy. \tag{19.54}$$

Note that the horizontal and vertical bonds have different dimensions. The horizontal bond dimension is d^2 if the vertical one is d.

Using these local tensors, the evolution operator becomes a zigzag MPS:

$$e^{-i\tau H} \;=\; \text{[zigzag MPS diagram]} \tag{19.55}$$

It could be further represented as an MPO by contracting the local tensors along the vertical direction:

$$e^{-i\tau H} \;=\; \cdots-\!\boxed{\tau_1}\!-\!\boxed{\tau_2}\!-\!\boxed{\tau_1}\!-\!\boxed{\tau_2}\!-\!\boxed{\tau_1}\!-\!\boxed{\tau_2}\!-\cdots \;, \tag{19.56}$$

where

$$-\!\boxed{\tau_1}\!- \;=\; \text{[diagram]}\;,\qquad -\!\boxed{\tau_2}\!- \;=\; \text{[diagram]}\;. \tag{19.57}$$

The time evolution of the wave function $|\Psi(t)\rangle$ is simply to apply successively these MPOs onto the initial MPS $|\Psi_0\rangle$, which spans a two-dimensional tensor network state:

$$|\Psi(t)\rangle \;=\; \text{[two-dimensional tensor network diagram]} \tag{19.58}$$

The contraction of this tensor network state can be done along the time direction, the same as in the TEBD calculation. It can also be done along the spatial direction. In this case, $\langle\Psi(t)|O|\Psi(t)\rangle$ can be expressed as

$$\langle\Psi(t)|O|\Psi(t)\rangle = \text{Tr}\!\left[(T_1 T_2)^{L-1} T_O T_2\right], \tag{19.59}$$

where T_1, T_2 and T_O are the column transfer matrices defined in Fig. 19.4.

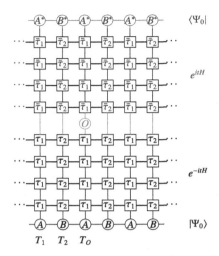

Figure 19.4 Tensor network representation of $\langle\Psi(t)|O|\Psi(t)\rangle$. $\bar{\tau}_\alpha$ ($\alpha = 1, 2$) is similarly defined as τ_α, but with negative ε; that is, with the local evolution $v_{ij} = \exp(-i\varepsilon H_{i,j})$. T_1, T_2, and T_O are the column transfer matrices.

The folding scheme uses a more efficient representation of the entanglement structure by folding the network in the time direction prior to contraction [97]. It suppresses the linear increase of the entanglement entropy with the evolution time t and significantly improves the accuracy of results. In this case, $\langle\Psi(t)|O|\Psi(t)\rangle$ is still determined by Eq. (19.59), but the transfer matrices are now defined by folding the upper half column transfer matrices downwards:

$$T_1 = \cdots \equiv \cdots , \quad T_O = \cdots \equiv \cdots , \tag{19.60}$$

where $\bar{\tau}_\alpha$ ($\alpha = 1, 2$) is similarly defined as τ_α, but with a negative ε:

$$-\!\!\!\bigcirc\!\!\!A_0\!\!\!-\; =\; -\!\!\!\bigcirc\!\!\!A\!\!\!-\!\!\!\bigcirc\!\!\!A^*\!\!\!- \quad,\qquad -\!\boxed{t_1}\!- \;=\; -\!\boxed{\tau_1}\!-\!\boxed{\bar{\tau}_1}\!- \quad, \tag{19.61}$$

and

$$-\!\boxed{\bar{t}_1}\!- \;=\; -\!\boxed{\tau_1}\!-\!\boxed{\bar{\tau}_1}\!- \quad,\qquad -\!\boxed{t_O}\!- \;=\; -\!\boxed{\tau_1}\!-\!\boxed{\bar{\tau}_1}\!\!\overset{\bigcirc\!\!O}{}- \quad. \tag{19.62}$$

T_2 is similarly defined. The contraction of $\langle\Psi(t)|O|\Psi(t)\rangle$ is done with these folded transfer matrices.

The folded transfer matrix method directly evaluates the expectation values of physical observables, instead of the time evolution of wave function. It reduces the linear buildup of entanglement in the wave function evolution with time and greatly extends the time range of simulation with a given precision desired.

20

Tangent-Space Approaches

20.1 Tangent Vectors of Uniform MPS

All uniform MPS with a given bond dimension span a manifold within the whole Hilbert space, defined by the map between the local tensor A and physical state $\Psi(A)$:

$$\Psi(A) = \quad \cdots -\!\!\underset{A}{\bigcirc}\!-\!\underset{A}{\bigcirc}\!-\!\underset{A}{\bigcirc}\!-\!\underset{A}{\bigcirc}\!-\!\underset{A}{\bigcirc}\!- \cdots \quad . \tag{20.1}$$

This is not a linear map because any sum of two MPS with a given bond dimension does not belong to the same manifold.

Associated with each MPS, a tangent vector is defined by taking the derivative of $\Psi(A)$ with respect to A and replacing each tensor A eliminated by the derivative with a tensor B. This tangent vector is equivalent to replacing tensor A at site 0 in $\Psi(A)$ by B and then boosting it into a translation-invariant state using the translation operator:

$$\Psi_\partial(B, A) \equiv \sum_n T^n \left(\cdots -\!\!\underset{A}{\bigcirc}\!-\!\underset{A}{\bigcirc}\!-\!\overset{\sigma_0}{\underset{B}{\bigcirc}}\!-\!\underset{A}{\bigcirc}\!-\!\underset{A}{\bigcirc}\!- \cdots \right)$$

$$= \sum_n \quad \cdots -\!\!\underset{A}{\bigcirc}\!-\!\underset{A}{\bigcirc}\!-\!\overset{\sigma_n}{\underset{B}{\bigcirc}}\!-\!\underset{A}{\bigcirc}\!-\!\underset{A}{\bigcirc}\!- \cdots \quad , \tag{20.2}$$

where T is the translation operator, which shifts the lattice over a single site. Ψ_∂ is a tangent vector because the derivative of $\Psi(A)$ with respect to parameter t is given by

$$\frac{\partial \Psi(A)}{\partial t} = \Psi_\partial(\partial_t A, A). \tag{20.3}$$

Here $B = \partial_t A$ is just a derivative of A. In general, B describes a linear combination of the partial derivatives. It presents a parametrization of the full tangent space of $\Psi(A)$. Unlike $\Psi(A)$, its tangent space is a linear subspace of the Hilbert space.

Using the translation invariance of the tangent vector $\Psi_\partial(B, A)$, it is simple to show its overlap with $\Psi(A)$:

$$\langle \Psi(A)|\Psi_\partial(B,A)\rangle = N \left(\quad \right)$$

$$= N \quad , \qquad (20.4)$$

where N is the lattice size, which diverges in the thermodynamic limit. l and r are the dominant left and right eigenvectors of the transfer matrix $\mathcal{E} = A \otimes A^*$, respectively.

20.1.1 Gauge Fixing

Given a local tensor A in $\Psi(A)$, the local tensor B in the tangent vector $\Psi_\partial(B, A)$ is not fixed. There is a gauge redundancy in this tangent vector introduced by its translation invariance.

From the defining expression (20.2), it is straightforward to show that $\Psi_\partial(B, A)$ is invariant under the following transformation:

$$\text{—}B\text{—} \rightarrow \text{—}B\text{—} + \text{—}A\text{—}G\text{—} - \text{—}G\text{—}A\text{—} , \qquad (20.5)$$

where G is an invertible $D \times D$ matrix. One way to remove the gauge redundancy in B is to find a gauge matrix G such that the following left gauge fixing condition is satisfied:

$$\quad = \quad = 0. \qquad (20.6)$$

To ensure that this condition is satisfied, we need to show that for an arbitrary B, we can always find a gauge matrix G such that

$$\quad + \quad - \quad = 0. \qquad (20.7)$$

Using the property of the eigenequation of $\mathcal{E}$, we can write Eq. (20.7) as

$$\equiv X_B. \tag{20.8}$$

This is a set of linear equations. $I - \mathcal{E}$ is a $D^2 \times D^2$ matrix, and X_B is a vector of dimension D^2.

To prove that Eq. (20.8) indeed has a solution, let us assume λ_k to be the eigenvalues, with l_k and r_k the corresponding left and right eigenvectors of $\mathcal{E}$:

$$\mathcal{E} = \sum_k \lambda_k |r_k\rangle\langle l_k| = |r\rangle\langle l| + \sum_{k>1} \lambda_k |r_k\rangle\langle l_k|. \tag{20.9}$$

The largest eigenvalue $\lambda_1 = 1$, and the corresponding eigenvectors are $l_1 = l$ and $r_1 = r$. For other eigenvalues, $\lambda_k < 1$ ($k > 1$). In this representation,

$$I - \mathcal{E} = \sum_{k>1} (1 - \lambda_k) |r_k\rangle\langle l_k| \tag{20.10}$$

is regularly defined and invertible. Similarly, we have

$$X_B = \sum_k X_B |r_k\rangle\langle l_k| = \sum_{k>1} X_B |r_k\rangle\langle l_k|. \tag{20.11}$$

This equation holds because

$$X_B |r\rangle = 0, \tag{20.12}$$

according to Eq. (20.8).

$I - \mathcal{E}$ is not a full rank matrix. However, it has the same rank as the augmented matrix $(I - \mathcal{E}, X_B^\dagger)$:

$$\text{rank}(I - \mathcal{E}) = \text{rank}(I - \mathcal{E}, X_B^\dagger) = D^2 - 1. \tag{20.13}$$

It implies that Eq. (20.8) has more than one solution. Thus we can indeed find a gauge matrix G such that Eq. (20.6) holds.

From Eq. (20.4), we immediately find that the left gauge fixing is to enforce the constraint that the tangent vector $\Psi_\partial(B, A)$ is orthogonal to its original MPS $\Psi(A)$:

$$\langle\Psi(A)|\Psi_\partial(B, A)\rangle = N \quad \text{} \quad = 0. \tag{20.14}$$

Hence we can always choose a gauge so that there is no overlap between $\Psi(A)$ and $\Psi_\partial(B, A)$.

As G is not uniquely determined, the gauge degrees of freedom are just partially fixed. To further reduce the variational parameters in B, we parameterize B by a product of four tensors:

$$-\!\!\!\!\!\overbrace{B}\!\!\!\!- \ = \ -\!\!\bigcirc\!\!-\!\!\overbrace{g}\!\!-\!\!\overbrace{V}\!\!-\!\!\bigcirc\!\!- \quad . \tag{20.15}$$
$$l^{-1/2} \qquad\qquad r^{-1/2}$$

Here g is the null space solution of the $D \times dD$ matrix:

$$M_{a,bc} \ = \ l^{1/2}\bigcirc\!\!\!\begin{array}{c}\overbrace{A^*}\!\!-a\\[4pt] c\\[4pt] b\end{array} \quad . \tag{20.16}$$

The nullity, or the number of null solutions, of this matrix is at least $dD - D$. Thus g is a $(d \times D) \times D_r$ matrix with $D_r \geq (d-1)D$, determined by the equation

$$l^{1/2}\bigcirc\!\!\!\begin{array}{c}\overbrace{A^*}\\[6pt] \overbrace{g}\end{array} \ = 0. \tag{20.17}$$

Furthermore, g can be left orthonormalized such that

$$\begin{array}{c}\overbrace{g^*}\\[6pt] \overbrace{g}\end{array} \ = I. \tag{20.18}$$

In Eq. (20.15), V is a $(d-1)D \times D$ matrix. It is the gauge-independent matrix that needs to be variationally optimized. This parametrization automatically enforces the left gauge fixing condition and eliminates the redundancy in B. Therefore, the tangent vector can be represented just using the V-matrix as

$$\Psi_\partial(V, A) = \Psi_\partial(B(V), A) \tag{20.19}$$

with

$$\Psi_\partial(V, A) = \sum_n \ \cdots\!\!-\!\!\overbrace{A}\!\!-\!\!\overbrace{A}\!\!-\!\!\bigcirc\!\!-\!\!\overset{\sigma_n}{\overbrace{g}}\!\!-\!\!\overbrace{V}\!\!-\!\!\bigcirc\!\!-\!\!\overbrace{A}\!\!-\!\!\overbrace{A}\!\!-\!\!\cdots \quad . \tag{20.20}$$
$$l^{-1/2} \qquad\qquad r^{-1/2}$$

20.1.2 Tangent Space Projection

In the application of the tangent-space approach, such as in the solution of the Schrödinger equation using the time-dependent variational principle (TDVP) method [95] to be introduced in §20.2, we often need to project a given quantum state $|\Phi\rangle$ (not necessary to be an MPS) in the Hilbert space onto the tangent space of $\Psi(A)$; namely, to find a matrix V such that

$$\Psi_\partial(V,A) = P_A|\Phi\rangle, \tag{20.21}$$

where P_A is a projector that orthogonally projects $|\Phi\rangle$ onto a tangent vector of $|\Psi(A)\rangle$. This projection is equivalent to optimizing a tangent vector so that the distance between $\Psi_\partial(V,A)$ and Φ, or the cost function

$$\begin{aligned}
f(V) &= \|\,|\Phi\rangle - |\Psi_\partial(V,A)\rangle\,\|^2 \\
&= \langle\Phi|\Phi\rangle - \langle\Phi|\Psi_\partial\rangle - \langle\Psi_\partial|\Phi\rangle + \langle\Psi_\partial|\Psi_\partial\rangle
\end{aligned} \tag{20.22}$$

is minimized.

The norm of the tangent vector $\langle\Psi_\partial(B,A)|\Psi_\partial(B',A)\rangle$ can be readily evaluated from the defining expression of $\Psi_\partial(V,A)$. It is straightforward to show using the left gauge fixing condition that

$$\langle\Psi_\partial(B,A)|\Psi_\partial(B',A)\rangle = N \quad \boxed{\;l\;} \overset{\displaystyle B^*}{\underset{\displaystyle B'}{\Big|}} \boxed{\;r\;} \quad . \tag{20.23}$$

In terms of variational matrices, it becomes

$$\langle\Psi_\partial(V,A)|\Psi_\partial(V',A)\rangle = N \quad \overset{\displaystyle V^*}{\underset{\displaystyle V'}{\Big[\;\;\Big]}} = N\,\mathrm{Tr}\left(V^\dagger V'\right). \tag{20.24}$$

The minimum of $f(V)$ is determined by the equation at which the V^*-derivative of $f(V)$ vanishes:

$$NV = \frac{\partial}{\partial V^*}\langle\Psi_\partial(V,A)|\Phi\rangle. \tag{20.25}$$

If Φ is translation invariant, the divergent factor N cancels out from the two sides of the equation, leading to the solution:

$$\tag{20.26}$$

Substituting it into Eq. (20.21), the tangent space projector is found to be

$$(20.27)$$

It does not depend on V, as expected.

20.2 Time-Dependent Variational Principle

Chapter 19 introduced a number of methods for solving time-dependent problems. However, these methods generally do not inherit all the symmetries of the Hamiltonian. The translation invariance, for example, is broken by the pace-keeping DMRG, the adaptive time-dependent DMRG, and other DMRG-related methods. TEBD uses the Trotter–Suzuki decomposition in performing the time-dependent evolution. This decomposition itself is symplectic. However, the truncation after each evolution step ruins this symmetry. In particular, for a time-independent Hamiltonian, the expectation value of the Hamiltonian is a constant of motion but will drift away in a simulation based on TEBD.

In 2011, Haegeman *et al.* proposed the TDVP method [95] to solve these problems. This method is formulated on an arbitrary variational manifold for a translation-invariant Hamiltonian, with short- and/or long-range interactions. It transforms the linear Schrödinger equation in the whole Hilbert space into a nonlinear set of symplectic differential equations in a restricted parameter space of the variational manifold. This approach is applicable to the imaginary time evolution (which does not maintain the symplectic symmetry).

TDVP does not need a Trotter–Suzuki decomposition. Instead, it assumes that the quantum state can evolve within the variational manifold. This assumption is the primary source of error in this method. In addition, numerical integration of the TDVP equation, which requires a discretization of the time variable, also induces an error in the final results.

TDVP is aiming to solve the Schrödinger equation:

$$i\frac{\partial}{\partial t}|\Psi(A)\rangle = H|\Psi(A)\rangle, \tag{20.28}$$

within the manifold of uniform MPS $|\Psi(A)\rangle$. The time derivative of $\Psi(A)$ is a tangent vector:

$$\frac{\partial}{\partial t}|\Psi(A)\rangle = |\Psi_\partial(\dot{A}, A)\rangle. \tag{20.29}$$

However, the right-hand side of (20.28), $H|\Psi(A)\rangle$, is in general not a vector in the tangent space of $|\Psi(A)\rangle$. To find the gradient of the local tensor $\dot{A}$ that dictates the direction of time evolving of the wave function, an approximation made by TDVP is to project $H|\Psi(A)\rangle$ onto the tangent space of $|\Psi(A)\rangle$ so that

$$i|\Psi_\partial(\dot{A}, A)\rangle = P_A H|\Psi(A)\rangle. \tag{20.30}$$

This, as discussed in §20.1.2, is equivalent to variationally minimizing the function:

$$f(\dot{A}) = \left\| H|\Psi(A)\rangle - i|\Psi_\partial(\dot{A}, A)\rangle \right\|^2. \tag{20.31}$$

The solution to this minimization problem is

$$V = \frac{1}{iN} \frac{\partial}{\partial V^*} \langle \Psi_\partial(V, A)|H|\Psi(A)\rangle, \tag{20.32}$$

where V is the variational matrix that parameterizes $\dot{A}$:

$$\tag{20.33}$$

Eq. (20.32) can also be expressed as

$$V = \frac{1}{iN} \frac{\partial \dot{A}^*}{\partial V^*} \left[\frac{\partial}{\partial \dot{A}^*} \langle \Psi_\partial(\dot{A}, A)|H|\Psi(A)\rangle \right]. \tag{20.34}$$

In the following discussion, we assume that H contains only nearest neighbor interactions:

$$H = \sum_i h_{i,i+1}. \tag{20.35}$$

In this case, the solution reduces to

$$V = -i \frac{\partial \dot{A}^*}{\partial V^*} \left[\frac{\partial}{\partial \dot{A}^*} \langle \Psi_\partial(\dot{A}, A)|h_{1,2}|\Psi(A)\rangle \right]. \tag{20.36}$$

The most consuming part of TDVP calculation is to compute the derivative $\partial_{\dot{A}^*} \langle \Psi_\partial(\dot{A}, A)|h_{1,2}|\Psi(A)\rangle$, which will be discussed in detail. Once this derivative is determined, we can calculate $\dot{A}$ using Eqs. (20.33) and (20.36). The differential equation of $A(t)$ is then solved simply using the Euler scheme:

$$A(t + \delta t) = A(t) + \delta t \dot{A}(t). \tag{20.37}$$

It yields a new local tensor at $t + \delta t$, which is used as an input for the next iteration.

20.2.1 $\langle \Psi_\partial(B,A)|h_{1,2}|\Psi(A) \rangle$ and Its Derivative

The matrix elements of the local Hamiltonian, $\langle \Psi_\partial(B,A)|h_{1,2}|\Psi(A) \rangle$, can be evaluated from the defining expression of $\Psi_\partial(B,A)$ and $\Psi(A)$. In the thermodynamic limit, it can be expressed as a sum of four terms:

$$\langle \Psi_\partial(B,A)|h_{1,2}|\Psi(A) \rangle$$

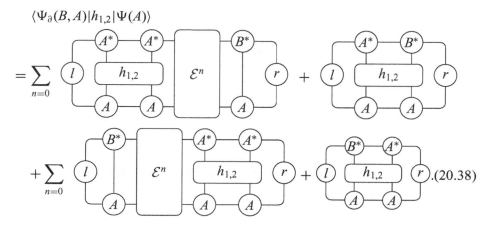

$$.(20.38)$$

To simplify this expression, we expand the transfer matrix $\mathcal{E}$ using its left and right eigenvectors and write it as a sum of terms:

$$\mathcal{E} = |r\rangle\langle l| + \bar{\mathcal{E}}, \tag{20.39}$$

where

$$\bar{\mathcal{E}} = \sum_{k>1} \lambda_k |r_k\rangle\langle l_k|. \tag{20.40}$$

The first term represents the contribution from the dominant eigenvectors of $\mathcal{E}$. The dominant eigenvalue λ_1 is normalized to 1 and is assumed to be non-degenerate. $\bar{\mathcal{E}}$ contains the contribution from all other eigenvectors of $\mathcal{E}$ whose eigenvalues are less than 1 in their absolute values. These terms are orthogonal to each other, so we have

$$\mathcal{E}^n = |r\rangle\langle l| + \bar{\mathcal{E}}^n = |r\rangle\langle l| + \sum_{k>1} \lambda_k^n |r_k\rangle\langle l_k|. \tag{20.41}$$

Using the left gauge fixing condition (20.14), it is straightforward to show that the projector $|r\rangle\langle l|$ has a null contribution to the geometric series of $\mathcal{E}$ in (20.38). Thus only the geometric series of $\bar{\mathcal{E}}$ needs to be considered. In the thermodynamic limit, this series converges:

$$\sum_{n=0} \bar{\mathcal{E}}^n = \frac{1}{I - \bar{\mathcal{E}}} = \sum_{k>1} \frac{1}{1 - \lambda_k} |r_k\rangle\langle l_k|. \tag{20.42}$$

Here $I - \bar{\mathcal{E}}$ is defined in the eigenspace of $\mathcal{E}$ excluding the subspace spanned by the largest eigenvectors; that is, $(|r\rangle, \langle l|)$. It allows us to rewrite (20.38) as

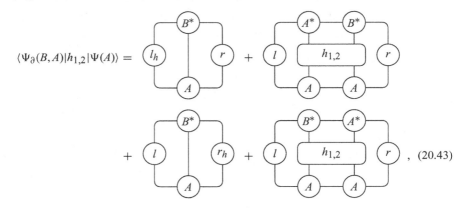

$$, \quad (20.43)$$

where $\langle l_h|$ is a partial contraction to all the terms with $h_{1,2}$ being on the left side of B^*:

$$(20.44)$$

Similarly, $|r_h\rangle$ is a partial contraction to all the terms where $h_{1,2}$ is on the right side of B^*:

$$(20.45)$$

Taking the derivative of $\langle \Psi_\partial(B, A)|h_{1,2}|\Psi(A)\rangle$ with respect to B^* is just to remove the B^*-tensor in $\langle \Psi_\partial(B, A)|h_{1,2}|\Psi(A)\rangle$. This yields

$$\frac{\partial}{\partial B^*}\langle \Psi_\partial(B, A)|h_{1,2}|\Psi(A)\rangle$$

$$+ \; \boxed{l} \; | \; \boxed{r_h} \; + \; \boxed{l} \; \boxed{h_{1,2}} \; \boxed{r} \; . \tag{20.46}$$

20.2.2 Determination of the Ground-State MPS

In a translation-invariant system without spontaneous symmetry breaking, the local tensors become identical on all the lattice sites, which dramatically reduces the number of variational parameters. In principle, all the approaches introduced in Chapter 12 can be applied to determine the local tensors in the ground-states. The simple update, for example, can be directly used to find the ground state wave function. The result can be further used as an input to refine the wave function using either the full update or the variational minimization. However, as all the local tensors evolve simultaneously in the variational optimization, the wave function becomes highly nonlinear. One has to solve the full update or variational minimization problem self-consistently by taking the local tensors obtained from the previous iteration as an input for this iteration. This nonlinearity may cause numerical instability in the evaluation of local tensors.

In a finite system, one can follow the steps given in §12.3.2 to perform a full update calculation. Each full update starts by evaluating the norm and the overlap matrices, $\mathcal{N}$ and $\mathcal{M}$. Equation (12.55) is then solved iteratively until the local tensors converge. For an infinite system, however, the full update can be done more efficiently by representing the evolution operator $\exp(-\varepsilon H)$ as a transfer matrix and solving the fixed-point equations that the converged local tensors satisfy. A detailed introduction to this kind of fixed-point approach is given in §14.3.

The ground-state optimization can also be implemented by applying the projection operator $\exp(-\tau H)$ onto a uniform MPS $|\Psi(A_0)\rangle$ with an arbitrarily initialized local tensor A_0 that is not orthogonal to the ground state. Again, the projection is performed within the manifold of MPS with a fixed bond dimension. In the infinite τ limit, the resultant MPS is just the solution of the ground state:

$$|\Psi(A)\rangle = \lim_{\tau \to \infty} e^{-\tau H} |\Psi(A_0)\rangle. \tag{20.47}$$

This projection can also be realized by taking an imaginary time evolution and solving the equation within the manifold of MPS:

$$-\frac{\partial}{\partial \tau} |\Psi(A)\rangle = H |\Psi(A)\rangle. \tag{20.48}$$

It can be regarded as the Schrödinger equation in the imaginary time space. By replacing i with -1, it is straightforward to solve this equation using the method introduced in §20.2. Again, we parameterize the derivative $\partial_\tau A$ so that it satisfies the left gauge fixing condition:

$$
\underset{\partial_\tau A}{-\!\!\!\bigcirc\!\!\!-} \;=\; \underset{l^{-1/2}}{-\!\!\!\bigcirc\!\!\!-}\!\!\underset{}{\textcircled{g}}\!\!\underset{}{\textcircled{V}}\!\!\underset{r^{-1/2}}{-\!\!\!\bigcirc\!\!\!-} \;. \tag{20.49}
$$

The solution of V is simply given by Eq. (20.36) with i replaced by -1

$$
V = \frac{\delta(\partial_\tau A^*)}{\delta V^*}\left[\frac{\delta}{\delta(\partial_\tau A^*)}\langle\Psi_\partial(\partial_\tau A, A)|h_{1,2}|\Psi(A)\rangle\right]. \tag{20.50}
$$

We then use the derivative $\partial_\tau A$ that is found by inserting the result of V back to Eq. (20.49) to update the local tensor from τ to $\tau + \delta\tau$:

$$
A(\tau + \delta\tau) = A(\tau) + \delta\tau(\partial_\tau A). \tag{20.51}
$$

Repeating these steps allows the local tensor to evolve quasi-continuously in the imaginary time space. The solution in the infinite-τ limit gives the ground-state MPS.

Alternatively, one can also determine the local tensor A in the uniform MPS $\Psi(A)$ by minimizing the energy density per bond:

$$
e = \frac{1}{N}\frac{\langle\Psi|H|\Psi\rangle}{\langle\Psi|\Psi\rangle} = \frac{\langle\Psi|h_{1,2}|\Psi\rangle}{\langle\Psi|\Psi\rangle}. \tag{20.52}
$$

This minimization problem can be solved by employing the steepest descent or other optimization approaches or by solving the Ritz–Rayleigh equation as defined by Eq. (12.6). In either case, we need to calculate the derivative of e with respect to the local tensors.

The derivative $\delta e/\delta A$ can be computed, for example, by simply employing the automatic differentiation approach [120]. In this case, the derivative is determined purely by running the computational codes that implement the automatic differentiation without explicitly evaluating the effective Hamiltonian as well as the norm matrices.

20.3 Single-Mode Excitations

The single-mode approximation is based on the assumption that the ground state of a bosonic system has a scarcity of long-wavelength excitations. It was initially introduced by Bijl et al. [40], and later explored by Feynman to accurately calculate the spectrum of phonons in superfluid ^{4}He [41, 42]. Feynman's general idea was that since the superfluid ground state is a Bose condensate, low-lying excitations just above it cannot be single-particle excitations but are necessarily collective excitations that should manifest as density waves of the superfluid, similar to elastic waves in a continuum. This approximation was first used in the context of spin systems by Arovas *et al.* [414].

The single-mode approximation was first applied to a periodic MPS by representing the excited state as a momentum superposition of a locally perturbed ground state by Östlund and Rommer in 1995 [20]. The perturbation was implemented through a bond matrix that is variationally optimized. Later, a slightly modified ansatz for the momentum superposition of locally perturbed MPS was introduced. In Ref. [43], a Fourier transform of some local operators was used to create excitations, which is closely related to the strategy first introduced by Feynman [41]. In Ref. [44], on the other hand, a momentum superposition is taken by replacing one of the local tensors in the ground-state MPS with a perturbed local tensor that is variationally optimized.

The extension of the single-mode approximation with MPS in the thermodynamic limit was first explored by Haegeman *et al.* [45]. The system is assumed to be translation invariant, and the ground state is represented by a uniform MPS $\Psi_0(A)$. The simplest ansatz of single-particle excitation is to replace tensor A at site 0 in Ψ_0 by a new tensor B and then boost it into a momentum eigenstate using the translation operator:

$$\Psi_k(B) = \sum_n e^{ik\cdot n} T^n \quad \cdots \underset{\substack{\big| \\ A}}{-}\underset{\substack{\sigma_{-2} \\ \big| \\ A}}{-}\underset{\substack{\sigma_{-1} \\ \big| \\ B}}{-}\underset{\substack{\sigma_0 \\ \big| \\ A}}{-}\underset{\substack{\sigma_1 \\ \big| \\ A}}{}\cdots$$

$$= \sum_n e^{ik\cdot n} \quad \cdots \underset{\substack{\big| \\ A}}{-}\underset{\substack{\sigma_{n-2} \\ \big| \\ A}}{-}\underset{\substack{\sigma_{n-1} \\ \big| \\ B}}{-}\underset{\substack{\sigma_n \\ \big| \\ A}}{-}\underset{\substack{\sigma_{n+1} \\ \big| \\ A}}{}\cdots \quad . \tag{20.53}$$

Tensor B perturbs the ground state over a finite region around it. It contains all the variational parameters for describing the single-particle excitation.

Clearly, $\Psi_k(B)$ is an eigenstate of the translation operator $T = \exp(-iP)$ with P the momentum operator:

$$T|\Psi_k(B)\rangle = e^{-ik}|\Psi_k(B)\rangle, \tag{20.54}$$

and k is the eigenvalue of the momentum operator:

$$P|\Psi_k(B)\rangle = k|\Psi_k(B)\rangle. \tag{20.55}$$

The single-mode approximated MPS wave function (20.53) works because it is believed that low-lying excited states also satisfy the area law of entanglement entropy, like the ground state, up to a possible logarithmic correction. In (20.53), we assume that the perturbed local tensor B acts only on one site. One can further improve the approximation by allowing the perturbed tensor B to act on a block containing more than one site [415]. In practical calculations, however, the size of the block cannot be too large due to the exponential growth of computational cost. To reduce the cost, one can replace each A-tensor in this block with a perturbed tensor that is variationally optimized.

Furthermore, by imposing symmetry to block diagonalize each local tensor, one can attach an extra quantum number to tensor B. This allows us to simulate an excitation with a particular quantum number, which can be either topologically trivial or nontrivial [415].

20.3.1 Gauge Fixing

It is straightforward to show that $\Psi_k(B)$ is invariant under an additive gauge transformation for tensor B:

$$-\!\!\fbox{B}\!\!- \rightarrow -\!\!\fbox{B}\!\!- + e^{ik} -\!\!\fbox{A}\!\!-\!\!\fbox{G}\!\!- - -\!\!\fbox{G}\!\!-\!\!\fbox{A}\!\!- , \tag{20.56}$$

where G is an invertible $D \times D$ matrix. This can be verified by substituting this expression into Eq. (20.53). As for the tangent vector of Ψ_0, we can also find a gauge matrix G so that the left gauge fixing condition is satisfied:

$$\left(\begin{array}{c} l \end{array}\right)\!\!\fbox{A^*}\!\!\fbox{B} = \left(\begin{array}{c} l \end{array}\right)\!\!\fbox{B^*}\!\!\fbox{A} = 0. \tag{20.57}$$

This gauge fixing removes the gauge redundancy in B. To prove this, we first assume that our gauge fixing equation is not satisfied. In that case, we can always find a gauge matrix G so that

$$\left(\begin{array}{c} l \end{array}\right)\!\!\fbox{A^*}\!\!\fbox{B} + e^{ik}\left(\begin{array}{c} l \end{array}\right)\!\!\fbox{A^*}\!\!\fbox{A}\!\!\fbox{G} - \left(\begin{array}{c} l \end{array}\right)\!\!\fbox{A^*}\!\!\fbox{G}\!\!\fbox{A} = 0. \tag{20.58}$$

Since $\langle l|$ is the maximal left eigenvector of the transfer matrix with eigenvalue 1, the equation can also be written as

$$\left(\begin{array}{c} l \\ G \end{array}\right)\!\!\fbox{$e^{ik} - \mathcal{E}$} = -\left(\begin{array}{c} l \end{array}\right)\!\!\fbox{A^*}\!\!\fbox{B} . \tag{20.59}$$

This is a set of linear equations. $e^{ik} - \mathcal{E}$ is a $D^2 \times D^2$ matrix. When $k = 0$, it reduces to Eq. (20.8). As already shown, this equation has a solution, and the left gauge fixing condition holds.

If $k \neq 0$, $e^{ik} - \mathcal{E}$ is a full rank matrix, the solution to the gauge matrix G is

$$(20.60)$$

This shows that one can indeed find a gauge matrix G so that Eq. (20.57) is fulfilled. The gauge degrees of freedom are completely fixed by Eq. (20.57) when $k \neq 0$, but only partially fixed when $k = 0$.

Again, we impose the left gauge fixing by parameterizing B as a product of four tensors, the same as in Eq. (20.15):

$$(20.61)$$

This separates the gauge redundancy and the independent variational parameters. V is the $(d - 1)D \times D$ gauge-independent matrix that needs to be variationally optimized.

From the momentum conservation, it is simple to show that $\Psi_k(B)$ is orthogonal to Ψ_0 for any finite k:

$$\langle \Psi_0 | \Psi_k(B) \rangle = 0 \qquad (k \neq 0). \tag{20.62}$$

For $k = 0$, the overlap between $\Psi_k(B)$ and Ψ_0 is

$$(20.63)$$

which is also zero when the left gauge fixing condition holds.

20.3.2 Variational Optimization

Given Ψ_0, tensor B or V should be determined by minimizing the cost function:

$$f[B] = \frac{\langle\Psi_k(B)|H - E_0|\Psi_k(B)\rangle}{\langle\Psi_k(B)|\Psi_k(B)\rangle}, \tag{20.64}$$

where E_0 is the ground-state energy. It is explicitly included in the above expression to remove the divergence in the excitation energy with the system size. By taking the derivative with respect to B, this minimization problem is reduced to solving the extreme equation:

$$\frac{\partial}{\partial B}\langle\Psi_k(B)|H - E_0|\Psi_k(B)\rangle = \omega_k \frac{\partial}{\partial B}\langle\Psi_k(B)|\Psi_k(B)\rangle, \tag{20.65}$$

where ω_k is the excitation energy at the extreme point:

$$\omega_k = \frac{\langle\Psi_k(B)|H - E_0|\Psi_k(B)\rangle}{\langle\Psi_k(B)|\Psi_k(B)\rangle}. \tag{20.66}$$

Both the numerator and the denominator of $f[B]$ are quadratic functions of B. If we convert B into a vector, we can represent them as

$$\langle\Psi_k(B)|H - E_0|\Psi_k(B)\rangle = B^\dagger \mathcal{H}_k B, \tag{20.67}$$

$$\langle\Psi_k(B)|\Psi_k(B)\rangle = B^\dagger \mathcal{N}_k B, \tag{20.68}$$

where $\mathcal{H}_k$ and $\mathcal{N}_k$ are the effective Hamiltonian and norm matrices, respectively. Substituting them into Eq. (20.65), we obtain the Ritz–Rayleigh equation that determines B:

$$\mathcal{H}_k B = \omega_k \mathcal{N}_k B. \tag{20.69}$$

This generalized eigenproblem determines the one-particle dynamics on top of the ground state. Both the excitation energy ω_k and B are determined by solving this eigenproblem.

If we parameterize B with Eq. (20.61) and take V as variational parameters, the expectation values could also be represented as

$$\langle\Psi_k(B)|H - E_0|\Psi_k(B)\rangle = V^\dagger \mathcal{H}_k^V V, \tag{20.70}$$

$$\langle\Psi_k(B)|\Psi_k(B)\rangle = V^\dagger \mathcal{N}_k^V V. \tag{20.71}$$

$\mathcal{H}_k^V$ and $\mathcal{N}_k^V$ are the effective Hamiltonian and norm matrices. The corresponding eigenequation becomes

$$\mathcal{H}_k^V V = \omega_k \mathcal{N}_k^V V. \tag{20.72}$$

The norm and Hamiltonian matrices, $\mathcal{N}_k$ and $\mathcal{H}_k$ (similarly $\mathcal{N}_k^V$ and $\mathcal{H}_k^V$), can be determined from the derivatives of $\langle\Psi_k(B)|\Psi_k(B)\rangle$ and $\langle\Psi_k(B)|H - E_0|\Psi_k(B)\rangle$, with respect to both B and B^*. This is equivalent to removing the B and B^* tensors from $\langle\Psi_k(B)|\Psi_k(B)\rangle$ and $\langle\Psi_k(B)|H - E_0|\Psi_k(B)\rangle$, respectively. Next we discuss the calculation of these two expectation values.

20.3.3 The Norm Matrix $\langle\Psi_k(B)|\Psi_k(B)\rangle$

The norm matrix $\langle\Psi_k(B)|\Psi_k(B)\rangle$ is a double infinite sum. It can be grouped according to the relative positions of B and B^* tensors into two terms:

$\langle\Psi_k(B)|\Psi_k(B)\rangle$

$$(20.73)$$

The last term vanishes because B satisfies the left gauge condition (20.57), so that

$$(20.74)$$

The last equality indicates that the effective normalization matrix $\mathcal{N}_k^V$ is an identity up to a constant:

$$\mathcal{N}_k^V = N. \qquad (20.75)$$

20.3.4 The Hamiltonian Matrix $\langle\Psi_k(B)|H - E_0|\Psi_k(B)\rangle$

The expectation value $\langle\Psi_k(B)|H - E_0|\Psi_k(B)\rangle$ depends on the Hamiltonian. We consider a system whose Hamiltonian contains just the nearest-neighbor interactions for simplicity. The discussion that follows, however, can be readily extended to more general cases. Using the fact that $\Psi_k(B)$ is the eigenvector of the translation operator, we obtain

$$\langle\Psi_k(B)|H - E_0|\Psi_k(B)\rangle = N\langle\Psi_k(B)|\Delta_{12}|\Psi_k(B)\rangle, \qquad (20.76)$$

where

$$\Delta_{12} = h_{12} - \bar{h}_0, \tag{20.77}$$

and $\bar{h}_0$ is the bond energy in the ground state:

$$\bar{h}_0 = \langle \Psi_0 | h_{12} | \Psi_0 \rangle = \quad \text{(diagram)} \tag{20.78}$$

$|\Psi_0\rangle$ is assumed to be normalized.

The expectation value $\langle \Psi_k(B) | \Delta_{12} | \Psi_k(B) \rangle$ is a quadratic function of B and a double infinite sum:

$$\langle \Psi_k(B) | \Delta_{12} | \Psi_k(B) \rangle = \sum_{m,n} e^{ik(m-n)} \langle \phi(B) | T^{-n} \Delta_{12} T^m | \phi(B) \rangle, \tag{20.79}$$

where

$$\phi(B) = \text{Tr} \left(\cdots A[\sigma_{-2}] A[\sigma_{-1}] B[\sigma_0] A[\sigma_1] A[\sigma_2] \cdots \right). \tag{20.80}$$

Depending on the relative positions of these two B-tensors, m and n, this expectation value can be grouped into a sum of two terms:

$$\langle \Psi_k(B) | \Delta_{12} | \Psi_k(B) \rangle = \Gamma_1 + \Gamma_2, \tag{20.81}$$

where

$$\Gamma_1 = \sum_m \langle \phi(B) | T^{-m} \Delta_{12} T^m | \phi(B) \rangle, \tag{20.82}$$

$$\Gamma_2 = \sum_{m \neq n} e^{ik(m-n)} \langle \phi(B) | T^{-n} \Delta_{12} T^m | \phi(B) \rangle. \tag{20.83}$$

Γ_1 contains the terms with $n = m$, and Γ_2 contains all other terms. We calculate these two terms separately.

The Γ_1-Term

The Γ_1-term can be further written as a sum of three terms, depending on the relative position of m with respect to Δ_{12}:

$$\Gamma_1 = \Sigma_1 + \Sigma_2 + \Sigma_3, \tag{20.84}$$

$$\Sigma_1 = \sum_{m < 1} \langle \phi(B) | T^{-m} \Delta_{12} T^m | \phi(B) \rangle, \tag{20.85}$$

$$\Sigma_2 = \sum_{1 \leq m \leq 2} \langle \phi(B) | T^{-m} \Delta_{12} T^m | \phi(B) \rangle, \tag{20.86}$$

$$\Sigma_3 = \sum_{m>2} \langle \phi(B)|T^{-m}\Delta_{12}T^m|\phi(B)\rangle. \tag{20.87}$$

We now derive the expression for these three terms.

(i) Using the left and right eigenvectors of the transfer matrix, it is simple to show that Σ_1 is

$$\Sigma_1 = \sum_{n\geq 0} \quad \text{(20.88)}$$

Since the expectation value of Δ_{12} vanishes in the ground state

$$= 0, \tag{20.89}$$

Σ_1 can be further simplified as

$$\Sigma_1 = \quad \text{(20.90)}$$

where

$$, \tag{20.91}$$

and $(I - \bar{\mathcal{E}})^{-1}$ is defined in the subspace spanned by the nonzero eigenvalues and eigenstates of $I - \bar{\mathcal{E}}$.

(ii) Using the property of the transfer matrix, Σ_2 is found to be

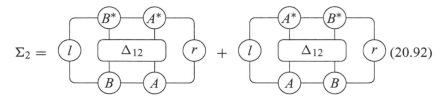

$$\hspace{10cm} (20.92)$$

(iii) Similar to Σ_1, Σ_3 is

$$\Sigma_3 = \hspace{6cm} , \hspace{2cm} (20.93)$$

where

$$\hspace{10cm} (20.94)$$

The Γ_2-Term

In the double summation of Γ_2, if m or n is less than 1, the corresponding term vanishes if the left gauge condition of B is imposed. Hence only the terms with both m and n larger than or equal to 1 need to be considered. Moreover, we only need to calculate the terms with $m > n$ since the other terms are equal to their complex conjugate. In this case, we can write Γ_2 as a sum of the following three terms and their complex conjugate:

$$\Gamma_2 = \Sigma_4 + \Sigma_5 + \Sigma_6 + c.c., \hspace{2cm} (20.95)$$

$$\Sigma_4 = e^{ik}\langle\phi(B)|T^{-1}\Delta_{12}T^2|\phi(B)\rangle, \hspace{2cm} (20.96)$$

$$\Sigma_5 = \sum_{1\leq n\leq 2, m>2} e^{ik(m-n)}\langle\phi(B)|T^{-n}\Delta_{12}T^m|\phi(B)\rangle, \hspace{2cm} (20.97)$$

$$\Sigma_6 = \sum_{2\leq n<m} e^{ik(m-n)}\langle\phi(B)|T^{-n}\Delta_{12}T^m|\phi(B)\rangle. \hspace{2cm} (20.98)$$

Again, the summations in these expressions can be explicitly evaluated using the left gauge condition of B and Eq. (20.89).

(i) Σ_4 is relatively simple to calculate. The result is

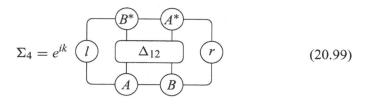

$$\Sigma_4 = e^{ik} \quad\quad\quad\quad (20.99)$$

(ii) In Σ_5, n takes two values, $n = 1$ and $n = 2$. Let us first consider the case $n = 1$. Using the left gauge condition, it is simple to show that

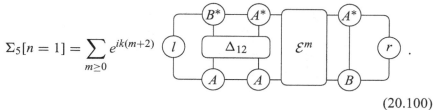

$$\Sigma_5[n = 1] = \sum_{m \geq 0} e^{ik(m+2)} \quad\quad\quad (20.100)$$

Using the left gauge condition, we can replace $\mathcal{E}$ by $\bar{\mathcal{E}}$ and simplify the expression as

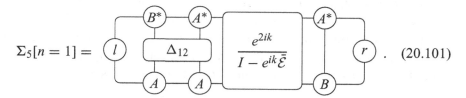

$$\Sigma_5[n = 1] = \quad\quad\quad (20.101)$$

The expression for the $n = 2$ term is similar. The final result of Σ_5 is

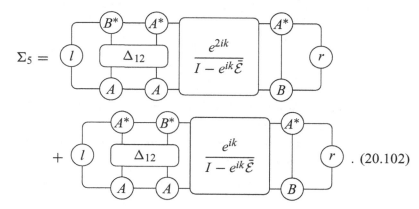

$$\Sigma_5 = \quad\quad + \quad\quad . \quad (20.102)$$

(iii) Σ_6 can be rewritten as a double infinite sum and simplified as

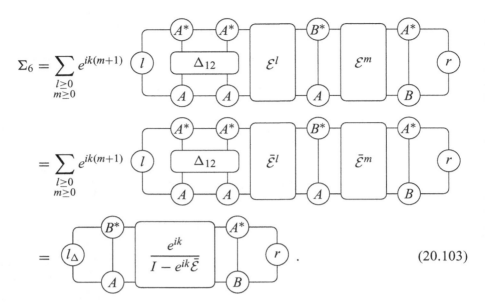

$$= \text{(1}_\Delta\text{)} \quad \frac{e^{ik}}{I - e^{ik}\bar{\mathcal{E}}} \quad \text{(r)} \quad . \tag{20.103}$$

20.4 Excitations Represented with PEPS

The single-mode approximation can also be implemented to evaluate excited states represented by PEPS. This approach was first exploited by Vanderstraeten *et al.* [121]. Similar to one dimension, the single-particle excitation is assumed to be a momentum superposition of PEPS by replacing one of the local tensors A from a translation-invariant ground state with a perturbed local tensor B:

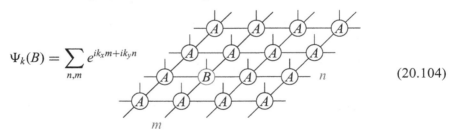

$$\Psi_k(B) = \sum_{n,m} e^{ik_x m + ik_y n} \tag{20.104}$$

where (m, n) is the lattice coordinate, and $k = (k_x, k_y)$ is the momentum of the state. Both A and B are five-leg tensors.

For finite k, the overlap between this excited state and the ground state is zero due to momentum conservation:

$$\langle \Psi_0 | \Psi_{k \neq 0}(B) \rangle = 0. \tag{20.105}$$

When $k = 0$, the overlap between Ψ_0 and $\Psi_{k=0}(B)$, from a top view of the double-layer tensor diagram, is

$$\langle \Psi_0 | \Psi_{k=0}(B) \rangle \;=\; N \quad \text{} \quad , \tag{20.106}$$

where N is the total number of lattice sites, and $\mathcal{E}$ and $\mathcal{B}$ are double-layer tensors, defined by

$$\text{} \tag{20.107}$$

By contracting all A tensors, eventually, we should be able to express this overlap as an inner product between a vector B, which is converted from the five-leg tensor B, and another vector $\mathcal{G}$:

$$\langle \Psi_0 | \Psi_{k=0}(B) \rangle = N \mathcal{G}^\dagger \cdot B. \tag{20.108}$$

Here we assume that the ground-state wave function is normalized to 1. $\mathcal{G}$ is obtained by contracting the whole tensor network excluding the B-tensor. This contraction can be done only approximately. As the excited state must be orthogonal to the ground state, Eq. (20.108) requires that B must be confined to the variational subspace orthogonal to $\mathcal{G}$; that is,

$$\mathcal{G}^\dagger \cdot B = 0. \tag{20.109}$$

Furthermore, it is simple to show that $\Psi_k(B)$ is invariant under the gauge transformation along the x-axis:

$$\text{} \tag{20.110}$$

as well as along the y-axis direction:

$$\text{} \tag{20.111}$$

In order words, if we set

$$\text{} \tag{20.112}$$

or

$$\text{\includegraphics{}} = e^{iky} \text{\includegraphics{}} - \text{\includegraphics{}} \quad , \tag{20.113}$$

this would yield a null state

$$\Psi_k(B_X) = \Psi_k(B_Y) = 0. \tag{20.114}$$

X and Y are two arbitrary $D \times D$ matrices. This suggests that in order to determine the physically allowed B-tensors in the tangent space, one should exclude the tensors in the subspace spanned by all linearly independent B_X- and B_Y-tensors. Hence the physically allowed vectorized B-tensors should be orthogonal to vectorized B_X and B_Y-tensors, namely

$$B_X^\dagger \cdot B = B_Y^\dagger \cdot B = 0. \tag{20.115}$$

This, together with Eq. (20.109), indicates that the physically allowed vectorized B-tensors in the tangent space should be the non-trivial solutions of the equation:

$$(\mathcal{G}, B_X, B_Y)^\dagger \cdot B = 0. \tag{20.116}$$

$(\mathcal{G}, B_X, B_Y)$ is a $(2D^2 + 1)$-column matrix. Its rank is at most $2D^2 + 1$. Thus there are at most $2D^2 + 1$ linearly independent vectorized B-basis tensors that should be excluded from the tangent space of $|\Psi\rangle$.

As the dimension of B-vector is dD^4, the homogeneous system of linear equations (20.116) should have at least $dD^4 - 2D^2 - 1$ basic solutions. Let us denote these basic solutions as B_α ($\alpha = 1, \ldots, N_B$), with N_B the total number of basic solutions.

To find the excited states, we start from all physically allowed and orthogonal B-tensors in the tangent space. The momentum eigenstates constructed from these B-tensors are also linearly independent, but they are usually not orthogonal to each other. From the wave function overlap

$$\langle \Psi_k(B_\alpha) | \Psi_k(B_\beta) \rangle = B_\alpha^\dagger \mathcal{N}_{k,\alpha\beta} B_\beta, \tag{20.117}$$

we can determine the norm matrix $\mathcal{N}_k$ in the space spanned by the basic solutions B_α. Similarly, the effective Hamiltonian matrix is determined by the overlap

$$\langle \Psi_k(B_\alpha) | H | \Psi_k(B_\beta) \rangle = B_\alpha^\dagger \mathcal{H}_{k,\alpha\beta} B_\beta. \tag{20.118}$$

Evaluation of these effective Hamiltonian and norm matrices is much more complicated than that in one dimension. A scheme for evaluating these matrices can be found in Ref. [416].

The physical allowed B-tensor in the tangent space is a linear superposition of the basic solutions of Eq. (20.116):

$$B = \sum_\alpha b_\alpha B_\alpha, \tag{20.119}$$

where the coefficients $b = (b_1, \cdots, b_{N_B})$ are determined by the generalized eigen-equation:

$$\mathcal{H}_k b = \omega_k \mathcal{N}_k b. \tag{20.120}$$

ω_k is the excitation spectrum.

In practical calculation, removing the exact null modes does not always guarantee that the equation has a stable solution. We often need to reduce the tangent subspace further by eliminating the modes with a small but finite norm. These small-norm modes may lead to spurious eigenvalues ω_k.

Using the eigenvectors of $\mathcal{N}_k$, we can define a projection operator P_k to exclude all the small eigenmodes of $\mathcal{N}_k$. Suppose $|l_m\rangle$ to be eigenvectors of $\mathcal{N}_k$ with the eigenvalues η_m, and η_t is a threshold below which the eigenmodes of $\mathcal{N}_k$ are ignored; then

$$P_k = \sum_{\eta_m > \eta_t} |l_m\rangle \langle l_m|. \tag{20.121}$$

Now the solution of the b-vector is confined to

$$b_k = P_k b, \tag{20.122}$$

and the generalized eigenequation for determining b_k becomes

$$(P_k \mathcal{H}_k P_k)\, b_k = \omega_k\, (P_k \mathcal{N}_k P_k)\, b_k. \tag{20.123}$$

As an example of application, we calculate the zero-temperature dynamical spectral function with the above method for the spin-1/2 antiferromagnetic Heisenberg model on the square lattice:

$$S(k, \omega) = \sum_\alpha S^\alpha(k, \omega), \tag{20.124}$$

where

$$\begin{aligned}
S^\alpha(k, \omega) &= \langle \Psi_0 | S^\alpha_{-k} \delta(\omega - H + E_0) S^\alpha_k | \Psi_0 \rangle \\
&= \sum_m \left| \langle \Psi_k(b_m) | S^\alpha_k | \Psi_0 \rangle \right|^2 \delta(\omega - E_m + E_0)
\end{aligned} \tag{20.125}$$

is the projected spectral functions along the three principal axes $\alpha = (x, y, z)$ and H is the Hamiltonian. $|\Psi_0\rangle$ is the ground state and E_0 is the corresponding energy. $|\Psi(b_m)\rangle$ and E_m are the PEPS excited states and the corresponding energies

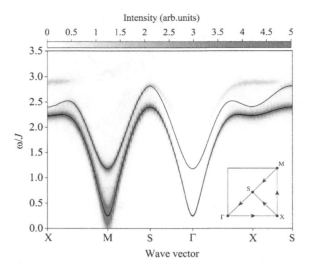

Figure 20.1 Dynamical spin structure factors in the ground state of the spin-1/2 Heisenberg model on the square lattice obtained from the single-mode approximation with $D = 4$. The energy dispersions for the two lowest excitation modes are shown by the solid curves. Inset: The first Brillouin zone and the momentum path (the arrowed lines) on which the spectral functions are shown.

obtained by solving the generalized eigenequation (20.120). The factor before the delta function,

$$w_k^\alpha(m) = \left| \langle \Psi_k(b_m) \left| S_k^\alpha \right| \Psi_0 \rangle \right|^2, \tag{20.126}$$

is the spectral weight at the excitation energy E_m. It is obtained by contracting the double layer tensor network $\langle \Psi(b_m) | S_k^\alpha | \Psi_0 \rangle$.

Figure 20.1 shows the intensity of the spin structural factor $S(k, \omega)$ calculated using the tensor network methods with the $D = 4$ PEPS along a representative path in the first Brillouin zone. In obtaining the spectral weight, the delta function in (20.124) is expanded by a Lorentzian broadening factor $\varepsilon = 0.05J$. The lowest excitation is the magnon mode. This is the transverse Goldstone mode, which should be gapless. It shows a finite excitation gap because the entanglement of the ground state is truncated by the finite bond dimension D of the PEPS used in the calculation. The gap value decreases with increasing D and tends to become zero in the infinite bond dimension limit [417]. The dispersion on the line from X (π, 0) to S ($\pi/2, \pi/2$) shows a small dip around X. This differs from a flat dispersion predicted by the conventional linear spin wave theory but is consistent with quantum Monte Carlo [418], as well as other PEPS calculations [416, 417].

The second-lowest energy excitation mode has a finite energy gap in the Brillouin zone. From the projected spectral functions along the three principal axes,

$S^x(k, \omega)$, $S^y(k, \omega)$, and $S^z(k, \omega)$, we find that this branch of excitations is the contribution of the longitudinal excitations, namely the contribution of the Higgs modes. In addition to these two lowest excitations, higher energy excitations also contribute to the spectral functions, but their intensities become significantly weaker.

21

Tree Tensor Network States

21.1 Canonical Representation

A tree tensor network state is defined on a Cayley tree [419] or Bethe lattice. It is a natural extension of MPS as well as a faithful representation and accurate approximation of the ground-state wave function on these lattices. A Bethe lattice is an infinite Cayley tree. It is possible to divide a tree tensor network state into two disconnected parts by cutting one bond like an MPS.

The boundary effect on a finite Cayley tree is strong since more than one-half of the total sites reside on the lattice edge. This effect may dramatically change some physical properties of the system. In some instances, the results obtained on a Cayley tree are qualitatively different from those obtained on the corresponding Bethe lattice. For example, the Ising model exhibits a phase transition on the Bethe lattice but not on a finite Cayley tree [420]. To avoid this boundary effect, we should directly investigate the tree tensor network states on a Bethe lattice [118].

A tree tensor network state is a product of local tensors as functions of local physical basis states on the vertices and the diagonal matrices, or bond vectors, defined on the links emitting from each lattice site [124, 118]. The bond vectors store the entanglement information, and the bond indices represent the quantum numbers of the virtual basis states. A bond links two sites.

A bipartite translation-invariant Bethe lattice is defined by dividing the system into two sublattices. All local tensors on the same sublattice are equal to each other. For example, when $z = 3$, a tree tensor network state on a $z = 3$ Bethe lattice can be expressed as

$$|\Psi\rangle = \sum_{\alpha\beta\gamma\sigma} \prod_{\substack{i \in A \\ j \in B}} \lambda_{x,\alpha_i} \lambda_{y,\beta_i} \lambda_{z,\gamma_i} A_{\alpha_i\beta_i\gamma_i}[\sigma_i] B_{\alpha_j\beta_j\gamma_j}[\sigma_j]|\sigma_i\sigma_j\rangle, \qquad (21.1)$$

or graphically,

$$|\Psi\rangle = \quad$$ 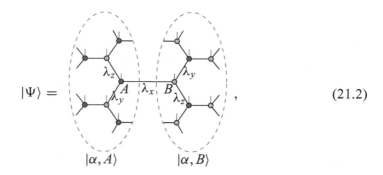   , $$\qquad (21.2)$$

where A and B are the local tensors defined on the A and B sublattice, respectively. $(\lambda_x, \lambda_y, \lambda_z)$ are three bond vectors. $\{\alpha\beta\gamma\sigma\}$ include all bond and physical indices. If the wave function is in the canonical form, the bond vector is just the singular vector whose elements are equal to the square roots of the eigenvalues of the reduced density matrix. The system becomes entirely translational invariant if $A = B$.

The tree tensor network state contains redundant gauge degrees of freedom on each bond. It is invariant if one inserts a product of two reciprocal matrices on a bond and absorbs each separately to a local tensor at the two ends of the bond. These gauge degrees of freedom can be fixed by introducing a canonical transformation to convert the tensor network state into a canonical form.

Like in an MPS, a bond index that connects two tensors in a tree tensor network state divides the whole tree into two subtrees. In the canonical representation, the Schmidt decomposition of the wave function reads

$$|\Psi\rangle = \sum_\alpha \lambda_{x,\alpha} |\alpha, A\rangle |\alpha, B\rangle, \qquad (21.3)$$

where $|\alpha, A\rangle$, as shown in Eq. (21.2), denotes a basis state in the semi-infinite block starting with a local tensor A. $|\alpha, B\rangle$ is similarly defined. Both $|\alpha, A\rangle$ and $|\alpha, B\rangle$ are orthonormalized:

$$\langle \alpha', A|\alpha, A\rangle = \delta_{\alpha',\alpha}, \qquad (21.4)$$
$$\langle \alpha', B|\alpha, B\rangle = \delta_{\alpha',\alpha}. \qquad (21.5)$$

The bond vectors $(\lambda_x, \lambda_y, \lambda_z)$ are the singular values of the wave function on the corresponding bonds. From the normalization of the wave function, we have

$$\sum_\alpha \lambda_{x,\alpha}^2 = \sum_\beta \lambda_{y,\beta}^2 = \sum_\gamma \lambda_{z,\gamma}^2 = 1. \qquad (21.6)$$

On a bipartite translation-invariant Bethe lattice, the basis state $|\alpha, A\rangle$ can be expressed using $|\alpha, B\rangle$ as

$$|\alpha, A\rangle = \sum_{\beta\gamma\sigma} A_{\alpha\beta\gamma} \lambda_{y,\beta} \lambda_{z,\gamma} |\gamma, B\rangle |\beta, B\rangle |\sigma\rangle, \tag{21.7}$$

or graphically

$$|\alpha, A\rangle \bigcirc\!\!- \quad = \quad \tag{21.8}$$

Here $|\sigma\rangle$ is a physical basis state of local tensor A. Substituting this expression into Eq. (21.3), the wave function becomes

$$|\Psi\rangle = \sum_{\alpha\beta\gamma} \lambda_{x\alpha} \lambda_{y\beta} \lambda_{z\gamma} A_{\alpha\beta\gamma} |\gamma B\rangle |\beta B\rangle |\alpha B\rangle |\sigma\rangle. \tag{21.9}$$

This is a canonical wave function defined around a site on the A-sublattice.

As both $|\alpha, A\rangle$ and $|\alpha, B\rangle$ are orthonormalized, it is straightforward to show that the local tensor with the associated bond vectors satisfies the canonical equation:

$$\langle \alpha', A | \alpha, A \rangle = \sum_{\beta\gamma\sigma} \lambda_{y,\beta}^2 \lambda_{z,\gamma}^2 A_{\alpha\beta\gamma}^* [\sigma] A_{\alpha'\beta\gamma} [\sigma] = \delta_{\alpha,\alpha'}. \tag{21.10}$$

Graphically, it can be represented as

$$= I. \tag{21.11}$$

By swapping the bond indices of the local tensors, it can be shown that the local tensors should also satisfy the other two canonical equations:

$$\sum_{\alpha\gamma\sigma} \lambda_{x,\alpha}^2 \lambda_{z,\gamma}^2 A_{\alpha\beta\gamma}^* [\sigma] A_{\alpha\beta'\gamma} [\sigma] = \delta_{\beta,\beta'}, \tag{21.12}$$

$$\sum_{\alpha\beta\sigma} \lambda_{x,\alpha}^2 \lambda_{y,\beta}^2 A_{\alpha\beta\gamma}^* [\sigma] A_{\alpha\beta\gamma'} [\sigma] = \delta_{\gamma,\gamma'}. \tag{21.13}$$

Replacing A by B, we obtain the canonical equations for tensor B.

The canonical decomposition, defined by Eq. (21.3), is equivalent to the canonical equations defined by Eqs. (21.10)–(21.13) for the local tensors. One can cut an arbitrary bond to divide the Bethe lattice into two sublattices and define the reduced density matrix by integrating all the tensors in one of the sublattices. It is simple to show that the square of $\lambda_{x,\alpha}$ in Eq. (21.3) is just the eigenvalue of the reduced density matrix, and $|\alpha, A\rangle$ or $|\alpha, B\rangle$ the corresponding eigenvector.

It is straightforward to generalize this derivation to obtain the canonical equations for local tensors on an arbitrary Bethe lattice with or without translation invariance. The canonical form of the tree tensor network state is useful in practical calculations. As the bond vector describes the entanglement spectrum between the two subblocks linked by this bond, selecting the virtual bond basis states according to this spectrum provides an optimal truncation scheme. Furthermore, the contribution of the environment tensors can be faithfully represented by the bond vectors on the dangling bonds surrounding the system tensors. As described in §21.4, this canonical representation leads to an efficient and nearly optimized scheme to evaluate the ground-state wave function by taking the imaginary time evolution.

In the canonical representation of the tree tensor network state, the expectation value of a local operator can be readily evaluated. For example, for a local operator $\hat{O}$ defined at a site on the A sublattice, its expectation value can be evaluated by utilizing the canonical wave function (21.9) as

$$\langle \hat{O} \rangle = \langle \Psi | \hat{O} | \Psi \rangle = \sum_{\alpha\beta\gamma\sigma\sigma'} A^*_{\alpha\beta\gamma}[\sigma'] A_{\alpha\beta\gamma}[\sigma] \lambda^2_{x\alpha} \lambda^2_{y\beta} \lambda^2_{z\gamma} \langle \sigma' | \hat{O} | \sigma \rangle. \tag{21.14}$$

The expectation values of other local operators, defined on a few lattice sites or a cluster, can be similarly determined.

21.2 Canonicalization

An arbitrarily given tree tensor network state (21.1) is generally not canonical. Nevertheless, it can always be canonicalized similarly to canonicalizing an MPS. To do this, we define a two-site matrix by tracing the local tensors on the neighboring sites and all the bond vectors connecting to these two sites. For example, for the two sites connected by a horizontal bond, that is, a bond along the x-axis, this two-site matrix is defined by

$$M_{ab\sigma_1, a'b'\sigma_2} = \quad\quad \quad\quad . \tag{21.15}$$

The rank of this matrix is upper bounded by the bond dimension of λ_x (i.e. D). Similarly, we can define the two-site tensor on the bond along the y- or z-axis.

To canonicalize the wave function, we factorize the two-site tensor (21.15) by SVD

$$M_{ab\sigma_1,a'b'\sigma_2} = \sum_l U_{ab\sigma_1,l}\lambda_l V_{a'b'\sigma_2,l}. \tag{21.16}$$

Since the rank of M is not more than the bond dimension D, λ_l has at most D nonzero singular values. Hence both U and V are isometric matrices. Now we update λ_x by λ, and local tensors A and B by the formula

$$c = \lambda_{y,a}^{-1}\lambda_{z,b}^{-1}U_{ab\sigma,c} \qquad c = \lambda_{y,a}^{-1}\lambda_{z,b}^{-1}V_{ab\sigma,c} \ . \tag{21.17}$$

After performing similar factorization and update for the bonds along the y- and z-axis, we complete a cycle of iterations. Repeating this cycle many times, we will eventually obtain the canonicalized local tensors, A and B, as well as the bond vectors, λ_x, λ_y, and λ_z.

21.3 Husimi Lattice

A Cayley tree or Bethe lattice is a graph without any closed loops. On the other hand, if a graph contains loops or cycles, but any two cycles have at most one vertex in common, this graph is called a cactus or a cactus tree. As it was Husimi [421] who first introduced cactus graphs in the study of statistical mechanics, a cactus is also a Husimi tree [422, 423]. A Husimi tree is a connected graph whose building blocks or lobes are p-polygons ($p \geq 2$) and where every edge or bond belongs to at most one polygon. For example, the 2-polygon is a bond, and the 3-polygon is a triangle. In other words, a Husimi tree is a connected graph in which every building block is either a polygon or an edge. On the other hand, if all lobes consist of only one type of p-polygon, the system is known as a pure Husimi tree, of which the simplest is the Cayley tree, whose lobes consist only of bonds (or 2-polygons).

A pure Husimi tree is characterized by two numbers, (p, z), where p is the number of edges of the p-polygon and z is the coordination number of each vertex. The size of the loop is p. A Husimi tree with $z = 4$ and $p > 2$ can be derived from the corresponding Cayley tree with coordination number p if each bond of the Cayley tree is replaced by a single vertex and each vertex by a single p-polygon. For example, the cactus graph shown in (21.18) is a pure triangular Husimi tree of $(p, z) = (3, 4)$:

$$\tag{21.18}$$

A Husimi lattice is an infinite pure Husimi tree, which differs from a pure Husimi tree in the same way that a Bethe lattice differs from a Cayley tree. The pure Husimi tree has a central polygon and a boundary, and the number of polygons grows exponentially with the distance from the center. Hence the number of polygons on the boundary is of the same order as in the interior, even in the thermodynamic limit. By contrast, the Husimi lattice preserves translational invariance because it has no center and no boundary, and all of its polygons are equivalent. Therefore, any model defined on the Husimi tree may show strong finite-size effects and behave very differently from the Husimi lattice. Consequently, a model defined on the Husimi lattice is much more suitable for simulating a translation-invariant system than a model on the Husimi tree.

The triangular Husimi lattice has the same local geometry as the Kagome lattice. However, the Husimi lattice has weaker geometrical frustration than the Kagome lattice because the triangles are not reconnected in the former case. Due to this feature, it is much easier to investigate a model defined on the Husimi lattice than on the Kagome lattice. Like in a regular lattice, one can use a PESS to represent a quantum state defined in the Husimi lattice [114, 254]. For example, the following 3-PESS

$$(21.19)$$

can be introduced to represent a quantum state on the triangular Husimi lattice. In this graph, the dashed lines represent the original triangular Husimi lattice, open circles the three-index simplex tensors S_{abc}, solid circles the three-index projection tensors $A_{aa'}[\sigma]$ at each lattice site, in-plane solid lines the virtual indices of all tensors, and vertical lines the physical degrees of freedom of each site:

$$(21.20)$$

This PESS representation defines a wave function on a decorated Bethe lattice. As there are no loops in this representation, an advantage in representing a quantum state on the Husimi lattice using PESS is that the wave function can be readily determined using the approach introduced in the preceding sections.

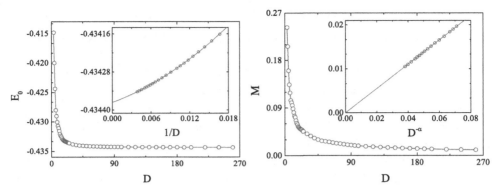

Figure 21.1 Ground-state energy per site E_0 (left panel) and local magnetization M (right panel) in the ground state of the spin-1/2 Heisenberg model on
the triangular Husimi lattice. Left panel: the inset shows the energy as a function of $1/D$, with a polynomial fit shown by the solid curve. The extrapolated
energy is $E_0(D \to \infty) = -0.43438(1)$. Right panel: the inset shows M as a
function of $D^{-\alpha}$ with $\alpha = 0.588(2)$. The intercept of the linear fit (solid line) is
$M(D \to \infty) = 0.00000(4)$. The plots in the left and right panels are reproduced
from figs. 7 and 8 in Ref. [254], respectively. Copyright by the American Physical
Society.

Figure 21.1 shows, as an example, how the ground-state energy per site E_0 and
the local magnetization $M = \sqrt{\langle S_{ix} \rangle^2 + \langle S_{iy} \rangle^2 + \langle S_{iz} \rangle^2}$ vary with the bond dimension D in the ground state of the spin-1/2 Heisenberg model on the triangular
Husimi lattice. Both E_0 and M converge algebraically with the bond dimension
in the large D limit. This implies that the ground state of this system is gapless
in the limit $D \to \infty$. Furthermore, the magnetization M varies as $D^{-0.588}$ in the
large D limit. An extrapolation of the numerical results to the limit $D \to \infty$ yields
$M(\infty) = 0.00000(4)$, which vanishes within numerical errors. It indicates that the
true ground state of the spin-1/2 Heisenberg model on the triangular Husimi lattice
is gapless and nonmagnetic; that is, an algebraic quantum spin liquid [254].

21.4 Determination of Tree Tensor Network State

Variational minimization is a commonly adopted approach for determining the
local tensors. It takes all the tensor elements as variational parameters and
determines them by minimizing the ground-state energy:

$$E = \frac{\langle \Psi | H | \Psi \rangle}{\langle \Psi | \Psi \rangle}. \tag{21.21}$$

However, the number of variational parameters the minimization can efficiently handle is limited, dramatically limiting the bond dimension that can be treated.

Similar to MPS, a tree tensor network state can be determined by taking the imaginary time evolution, namely by applying the projection operator $\exp(-\tau H)$ onto an arbitrary initial state that is not orthogonal to the ground state [118]. In the limit $\tau \to \infty$, the resulting wave function will converge to the ground state. However, this projection cannot be made in a single step since the interaction terms in the Hamiltonian do not commute with each other. Instead, one needs to use a small τ and apply this projection operator to the wave function recursively for sufficiently many times.

Let us use the tree-tensor network state of $z = 3$ to demonstrate how this method works. We assume the Hamiltonian is translation-invariant and contains only nearest-neighboring interactions:

$$H = \sum_{\langle ij \rangle} H_{ij}. \tag{21.22}$$

It can be further divided into three parts:

$$H = H_x + H_y + H_z, \tag{21.23}$$
$$H_\alpha = \sum_{i \in A} H_{i,i+\alpha} \quad (\alpha = x, y, z), \tag{21.24}$$

where H_α ($\alpha = x, y, z$) contains all the interaction terms along the α-direction. The lattice is divided into two sublattices. A is a collection of all the lattice sites in one of the sublattices. For small τ, one can use the Trotter–Suzuki formula to decouple approximately $\exp(-\tau H)$ into a product of three terms:

$$e^{-\tau H} \approx e^{-\tau H_z} e^{-\tau H_y} e^{-\tau H_x} + o(\tau^2). \tag{21.25}$$

A projection of H can then be readily performed using $\exp(-\tau H_\alpha)$ ($\alpha = x, y, z$) in three steps.

In the first step, the projection is made with H_x. As H_x contains only the interaction terms between two neighboring spins connected by horizontal bonds, this projection generates a new wave function:

$$e^{-\tau H_x} |\Psi\rangle = \text{Tr} \prod_{i \in A, j=i+\hat{x}} \sum_{\sigma_i \sigma_j} \lambda_{y,\beta_i} \lambda_{z,\gamma_i} M_{\beta_i \gamma_i, \beta_j \gamma_j}[\sigma_i, \sigma_j] |\sigma_i \sigma_j\rangle, \tag{21.26}$$

where Tr is to sum over all virtual bond indices and $|\sigma_i\rangle$ is the local basis states of dimension d. The local tensors A and B are coupled by the matrix elements of the local projection operator $\exp(-\tau H_{ij})$, and M is a tensor defined on two neighboring sites, i and j:

$$M_{\beta_i\gamma_i,\beta_j\gamma_j}[\sigma_i,\sigma_j] = \qquad \text{(21.27)}$$

Tensor M can be decoupled into a product of two updated local tensors through SVD so that the resulting tensor network state can return to its original form. However, the dimension of the bond linking the two local tensors is increased by a factor of d^2, which is equal to the dimension of the local projection operator $\exp(-\tau H_{ij})$. This bond dimension needs to be truncated by adequately considering the renormalization effect of the environment tensors. Bearing in mind the canonical form of the tree tensor network state, one can explicitly perform the canonical transformations so that the local tensors satisfy the canonical equations. The canonicalization provides an optimized scheme to truncate the virtual bond basis space. However, the computational cost of the canonicalization is generally very high.

Jiang *et al.* [116] proposed a simple approach, later called the "simple update" approach, to resolve this difficulty. As mentioned before, the bond vector λ in Eq. (21.1) measures the entanglement between two parts linked by this bond. Since each bond connects with two tensors, one can regard each λ as a product of two $\lambda^{1/2}$ and associate each of them to one of the tensors. It implies that one can take $\lambda^{1/2}$ as an effective entanglement mean field to simulate the contribution of environment tensors.

Thus if the tree tensor network state defined by Eq. (21.1) is already canonical or approximately canonical, the environment tensor can be simulated by the product of four bond vectors, $(\lambda_{y,\beta_i}, \lambda_{z,\gamma_i}, \lambda_{y,\beta_j}, \lambda_{z,\gamma_j})$, on the four links connecting M with the environment, similar as for an MPS in one dimension. This defines a renormalized tensor $\tilde{M}$:

$$\tilde{M}_{\beta_i\gamma_i\sigma_i,\beta_j\gamma_j\sigma_j} = \lambda_{y,\beta_i}\lambda_{z,\gamma_i}M_{\beta_i\gamma_i,\beta_j\gamma_j}[\sigma_i,\sigma_j]\lambda_{y,\beta_j}\lambda_{z,\gamma_j} \qquad \text{(21.28)}$$

In this definition, the four bonds connected to the environment are weighted by the corresponding bond vector λ, rather than $\lambda^{1/2}$. These extra $\lambda^{1/2}$ are included to simulate the renormalization effect from the environment. The dimension of $\tilde{M}$ is D^2d. Nevertheless, as already mentioned, the rank of $\tilde{M}$ is equal to Dd^2, less than D^2d.

Now an SVD is taken to decompose $\tilde{M}$ into a product of two tensors at sites i and j:

$$\tilde{M}_{\beta_i\gamma_i\sigma_i,\beta_j\gamma_j\sigma_j} = \sum_{\alpha} U_{\beta_i\gamma_i\sigma_i,\alpha}\tilde{\lambda}_{x,\alpha}V_{\beta_j\gamma_j\sigma_j,\alpha}, \qquad (21.29)$$

where U and V are two isometric matrices and $\tilde{\lambda}_x$ is the semi-positive singular matrix. $\tilde{\lambda}_x$ measures the entanglement between U and V tensors. The virtual basis space is truncated by keeping only the D largest singular values of $\tilde{\lambda}_x$. The truncated $\tilde{\lambda}_x$ is then used to update λ_x. After this, tensors A and B are updated using the formula

$$A_{\alpha\beta\gamma}[\sigma] = \lambda_{y,\beta}^{-1}\lambda_{z,\gamma}^{-1}U_{\beta\gamma\sigma,\alpha}, \qquad (21.30)$$

$$B_{\alpha\beta\gamma}[\sigma] = \lambda_{y,\beta}^{-1}\lambda_{z,\gamma}^{-1}V_{\beta\gamma\sigma,\alpha}. \qquad (21.31)$$

The local tensor $A[\sigma]$ such defined satisfies the canonical condition (21.10):

$$\sum_{\beta\gamma\sigma}\lambda_{y,\beta}^2\lambda_{z,\gamma}^2 A_{\alpha\beta\gamma}^*[\sigma]A_{\alpha'\beta\gamma}[\sigma] = \sum_{\beta\gamma\sigma}U_{\beta\gamma\sigma,\alpha'}^*U_{\beta\gamma\sigma,\alpha} = \delta_{\alpha\alpha'}. \qquad (21.32)$$

But it may not satisfy Eqs. (21.12)–(21.13).

This completes the projection for all horizontal bonds. The projection for the bonds along the other two directions can be made in the same way. By repeating this procedure, the local tensors, $A[\sigma]$ and $B[\sigma]$, and the bond vectors, λ_x, λ_y, and λ_z, will eventually converge. The converged tensor network state gives an accurate description of the ground state.

Equation (21.32) suggests that the converged local tensors obtained from this scheme satisfy approximately the three canonical equations defined by Eqs. (21.10)–(21.13). The deviation of these tensors from the rigorous canonical conditions is determined by the truncation error.

This truncation scheme is not optimal since the local tensors do not rigorously satisfy the canonical equations after each projection step. Nevertheless, it provides an efficient and accurate approach for determining the ground-state wave function. To minimize the truncation error, one should canonicalize all local tensors at each projection step. However, the computational cost for performing this canonicalization step is again very high.

Our self-consistent projection scheme takes the bond vectors as an effective field acting on the system tensor from the environment. This is, in fact, an entanglement mean-field approach, resembling the famous Bethe approximation first proposed by Bethe in 1935 [424]. It is also intimately connected with the dynamical mean-field theory [425, 426]. The key idea is to treat the inner tensors exactly and to use an effective mean field to approximate the interactions between the inner tensors and their environment tensors. This approximation converts the complicated global-optimization problem into a local one. It becomes rigorous if the tensor network state is already canonicalized, satisfying Eqs. (21.10)–(21.13).

Now let us consider how to implement the scheme efficiently. A simple approach is taking an SVD for $\tilde{M}$. However, as $\tilde{M}$ is a matrix of $dD^2 \times dD^2$, the cost scales as $O(d^3D^6)$. This cost can be reduced to $O(dD^4) + O(d^3D^3)$ if the SVD for the projection on the x bond (similarly for the other two bonds) is done in the following three steps:

(i) Factorize the following two $D^2 \times Dd$ matrices by QR decompositions:

$$A'_{\beta\gamma,\alpha\sigma} = \lambda_{y,\beta}\lambda_{z,\gamma}A_{\alpha\beta\gamma}[\sigma] = \sum_k Q^a_{\beta\gamma,k}R^a_{k,\alpha\sigma}, \qquad (21.33)$$

$$B'_{\beta\gamma,\alpha\sigma} = \lambda_{y,\beta}\lambda_{z,\gamma}B_{\alpha\beta\gamma}[\sigma] = \sum_k Q^b_{\beta\gamma,k}R^b_{k,\alpha\sigma}, \qquad (21.34)$$

where Q^μ ($\mu = a$ or b) is a $D^2 \times Dd$ column orthonormal matrix and R^μ is a $Dd \times Dd$ upper triangular matrix. The cost for taking the QR decomposition scales as $O(d^3D^3)$.

(ii) Define the gate matrix by applying the bond projection operator to the system:

$$G_{m\sigma_1,n\sigma_2} = \sum_{\alpha\sigma'_1\sigma'_2} \langle \sigma_1\sigma_2|e^{-H_{ij}\tau}|\sigma'_1\sigma'_2\rangle R^a_{m;\alpha\sigma'_1}\lambda_{x,\alpha}R^b_{n;\alpha\sigma'_2} \qquad (21.35)$$

and factorize it by SVD:

$$G_{m\sigma_1,n\sigma_2} = \sum_\alpha U_{m\sigma_1,\alpha}\lambda'_{x,\alpha}V_{n\sigma_2,\alpha}, \qquad (21.36)$$

where U and V are two $Dd \times Dd$ unitary matrices, and $\lambda'_{x,\alpha}$ is a semi-positive diagonal matrix. Again the cost scales as $O(d^3D^3)$.

(iii) Truncate the inner bond dimension by keeping the largest D matrix elements of λ'_x, and update the local tensors by the formula

$$A_{\alpha\beta\gamma}[\sigma] = \sum_m \lambda^{-1}_{y,\beta}\lambda^{-1}_{z,\gamma}Q^a_{\beta\gamma,m}U_{m\sigma,\alpha}, \qquad (21.37)$$

$$B_{\alpha\beta\gamma}[\sigma] = \sum_m \lambda^{-1}_{y,\beta}\lambda^{-1}_{z,\gamma}Q^b_{\beta\gamma,l}V_{m\sigma,\alpha}. \qquad (21.38)$$

The cost in this step scales as $O(dD^4)$.

21.5 Upper Bound of the Correlation Length

On a Bethe lattice, the lattice sites are not uniformly distributed. Particularly, the number of sites on the boundary of a finite connected region is roughly equal to the number of internal sites, different than in a regular lattice. Thus the boundary effect strongly affects the physical behaviors of correlation functions. A peculiar feature revealed is that the correlation function is always short-ranged on the Bethe lattice and the correlation length ξ is finite even at a critical point [427, 428, 118].

The correlation length is actually upper bounded by a number that depends on the coordination number z even at a critical point. This upper bound does not exist on a regular lattice where the correlation length is always divergent at critical points. This finite correlation length at the critical point is due to the peculiar topology of the Bethe lattice. To understand this, let us consider the uniform magnetic susceptibility χ in a spin system:

$$\chi = \sum_i \langle S_0^z S_i^z \rangle - \langle S_0^z \rangle \langle S_i^z \rangle, \tag{21.39}$$

where S_0^z is the spin at a reference site, and i runs over all lattice sites. $\langle S \rangle$ is the thermal average of operator S. For simplicity, we assume the system is rotationally symmetric around the reference site. In this case, the magnetic susceptibility can be expressed as

$$\chi = \sum_r n(r)\chi(r), \tag{21.40}$$

where r is the layer number at which site i resides. r equals the distance of site i from the reference site. $n(r)$ is the number of spins on the rth layer and

$$\chi(r) = \langle S_0^z S_r^z \rangle - \langle S_0^z \rangle \langle S_r^z \rangle. \tag{21.41}$$

On the Bethe lattice, $n(r) \propto (z-1)^r$. If $\chi(r) \propto \exp(-r/\xi)$ is an exponentially decay function of r in the large r limit, then

$$\chi \propto \sum_r (z-1)^r e^{-r/\xi} = \sum_r e^{r[\ln(z-1)-\xi^{-1}]}. \tag{21.42}$$

χ diverges if ξ approaches $1/\ln(z-1)$. This shows that the susceptibility diverges even if $\chi(r)$ decays exponentially in space. The critical point occurs at

$$\xi = \frac{1}{\ln(z-1)}, \tag{21.43}$$

which is an upper bound of the correlation length ξ on the Bethe lattice with rotational symmetry [429, 118]. On a regular lattice, the susceptibility is non-critical if the spin-spin correlation function $\chi(r)$ decays exponentially with r, since $n(r) \propto r^{d-1}$ and d is the spatial dimension of the lattice.

21.6 Thermodynamics

The correlation length of a physical quantity is finite, less than an upper bound. This finite correlation length is an inherent property of the Bethe lattice, irrespective of temperatures and interactions. It implies that the entanglement entropy also has an upper bound at finite temperatures and that a tree tensor network state presents a good representation of the ground state and the thermal density matrix on the Bethe lattice [257].

To evaluate thermodynamic quantities, we write the thermal density matrix

$$\rho(\beta) = e^{-\beta H} \tag{21.44}$$

as a product of two half-density matrices $\rho(\beta/2)$,

$$\rho(\beta) = \rho(\beta/2)\rho(\beta/2) = \rho^\dagger(\beta/2)\rho(\beta/2). \tag{21.45}$$

To evaluate thermodynamic quantities, we represent $\rho(\beta/2)$ as a tree tensor network operator:

$$\rho(\beta/2) = \qquad\qquad\qquad . \tag{21.46}$$

This tree tensor network operator can also be regarded as a tree tensor network state $|\rho(\beta/2)\rangle$, defined in the product space of the initial and final physical basis states, namely by grouping the initial and final basis states at each site as one set of basis states. In this representation, the partition function could be expressed as an inner product of the tree tensor network state:

$$Z(\beta) = \mathrm{Tr}\rho(\beta) = \langle\rho(\beta/2)|\rho(\beta/2)\rangle. \tag{21.47}$$

Hence, the partition function can be determined by contracting a bilayer tree-tensor network state.

We determine the tree tensor network state $|\rho(\beta/2)\rangle$ using the imaginary evolution approach introduced in §21.4. Initially, the tree tensor network state is the density matrix at infinite temperature, $\beta = 0$, where $|\rho(0)\rangle$ is simply an identity operator. At each step of evolution, the projection operator $\exp(-\tau H/2)$ with a small increment τ is applied to the tree tensor network state $|\rho(\beta/2)\rangle$. This generates a new tree tensor network state of the half-density matrix $|\rho((\beta + \tau)/2)\rangle$ but with higher bond dimensions. To reduce the truncation error, the canonicalization introduced in §21.2, instead of the simple update approach, should be imposed to canonicalize the tree tensor network state before cutting the bond dimension.

As an example, let us calculate the thermodynamic quantities of the antiferromagnetic Heisenberg model (4.1) with $J=1$ and $B=0$ on the $z=3$ Bethe lattice [257]. Figure 21.2 shows the temperature dependence of the entanglement entropy, the specific heat, and the magnetization. A λ-type jump is observed in the

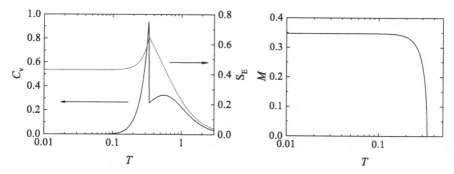

Figure 21.2 Temperature dependence of (left) the entanglement entropy and the specific heat and (right) the magnetization for the Heisenberg model on the $z = 3$ Bethe lattice. $D = 60$ is used in the imaginary time evolution.

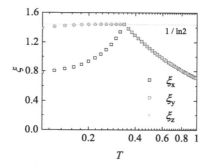

Figure 21.3 Temperature dependence of the correlation lengths along three principal axes for the spin-1/2 Heisenberg model on the $z = 3$ Bethe lattice obtained with the $D = 60$ tree tensor network state. The SU(2) spin rotation symmetry is spontaneously broken along the z-axis.

specific heat at the critical point $T_c \sim 0.34$. The magnetization is antiferromagnetically ordered at low temperatures and drops to zero continuously at the transition point, indicating that the transition from the paramagnetic to the antiferromagnetic phase is continuous. On the other hand, the entanglement entropy exhibits a cusp and gradually drops with decreasing temperature after crossing the critical point. This cusp is a natural consequence of the finite correlation length, allowing us to perform accurate thermal simulations with tree tensor network states for this model system.

Figure 21.3 shows how the correlation lengths along the three principal axes vary with temperature. Like the entanglement entropy, the correlation lengths exhibit cusps at the critical temperature. The correlation lengths along all three directions are equal in the paramagnetic phase. They increase with decreasing temperature and reach the upper bound on the $z = 3$ Bethe lattice (i.e. $\xi_c = 1/\ln 2$), at the critical point.

Below the critical temperature, the correlation lengths of the transverse and longitudinal modes behave differently. The correlation length of the longitudinal mode drops immediately below T_c because of the spontaneous symmetry breaking along the z-direction. On the other hand, the two transverse excitation modes remain critical in the antiferromagnetic ordered phase. Therefore, the correlation lengths of the transverse excitation modes are expected not to change with temperature below T_c. However, our tensor network calculation finds that the correlation lengths show a tiny drop at low temperatures. This drop results from the truncation error introduced by the tree tensor network states with a finite bond dimension D. By increasing the value of D, we find that this drop tends to approach zero.

22

Two-Dimensional Tensor Network States

22.1 PEPS

PEPS provides a variational ansatz for the ground states in two or higher dimensions. There are a number of methods that can be used to determine the local tensors. One can, for example, take each tensor element as a variational parameter and determine it by minimizing the energy. One can also take a projection approach to determine the ground-state PEPS by applying successively the imaginary time evolution operator $\exp(-\tau H)$ onto an arbitrary initial PEPS. At each step of evolution, the local tensors are updated. The converged PEPS in the infinite imaginary time limit ($\tau \to \infty$) is expected to be a good approximation to the true ground state.

There are two kinds of update schemes, named as the simple update [116] and the full update [117], respectively. The simple update is based on an entanglement mean-field approximation for the environment tensor. It provides a good approximation to the ground-state wave function. The full update takes account of the environment effect more accurately. It evaluates the environment tensor by contracting all local tensors in the environment using TMRG, boundary MPS or other RG approaches. The local tensors are updated by solving a set of linear equations.

Below we take a PEPS defined on a square lattice:

$$|\Psi\rangle = \mathrm{Tr} \prod_i A^i_{x_i, \bar{x}_i, y_i, \bar{y}_i}[\sigma_i]|\sigma_i\rangle, \tag{22.1}$$

as an example to discuss how these methods work. In this expression, $(x_i, \bar{x}_i, y_i, \bar{y}_i)$ are the virtual basis indices on the bonds emitted from site i along $(+\hat{x}, -\hat{x}, +\hat{y}, -\hat{y})$ directions, the trace is to sum over all internal bond indices as well as all physical indices, and

$$A^i_{x_i, \bar{x}_i, y_i, \bar{y}_i}[\sigma_i] = \quad \bar{x}_i \overset{\sigma_i \quad y_i}{\underset{\bar{y}_i \quad A^i}{\longleftrightarrow}} x_i \tag{22.2}$$

is the local tensor.

22.2 Variational Optimization

The variational minimization is to take all local tensor elements as variational parameters and determine them by minimizing the ground-state energy:

$$E = \frac{\langle \Psi | H | \Psi \rangle}{\langle \Psi | \Psi \rangle}. \tag{22.3}$$

This is an extension of the method introduced in §12.1 for determining an MPS in one dimension. Generally, it is difficult to determine all local tensors simultaneously. One can, however, determine them iteratively. At each step of the iteration, we take just one local tensor, for example, the local tensor at site i, A^i, as a variational tensor and fix all other local tensors. In this case, the optimization is to find a local tensor A^i such that $\langle \Psi | H | \Psi \rangle$ is minimized, subject to the constraint that the wave function norm, $\langle \Psi | \Psi \rangle$, is a fixed constant. This approach is equivalent to introducing a Lagrange multiplier, λ, and minimizing the functional

$$E[A^i] = \langle \Psi | H | \Psi \rangle - \lambda \left(\langle \Psi | \Psi \rangle - 1 \right). \tag{22.4}$$

Here $|\Psi\rangle$ is a linear function of A^i. Thus both $\langle \Psi | H | \Psi \rangle$ and $\langle \Psi | \Psi \rangle$ are quadratic functions of A^i, and can be generally expressed as

$$\langle \Psi | H | \Psi \rangle = \left(A^i \right)^\dagger \mathcal{H}^i A^i, \tag{22.5}$$

$$\langle \Psi | \Psi \rangle = \left(A^i \right)^\dagger \mathcal{N}^i A^i. \tag{22.6}$$

In these expressions, A^i is a vectorized tensor, or a five-leg vector of dimension $D^4 d$. $\mathcal{H}^i$ and $\mathcal{N}^i$ are the effective Hamiltonian and norm matrices, respectively. Both are $D^4 d \times D^4 d$ matrices. They are obtained by contracting all the local tensors in $\langle \Psi | H | \Psi \rangle$ and $\langle \Psi | \Psi \rangle$, excluding A^i and $(A^i)^\dagger$. These contractions are equivalent to taking derivatives for the corresponding expectation values with respect to A^i and $(A^i)^\dagger$:

$$\mathcal{H}^i = \frac{\delta^2 \langle \Psi | H | \Psi \rangle}{\delta (A^i)^\dagger \delta A_i}, \qquad \mathcal{N}^i = \frac{\delta^2 \langle \Psi | \Psi \rangle}{\delta (A^i)^\dagger \delta A_i}. \tag{22.7}$$

Taking the derivative of $E[A^i]$ with respect to A^i, we obtain the equation that determines A^i:

$$\mathcal{H}^i A^i = \lambda \mathcal{N}^i A^i. \tag{22.8}$$

This is a generalized eigenequation. We do not need to find all the eigenvalues and eigenvectors from this eigenequation. Instead, only the minimal eigenvalue with the corresponding eigenvector needs to be determined. λ at the extreme point where all local tensors become converged is nothing but the ground-state energy,

$$\lambda = \frac{\left(A^i\right)^\dagger \mathcal{H}^i A^i}{\left(A^i\right)^\dagger \mathcal{N}^i A^i} = \frac{\langle \Psi | H | \Psi \rangle}{\langle \Psi | \Psi \rangle}, \tag{22.9}$$

and $|\Psi\rangle$ is just the ground state.

This discussion indicates that determining a PEPS by variationally minimizing the ground-state energy is equivalent to solving a generalized eigenequation. This framework holds generally. However, to implement it in practical calculations, we need to accurately calculate the matrix elements of $\mathcal{H}^i$ and $\mathcal{N}^i$, which is technically challenging.

To determine $\mathcal{N}^i$, one needs to contract a double-layer tensor network $\langle \Psi | \Psi \rangle$, excluding the local tensor A^i and its Hermitian conjugate. $\mathcal{H}^i$ is a sum for all interaction terms in H. Each term in $\mathcal{H}^i$ is also a double-layer tensor network. The total number of these double-layer tensor network operators equals the number of interaction terms in H.

Both $\mathcal{N}^i$ and $\mathcal{H}^i$ can, in principle, be contracted out using, for example, TMRG or CTMRG introduced in Chapter 14, or the SRG method to be introduced in §23.3. To accurately contract these double-layer tensor network states, however, one has to keep a large number of basis states (or a large bond dimension) in the RG contraction. The cost increases quickly with the number of states retained, which limits the bond dimension of PEPS to a small value (generally less than 7) in practical calculations.

Furthermore, the solution of Eq. (22.8) depends strongly on the condition number of $\mathcal{N}^i$. If the singular value matrix (by normalizing the maximal eigenvalue to 1) of $\mathcal{N}^i$ is close to the unit matrix, the eigenvector can be accurately determined. On the other hand, if the singular value matrix deviates strongly from the unit matrix such that some of the singular values are much smaller than 1, the condition number is low. In this case, it is difficult to determine the eigenvector accurately, and the minimization may converge very slowly or, in some cases, even not converge.

22.3 Imaginary Time Evolution

We can also determine a PEPS wave function by taking imaginary time evolution, namely by applying the thermal density operator $\exp(-\tau H)$ to an arbitrary initial state $|\Psi\rangle$ not orthogonal to the true ground state. In the limit $\tau \to \infty$, the converged PEPS should be a good approximation to the ground state of H.

Next we take a Hamiltonian with only nearest-neighbor interactions as an example to show how this method works. To carry out the projection, we divide the Hamiltonian into four parts:

$$H = H_{\hat{x}} + H_{-\hat{x}} + H_{\hat{y}} + H_{-\hat{y}}, \tag{22.10}$$

$$H_\alpha = \sum_{i \in A} H_{i,i+\alpha}, \quad (\alpha = \hat{x}, -\hat{x}, \hat{y}, -\hat{y}). \tag{22.11}$$

Each part contains the terms that commute with each other. Specifically, H_α contains only the interaction terms acting on the bonds emitting from the sites on the A sublattice along the α direction.

For small τ, we use the Trotter–Suzuki formula to decouple approximately $\exp(-\tau H)$ into a product of four terms:

$$e^{-\tau H} \approx e^{-\tau H_{-\hat{y}}} e^{-\tau H_{\hat{y}}} e^{-\tau H_{-\hat{x}}} e^{-\tau H_{\hat{x}}} + o(\tau^2). \tag{22.12}$$

A projection of H is then performed successively using $\exp(-\tau H_\alpha)$ in four steps.

In the first step, the projection operator $\exp(-\tau H_{\hat{x}})$ is applied to the PEPS, which generates a new state

$$|\Psi'\rangle = e^{-\tau H_{\hat{x}}} |\Psi\rangle. \tag{22.13}$$

This newly generated state can also be represented as a PEPS. However, the dimension of the bonds to which the interaction H_x is applied is higher than the original one. We aim to truncate these higher bond dimensions so that $|\Psi'\rangle$ can be accurately updated by a new PEPS whose bond dimensions are all equal to D.

The projection for the bonds along the other three directions can be similarly made. To find a converged ground-state wave function, we need to repeat the update procedure many times. The update can be done in two ways, depending on the approximation. Here we give an introduction to these two update approaches.

22.3.1 Full Update

The full update is an optimization approach. It is to find a new PEPS $|\Psi\rangle$, whose bond dimensions are all equal to D, to accurately represent $|\Psi'\rangle$ so that the functional

$$f(\Psi) = \big\| |\Psi\rangle - |\Psi'\rangle \big\|^2 = \langle \Psi|\Psi\rangle + \langle \Psi'|\Psi'\rangle - 2\mathrm{Re}\,\langle \Psi|\Psi'\rangle \tag{22.14}$$

is minimized. This is a two-dimensional generalization of the variational optimization method introduced in §12.3.2. This minimization problem is solved by taking all local tensors A^i of $|\Psi\rangle$ as variational parameters and determining them by variationally minimizing $f(\Psi)$. Since $\langle \Psi'|\Psi'\rangle$ does not contain any variational parameters, minimizing $f(\Psi)$ is equivalent to minimizing the cost function:

$$g(\Psi) = \langle\Psi|\Psi\rangle - 2\mathrm{Re}\langle\Psi|\Psi'\rangle. \tag{22.15}$$

To avoid finding too many variational parameters at one time, again we determine the local tensors iteratively. At each iteration, we take just one local tensor, say A^i, as a free variational tensor and fix all other tensors. In this case, $\langle\Psi|\Psi\rangle$ is a quadratic function of A^i, whose expression is given by Eq. (22.6). $\langle\Psi|\Psi'\rangle$ is a linear function of A^i:

$$\langle\Psi|\Psi'\rangle = (A^i)^\dagger \mathcal{M}^i. \tag{22.16}$$

Again A^i is vectorized. $\mathcal{M}^i$ is a vector obtained by contracting out all local tensors in $\langle\Psi|\Psi'\rangle$ excluding A^i. It is equal to the derivative of $\langle\Psi|\Psi'\rangle$ with respect to A^i:

$$\mathcal{M}^i = \frac{\delta\langle\Psi|\Psi'\rangle}{\delta(A^i)^\dagger}. \tag{22.17}$$

The optimal A^i is determined by the extreme condition of $g(\Psi)$:

$$\mathcal{N}^i A^i = \mathcal{M}^i. \tag{22.18}$$

This is a set of linear equations. In case the norm matrix $\mathcal{N}^i$ is well conditioned, the solution is

$$A^i = (\mathcal{N}^i)^{-1} \mathcal{M}^i. \tag{22.19}$$

Other local tensors can be similarly calculated.

22.3.2 Simple Update

The simple update is based on an entanglement mean-field approximation. It assumes that the renormalization effect of environment tensors can be approximately described by certain vectors defined on the bonds separating the system and environment blocks. These bond vectors are introduced to mimic the entanglement spectra on the corresponding bonds. If there is no closed loop in the lattice, for example, in a Bethe lattice, the simple update, as discussed in Chapter 21, is almost optimal.

In order to understand this method, let us consider a bipartite translation invariant PEPS. We divide the lattice into two intercalated sublattices and denote them as sublattices A and B, respectively. The simple update is more convenient to implement in the $\Gamma - \lambda$ representation of PEPS. In this case, the wave function reads

$$|\Psi\rangle = \mathrm{Tr} \prod_{i\in A, j\in B} \lambda_{1,x_i}\lambda_{2,\bar{x}_i}\lambda_{3,y_i}\lambda_{4,\bar{y}_i}\Gamma^a_{x_i\bar{x}_iy_i\bar{y}_i}[m_i]\Gamma^b_{x_j\bar{x}_jy_j\bar{y}_j}[m_j]|m_im_j\rangle, \tag{22.20}$$

where Γ^a and Γ^b are the local tensors defined on sublattices A and B, respectively. λ_α ($\alpha = 1, 2, 3, 4$) is a positive vector defined on one of the four bonds emitted from a lattice site. The trace is to sum up all physical as well as virtual bond variables. Graphically, this wave function can be represented as

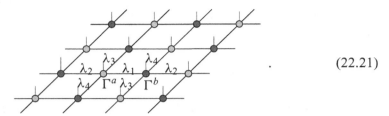

$$(22.21)$$

In Eq. (22.20), a bond vector is included explicitly at each bond. This bond vector provides an approximate measure of the entanglement spectrum on the corresponding bond. Since each bond connects two tensors, one can regard each bond vector λ as a product of two $\lambda^{1/2}$ and associate each of them to one of the tensors. Effectively, one can also take $\lambda^{1/2}$ as an effective entanglement field acting from one tensor to another.

In order to determine the local tensors, the renormalization effect of the environment tensors needs to be considered. This effect is approximately mimicked by these effective entanglement fields.

Taking the projection with $\exp(-\tau H_x)$, the wave function evolves to

$$
\begin{aligned}
|\Psi'\rangle &= e^{-\tau H_x}|\Psi\rangle \\
&= \mathrm{Tr} \prod_{i\in A, j=i+\hat{x}} \sum_{m'_i m'_j} |m_i m_j\rangle \langle m_i m_j | e^{-H_{ij}\tau} | m'_i m'_j \rangle \\
&\quad \lambda_{1,x_i} \lambda_{2,\bar{x}_i} \lambda_{3,y_i} \lambda_{4,\bar{y}_i} \Gamma^a_{x_i\bar{x}_iy_i\bar{y}_i}[m'_i] \Gamma^b_{x_j\bar{x}_jy_j\bar{y}_j}[m'_j].
\end{aligned}
$$

$$(22.22)$$

In this equation, two neighboring local tensors, Γ^a and Γ^b, at site i and $j = i + \hat{x}$, are coupled together by the matrix elements of $\exp\left(-H_{ij}\tau\right)$. In order to perform the projection for the next step, one needs to separate these two tensors so that the wave function returns to its original form.

To do this, let us define a $(D^3 d) \times (D^3 d)$ matrix using Γ^a and Γ^b and the bond vectors connected to them:

$$
S_{x_i y_i \bar{y}_i \sigma_i, \bar{x}_j y_j \bar{y}_j \sigma_j} =
$$

$$(22.23)$$

where

$$\text{(22.24)}$$

$$\text{(22.25)}$$

The three bonds connected to the three local tensors in the environment are weighted by the corresponding bond vectors λ_μ, rather than $\lambda_\mu^{1/2}$. These extra $\lambda_\mu^{1/2}$'s are included to account for the renormalization effect of the environment tensors.

To update Γ^a and Γ^b, we perform an SVD for S. The cost for directly singular-value decomposing S is of the order $O(D^9 d^3)$. This cost, nevertheless, can be reduced by first taking a QR decomposition for A and an LQ decomposition for B:

$$\text{(22.26)}$$

$$\text{(22.27)}$$

Q_a and Q_b are isometric matrices. The thick bond has a dimension of dD. The cost for taking this QR or LQ decomposition is of the order $O(D^5 d^2)$. Matrix S then becomes

$$\text{(22.28)}$$

As both Q_a and Q_b are isometric, S and the matrices enclosed by the dashed square in Eq. (22.28) should have identical singular spectra. Now we diagonalize this matrix, whose dimension is Dd^2, by SVD:

$$\text{(22.29)}$$

where U and V are two unitary matrices and λ is a positive diagonal matrix. The cost for taking this SVD is of the order $O(D^3d^3)$.

The dimension of λ is Dd^2. We use it to update λ_1 by keeping its largest D singular values. We then update Γ^a and Γ^b by the formulas:

$$\begin{array}{ccc} \rotatebox{0}{\Gamma^a} & = & \lambda_2^{-1}\;\boxed{Q_a}\;\lambda_3^{-1}\;\boxed{U}\;\lambda_4^{-1} \end{array} \tag{22.30}$$

$$\begin{array}{ccc} \rotatebox{0}{\Gamma^b} & = & \boxed{V}\;\boxed{Q_b}\;\lambda_4^{-1}\;\lambda_2^{-1}\;\lambda_3^{-1} \end{array} \tag{22.31}$$

The cost of this decompostion step is of the order $O(D^5d)$.

This completes the projection of $H_{\hat{x}}$. The projection for the bonds along the other three directions can be similarly implemented. An accurate ground-state wave function can be obtained by repeating this procedure sufficiently many times.

In these iteration steps, the renormalization effect of the environment is taken into account by a mean-field approximation. This method does not minimize the truncation error at each step of projection. Nevertheless, it is computationally stable and accurate because the Trotter and truncation errors are not accumulated in the projection. This method works very efficiently. It allows a tensor network state with D as large as $10^2 - 10^3$ to be determined in two dimensions.

22.4 Tensor Derivatives by Automatic Differentiation

From the discussions presented in §22.2 and §22.3.1, we know that $\mathcal{H}^i$, $\mathcal{N}^i$, and $\mathcal{M}^i$ are determined by the first or second derivatives of $\langle\Psi|H|\Psi\rangle$, $\langle\Psi|\Psi\rangle$, and $\langle\Psi|\Psi'\rangle$ with respect to A^i. If the system is translation-invariant and the local tensor is site-independent, each derivative involves a summation over all the lattice sites. In this case, it is more convenient to use the automatic differentiation to evaluate these derivatives.

We now use CTMRG and the tensor network state $|\Psi\rangle$ defined by Eq. (22.39) as an example to show how the automatic differentiation is imposed to evaluate the derivative of the scalar product $\mathcal{N} = \langle\Psi|\Psi\rangle$ with respect to the local tensor A (i.e. $\delta\mathcal{N}/\delta A$). For doing this, we first trace out the physical variables in $\mathcal{N}$ and convert it into a single-layer one whose local tensor B is defined by Eq. (22.41). B is a function of A, so we have

$$\frac{\delta N}{\delta A} = \frac{\delta N}{\delta B} \frac{\partial B}{\partial A}. \tag{22.32}$$

$\partial B / \partial A$ is simply to remove an A-tensor in B.

We evaluate the derivative $\delta N / \delta B$ using the automatic differentiation in three steps:

(i) Perform the CTMRG iteration, using the method introduced in §14.5.2, to find the converged corner and edge tensors C_0 and E_0. The corner tensors at the four corners may not equal each other. Here C_0 is a collection of the four corner tensors. Similarly, E_0 is a collection of the four edge tensors. For simplicity, we use $X_0 = (C_0, E_0)$ to represent all these corner and edge tensors.

(ii) Take X_0 previously obtained as the initial corner and edge tensors, irrespective of B, and repeat the CTMRG iteration for n more times. This generates a set of new corner and edge tensors, C_i and E_i $(i = 1, \cdots n)$. Again we use $X_i = (C_i, E_i)$ to represent these tensors. Finally, we obtain the norm scalar by contracting all the corner and edge tensors with the middle local tensor B. The graphical representation of this forward propagation is

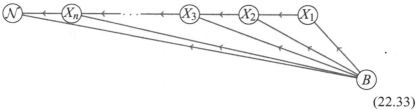

$$\tag{22.33}$$

X_i depends not just on X_{i-1}, but also on B:

$$X_i = X_i(X_{i-1}, B). \tag{22.34}$$

(iii) Define the adjoint variable of X_i:

$$\bar{X}_i = \frac{\partial N}{\partial X_i} \equiv \frac{\partial N}{\partial X_n} \frac{\partial X_n}{\partial X_{n-1}} \cdots \frac{\partial X_{i+1}}{\partial X_i} = \bar{X}_{i+1} \frac{\partial X_{i+1}}{\partial X_i}, \tag{22.35}$$

where $\bar{N} = 1$. From the backward propagation of the adjoint variables [339]

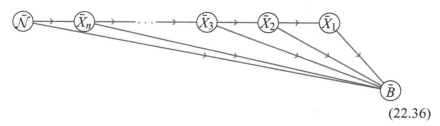

$$\tag{22.36}$$

we then obtain the desired result:

$$\frac{\delta \mathcal{N}}{\delta B} = \bar{B} = \sum_{i=1}^{n} \bar{X}_i \frac{\partial X_i}{\partial B} + \frac{\partial \mathcal{N}}{\partial B}. \tag{22.37}$$

There are $(n + 1)$ independent paths that one can go from $\mathcal{N}$ to $\bar{B}$. This is the reason why the above derivative is a sum of $(n + 1)$ terms.

As the cost for saving the derivative tensors in the intermediate steps is very high, we have to use a relatively small n in the forward and backward propagations.

22.5 Contraction of Double-Layer Tensor Networks

In the determination of local tensors using either the variational minimization or the full update approach, we need to calculate the effective Hamiltonian and norm matrices. These matrices are formed by contracting nearly all virtual and physical bond variables in the initial and final ground state represented by PEPS. In other words, each of these matrices can be represented as a double-layer tensor network with open legs associated with the tensors to be optimized. One of the layers is associated with the initial state $|\Psi\rangle$, and the other is associated with the final state $\langle\Psi|$. They are connected by the physical variables. Similar double-layer tensor networks appear, for example, in the evaluation of the norm, $\langle\Psi|\Psi\rangle$, or more generally in the evaluation of the expectation value of a physical variable O:

$$\langle O \rangle = \frac{\langle \Psi | O | \Psi \rangle}{\langle \Psi | \Psi \rangle}. \tag{22.38}$$

Both $\langle\Psi|\Psi\rangle$ and $\langle\Psi|O|\Psi\rangle$ are double-layer tensor networks.

To evaluate these double-layer tensor networks, one needs to contract all the internal bond variables. These contractions are difficult to handle rigorously. First, unlike MPS, there is no rigorous method to do these contractions. One has to use an approximate approach to reduce this two-dimensional contraction problem into a one-dimensional one. Second, the computational cost for carrying out these contractions is generally very high, which limits the bond dimension of PEPS or other two-dimensional tensor networks that can be reliably handled to a relatively small value.

Here we take a uniform PEPS of bond dimension D:

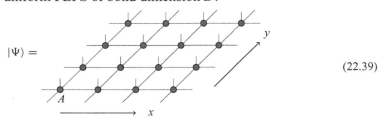

$$\tag{22.39}$$

as an example to elaborate the methods for evaluating the double-layer scalar product:

$$\langle \Psi | \Psi \rangle = \hspace{3cm} \tag{22.40}$$

Nevertheless, the idea can be generalized to an arbitrary double-layer tensor network with or without external bond indices.

22.5.1 Double-Layer Approach

A commonly adopted approach is to first trace the physical indices, reducing a double-layer tensor network into a single-layer one. By doing this, two local tensors at the same site are merged to form a reduced tensor with a higher bond dimension D^2;

$$-\!\!\boxed{B}\!\!- \;=\; \begin{matrix} A^* \\ A \end{matrix} \quad . \tag{22.41}$$

After tracing out the physical degrees of freedom, the scalar product $\langle \Psi | \Psi \rangle$ becomes

$$\langle \Psi | \Psi \rangle = \hspace{3cm} \tag{22.42}$$

This scalar tensor network could be traced out, for example, using CTMRG introduced in §14.5. However, the local tensor, B, is generally not symmetric. The transformation matrix cannot be determined simply by diagonalizing a symmetric reduced density matrix. Instead, it should generally be determined using the methods introduced in §15.

Besides CTMRG, one can also use the boundary MPS approach to contract the above scalar tensor network. It is basically to contract the whole tensor network from one row to another, starting from the bottom edge of the lattice at which all the edge tensors form an MPS. This contraction is equivalent to taking the evolution of the boundary MPS along the vertical direction. At each step of evolution, an approximation has to be used to truncate the bond dimension of MPS. If B is symmetric with respect to the two vertical indices, the methods introduced in §12.3 could be used. On the other hand, if B is not reflection symmetric along the horizontal axis, the method introduced in §16.3 should generally be used. After arriving at the top edge, the MPS becomes a simple product of matrices that can be readily contracted from left to right.

For an infinite lattice system, one can take each row of the two-dimensional tensor network (22.42) as a transfer matrix:

$$T = \quad \cdots \; -\!\!\!\text{\textcircled{B}}\!-\!\!\text{\textcircled{B}}\!-\!\!\text{\textcircled{B}}\!-\!\!\text{\textcircled{B}}\!- \; \cdots \tag{22.43}$$

and express $\langle \Psi | \Psi \rangle$ as a product of these transfer matrices:

$$\langle \Psi | \Psi \rangle = \lim_{N_y \to \infty} \operatorname{Tr} T^{N_y}, \tag{22.44}$$

where N_y is the lattice length along the y-axis. In the thermodynamic limit, $\langle \Psi | \Psi \rangle$ is determined purely by the largest eigenvalue of the transfer matrix. This eigenvalue and the corresponding eigenvector can be found by solving the equations that determine the fixed-point tensors using the method introduced in §14.3 if B is symmetric, or more generally the method introduced in §17.3 or §17.4 if B is not symmetric.

The accuracy of CTMRG or boundary-MPS depends on the bond dimension of local tensor B. To obtain a converged result, the bond dimension of the edge tensors should be at least of the order of the bond dimension of B (i.e. D^2), and consequently, the computational cost scales roughly as $(D^2 \times D^2)^3 = D^{12}$. This scaling power is very high, imposing a strong constraint on the bond dimension of $|\Psi\rangle$.

22.5.2 Nested Tensor Network Approach

A more efficient strategy is to sketch the tensor network representation of $\langle \Psi | \Psi \rangle$ in a single-layer picture without tracing over physical index [107, 430]. For this, we first shift all the top-layer tensors in $\langle \Psi |$ along the diagonal direction by a half unit cell and then press them down to the bottom layer. It results in an intertwined or nested single-layer tensor network:

$$\langle \Psi | \Psi \rangle = \quad \text{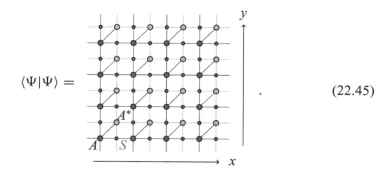} \quad . \tag{22.45}$$

The total number of local tensors is quadrupled. Besides the original A-tensor, a swap gate S is inserted at each intersect point between two bonds in the original top and bottom layers, respectively. S is simply a direct product of two identity operators:

$$x - \underset{y'}{\overset{y}{S}} - x' \; = \; \delta_{x,x'} \delta_{y,y'} \; . \tag{22.46}$$

The bond dimension of local tensors in this effective single-layer tensor network is equal to D or d, which is generally much smaller than D^2. Again this scalar tensor network can be renormalized with CTMRG or boundary-MPS. The leading computational cost scales as $(D \times D^2)^3 = D^9$, which is significantly smaller than D^{12}. This approach dramatically improves the contraction method, allowing PEPS or other two-dimensional tensor networks with much higher bond dimensions to be evaluated efficiently and reliably.

Alternatively, one can use the coarse-graining tensor renormalization group methods introduced in Chapter 23 to contract the scalar tensor networks. These methods contract the lattice iteratively. Each time two or more local tensors are merged and renormalized. As a result, the lattice size grows exponentially with the iteration steps and approaches the thermodynamic limit quickly.

23

Coarse-Graining Tensor Renormalization

23.1 Coarse-Graining Approaches

As discussed in Chapter 3, all classical or quantum lattice models with local interactions, such as the quantum Ising model, and the expectation value of a quantum tensor network state can be written as a product of local tensors with finite bond dimension, called a tensor network model. One has to trace out all local tensors to evaluate physical quantities from these tensor network models. For a classical statistical model in two dimensions, or an equivalent quantum lattice model in one dimension, the corresponding tensor network model can always be represented as a product of transfer matrices if the system is translation invariant. This tensor network model can be solved by diagonalizing the transfer matrix, for example, using TMRG introduced in Chapters 14 and 16 or boundary-MPS introduced in Chapter 12. However, extending these approaches to three or higher dimensions or a system without translation invariance, such as a spin glass system, is challenging.

The coarse-graining tensor renormalization group (TRG) is a method that can be used to overcome this difficulty. In two dimensions, this approach was first introduced by Levin and Nave in 2007 [128]. It is essentially a truncation scheme based on the singular value spectra of local tensors. It produces qualitatively correct results when the bond dimension of tensors is not too small. However, it converges slowly with the increase of bond dimension. This approach minimizes the error in the truncation of local tensors without considering the influence of the environment tensors and is a local optimization scheme.

A more sophisticated and globally optimized TRG method, called the second renormalization group (SRG), was proposed by Xie et al. in 2009 [129], whose method takes account of the interplay between the system (the local tensor here) and environment tensors. It includes the correction from the environment tensors in the renormalization of local tensors and significantly improves the accuracy of

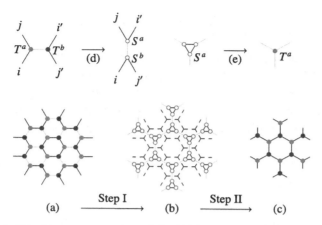

Figure 23.1 Graphic representation of the TRG steps. Step I (a→b): The tensor network model defined on the honeycomb lattice (a) is rewired into a tensor network model on a triangle-honeycomb lattice (b) by taking SVD for the product of two local tensors in (d). T_a and T_b are the local tensors defined on the two sublattices of the honeycomb lattice. S_a and S_b are defined by Eqs. (23.3) and (23.4), respectively. Step II (b→c): (e) an updated local tensor T_a is defined by contracting the three internal bonds on each triangle, leading to a new tensor network model on the squeezed honeycomb lattice (c).

results. This method is named SRG because it includes one more renormalization step than TRG.

Both TRG and SRG rely on the singular-value decomposition of tensors in the matrix representation. It is difficult to extend these methods to three or higher dimensions due to the increase in the order of local tensors and the change of lattice topology in the coarse-graining process. In 2012, a modified TRG, called HOTRG, was introduced to overcome this difficulty [131]. HOTRG is a coarse-graining TRG method that uses the higher-order singular value decomposition (HOSVD) of tensors. Again, HOTRG is a local optimization method. The corresponding SRG method is called HOSRG.

23.2 TRG

TRG is an iterative algorithm. It offers an approximate scheme to truncate the basis space according to the singular values of local tensors. Let us take the Ising model on a triangular lattice as an example of how the method works.

As discussed in Chapter 3, the triangular Ising model can be represented as a tensor network model, defined by Eq. (3.9), on a honeycomb lattice. A graphical representation of the corresponding tensor network model is shown in Fig. 23.1(a). The initial bond dimension of local tensors, T_a and T_b, is 2.

TRG is an iterative approach. It proceeds in two steps at each iteration:

Step I: Rewire a honeycomb lattice, Fig. 23.1(a), into a triangle-honeycomb one, Fig. 23.1(b), by transforming a pair of neighboring tensors, T^a and T^b, into two new tensors, S^a and S^b, defined in the rewired lattice by SVD. A schematic representation of this decomposition is shown in Fig. 23.1(d).

More specifically, we first contract the common bond index of T_a and T_b to define a matrix:

$$
\begin{array}{c}
\underset{j}{\overset{i'}{\rule{0pt}{0pt}}}\!\!=\!\!\boxed{M}\!\!=\!\!\underset{i}{\overset{j'}{\rule{0pt}{0pt}}}
\end{array}
\;=\;
\begin{array}{c}
\overset{j\;\;\;i'}{\underset{i\;\;\;j'}{\boxed{M}}}
\end{array}
\;=\;
\begin{array}{c}
\overset{j}{T^a}\!\!\!\overset{i'}{\rule{0pt}{0pt}}\;\;\overset{j}{\rule{0pt}{0pt}}\overset{i'}{T^b}\\[-2pt]
\underset{i}{\rule{0pt}{0pt}}\;\;\;\;\;\;\underset{j'}{\rule{0pt}{0pt}}
\end{array}
\;,
\tag{23.1}
$$

where (i',j) and (i,j') are the row and column indices of this matrix, respectively. If the bond dimension of T^a and T^b is D, then the dimension of M is D^2. Initially, $D = 2$, hence the dimension of M is 4.

We now factorize M into a product of three matrices (or tensors) by SVD:

$$
M_{i'j,ij'} = \sum_{n} U^*_{n,i'j}\lambda_n V_{n,ij'},
\tag{23.2}
$$

where λ is a semipositive diagonal matrix. The singular value λ_n measures the entanglement between two singular vectors U_n and V_n.

Two new local tensors defined in the rewired triangle-honeycomb lattice, S^a and S^b, are defined by

$$
S^a_{n,i'j} = U_{n,i'j}\sqrt{\lambda_n},
\tag{23.3}
$$

$$
S^b_{n,ij'} = V_{n,ij'}\sqrt{\lambda_n}.
\tag{23.4}
$$

In these matrices, M reads

$$
M = (S^a)^{\dagger} S^b.
\tag{23.5}
$$

The dimension of λ equals D^2. If it is larger than the maximal bond dimension allowed to retain, the singular spectra are truncated.

Step II: Contract all internal bond indices within each small triangle in the triangle-honeycomb lattice. This defines two updated local tensors in the squeezed honeycomb lattice. For example, the updated local tensor T^a is given in Fig. 23.1(e). T^b is similarly updated. It leads to a coarse-grained honeycomb tensor network model, Fig. 23.1(c), whose lattice size is reduced by a factor of 3 in comparison with the original one.

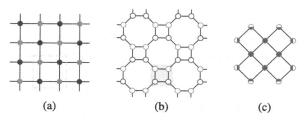

Figure 23.2 TRG steps on a square lattice: Step I (a→b): The tensor network model defined on a square lattice (a) is rewired into a tensor network model on a square-octagon lattice (b) by factorizing each local tensor with SVD. Step II (b→c): The four tensors on each small square in (b) are contracted to produce a new local tensor on a squeezed square lattice (c).

By repeating these steps iteratively, one can finally reduce the lattice to just two sites. Then, the partition function is obtained by tracing all bond indices of the tensors on these two sites, assuming a central symmetric boundary condition.

23.2.1 Other Kinds of Lattices

TRG relies on an approximate factorization scheme of local tensors, achieved by SVD. The idea of TRG, nevertheless, works very generally. Moreover, it can be readily extended to other kinds of lattices in two dimensions [128, 431, 308]. As an example, let us consider how TRG works on a square lattice.

A TRG iteration on a square lattice also takes two steps. The first is to take SVD for each local tensor to rewire the square lattice, shown in Fig. 23.2(a), to a square-octagon one, shown in Fig. 23.2(b). Unlike in the honeycomb lattice, there is no need to combine two local tensors before decomposition. For the local tensors on the two sublattices in Fig. 23.2(a), we factorize them by SVD along two different diagonal directions:

$$\underset{M}{\vdash} \xrightarrow{\text{SVD}} \curlyvee \quad , \quad \underset{M}{\vdash} \xrightarrow{\text{SVD}} \curlywedge \quad . \tag{23.6}$$

This factorization converts a tensor network model from a square lattice onto a square-octagon lattice. The bond dimension linking two newly generated tensors is D^2 if the bond dimension of the original tensor is D. It should be truncated according to the singular values of M.

The second step is to sum all the internal indices on each small square in the square-octagon lattice, Fig. 23.2(b). It renormalizes the four tensors on four corners of the square into a single tensor:

$$\times \; = \; \boxtimes \, , \tag{23.7}$$

and squeezes the square-octagon lattice into a square lattice shown in Fig. 23.2(c). This step is rigorous, without taking any approximation. The newly generated square lattice has half the size of the original square lattice. Its axes point in diagonal directions.

Repeating these above two steps will squeeze the lattice into one site whose local tensor is readily contracted, yielding the value of the partition function.

23.3 Second Renormalized TRG

TRG is a local rather than a global optimization approach because it does not consider the renormalization effect of the environment tensors in the truncation of local tensor M. On the honeycomb lattice, the environment tensor, M^{env}, as shown below,

$$\tag{23.8}$$

contains all the lattice sites except the two on which M is defined. TRG minimizes the truncation error for the local tensor M. However, in the calculation of thermodynamic quantities, it is the error of the partition function that should be minimized.

The renormalization effect of M^{env} to M, which is not considered in TRG, should be included to minimize the truncation error globally. We will discuss how to determine M^{env} in §23.4. Before doing that, let us consider how TRG is modified by M^{env}: a second renormalization group (SRG) method [129].

Given a local matrix M and its corresponding environment tensor M^{env}, the partition function is simply the scalar product of these two tensors:

$$Z = \text{Tr} M \cdot M^{\text{env}}. \tag{23.9}$$

Using the matrix representation of M^{env},

$$\tag{23.10}$$

and Eq. (23.5), we can also represent the partition function as a product of three matrices:

$$Z = \text{Tr} M^{\text{env}} \left(S^a\right)^\dagger S^b = \text{Tr} S^b M^{\text{env}} \left(S^a\right)^\dagger. \tag{23.11}$$

It motivates us to define a density matrix on the bond that separates S_a and S_b:

$$\rho = S^b M^{\text{env}} (S^a)^\dagger. \tag{23.12}$$

Graphically, it can be represented as

$$\rho = \quad \overset{(S^a)^\dagger}{\underset{S^b}{}} \quad M^{\text{env}} . \tag{23.13}$$

In order to determine the optimized truncation scheme in the SRG, we diagonalize the bond density matrix by a pair of canonical matrices P and Q:

$$\rho = P \Lambda Q, \tag{23.14}$$

where Λ is the diagonal eigenvalue matrix. Q is the inverse of P, $Q = P^{-1}$. The eigenvalue of ρ is generally a complex number. If the bond density matrix is real, $\rho^* = \rho$, the complex eigenvalues with the corresponding eigenvectors should appear in pairs. We assume that the eigenvalues are arranged in descending order according to their absolute values. To truncate the basis space, we keep the first D eigenvalues whose absolute values are the largest. After truncation, P becomes a $D_0 \times D$ matrix and Q becomes a $D \times D_0$ matrix, with D_0 the bond dimension before truncation.

In the truncated space, the column vectors of P form a set of biorthonormal basis states with the row vectors of Q:

$$QP = I_D. \tag{23.15}$$

P and Q form a pair of transformation matrices. They transform the bond density matrix to a $D \times D$ diagonal matrix:

$$Q\rho P = QS^b M^{\text{env}} (S^a)^\dagger P = \Lambda_D, \tag{23.16}$$

where Λ_D is the eigenvalue matrix of ρ after truncation.

M is now approximately given by

$$M \approx (S^a)^\dagger PQS^b. \tag{23.17}$$

Under this approximation, the partition function is given by the sum of all retained eigenvalues:

$$Z = \text{Tr} M M^{\text{env}} \approx \text{Tr} \Lambda_D. \tag{23.18}$$

This suggests that the optimized local tensors, $\tilde{S}^a$ and $\tilde{S}^b$, in the rewired lattice, should be defined as

$$\tilde{S}^a = P^\dagger S^a, \qquad \tilde{S}^b = QS^b. \tag{23.19}$$

Once the local tensors in the rewired lattice are determined, we can update the local tensors in the squeezed lattice by contracting all internal bond indices on each triangle. Again, we can obtain the value of the partition function by repeating the procedure for sufficiently many steps.

The difference between TRG and SRG is similar to that between NRG [9] and DMRG [18]. In NRG, the block Hamiltonian is optimized, but the interaction between different blocks is ignored. In DMRG, on the other hand, the basis states of the system are optimized by fully considering the interplay between the system and the environment blocks through the reduced density matrix.

23.4 Determination of the Environment Tensor

In DMRG, a superblock can be separated into a system block and an environment block by cutting just one bond. Thus the environment (block) can be readily identified and integrated out. However, in TRG, it is highly nontrivial to handle the environment because to separate the environment from the system, one needs to cut four bonds. The system contains only two sites, but the environment contains all other sites. In this case, evaluating the environment tensor is even more complicated than the partition function.

There are at least three approaches that can be used to determine the environment tensor:

Approach I: Take the entanglement spectra on the bonds connecting the system with the environment as an effective field to simulate the environment contribution. This is an entanglement mean-field approximation (or cavity approximation). This approach is physically intuitive and easy to implement.

Approach II: Evaluate the environment tensor by taking a combined forward and backward iteration. The forward iteration is to repeat all the steps of TRG. The backward iteration, on the other hand, is used to determine the environment tensor. This approach is more accurate than the first but has a higher computational cost.

Approach III: Evaluate the environment tensor by employing the automatic differentiation with back propagation. This approach is equivalent to the second one but implemented differently.

Below we take the honeycomb lattice as an example to introduce these approaches.

23.4.1 Entanglement Mean-Field Approach

For the M-tensor defined in Eq. (23.8), instead of factorizing it along the vertical direction, one can also factorize it along the horizontal direction by SVD:

$$\tag{23.20}$$

An essential property of the singular value matrix λ, is that it effectively measures the entanglement or interaction between the left and right unitary matrices, U and V. This singular value matrix also lives on all four bonds connecting the system tensor M with the environment tensor M^{env}. They serve as an entanglement field acting on all four dangling bonds (i, j, i', j'). Since each λ connects two tensors on the two ends of the bond, we can attribute half of it, that is, $\lambda^{1/2}$, to the system tensor M and the other half to the environment tensor M^{env}. Hence we can take the singular value matrix as an entanglement mean-field to approximate the environment tensor by

$$M^{\text{env}}_{i'j,ij'} \approx \sqrt{\lambda_j \lambda_i \lambda_{j'} \lambda_{i'}}, \tag{23.21}$$

In this discussion, the singular bond spectrum λ_i is determined from M by SVD without considering the environment contribution. A self-consistent treatment for the effective mean-field at each bond should take into account the environment contribution in obtaining the bond singular value spectrum by SVD. Hence λ is determined by the self-consistent equation

$$\tag{23.22}$$

where a is a normalization constant. λ on the right-hand side equals that on the left-hand side. It is not easy to solve this self-consistent equation. Nevertheless, it can be solved iteratively around a critical point.

23.4.2 TRG Approach

The environment tensor M^{env} can also be determined by tracing out all the tensors in the environment using TRG [129, 130]. This approach is more accurate but also more complicated than the mean-field approach. In order to understand how this approach works, let us first consider how the environment tensor changes under the transformation of TRG.

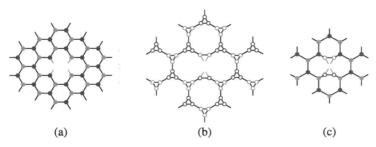

(a)   (b)   (c)

Figure 23.3 Evolution of the environment tensor under the TRG transformation.

Figure 23.3 shows how the environment tensor evolves under one iteration of TRG. At each iteration, for example, the nth iteration of TRG, a product of two local tensors linked by a horizontal bond in the original lattice [Fig. 23.3(a)], is first factorized into two local tensors, $S^{a,n}$ and $S^{b,n}$, defined by Eqs. (23.3)–(23.4) in the rewired lattice (Fig. 23.3(b)). Here a superscript n is added to these tensors to emphasize that they are the tensors obtained in the nth TRG iteration. After the truncation, a decimated environment lattice, shown in Fig. 23.3(c), is obtained by tracing out the three internal bonds on each triangle and replacing them with a single tensor. Therefore, the environment tensor at the nth iteration, M_n^{env}, can be approximately represented by Fig. 23.3(c). This decimated environment tensor, which is redrawn in Fig. 23.4(a), can be factorized as a product of the environment tensor at the $(n + 1)$th iteration, shown in 23.4(b), and four S-tensors shown in 23.4(c); namely,

$$\left(M_n^{\mathrm{env}}\right)_{ijkl} = \sum_{l'i'j'k'} \sum_{pq} \left(M_{n+1}^{\mathrm{env}}\right)_{l'i'j'k'} S_{i'ip}^{a,n} S_{l'pl}^{a,n} S_{j'qj}^{b,n} S_{k'kq}^{b,n}. \tag{23.23}$$

An important property revealed by the above recursion formula is that the environment tensor at the nth iteration, M_n^{env}, is determined by the environment tensor at the $(n + 1)$th iteration, M_{n+1}^{env}. It suggests that in order to determine M^{env}, which is the environment tensor at the first iteration, $M_1^{\mathrm{env}} = M^{\mathrm{env}}$, we should take a backward propagation to go from the last iteration to the first one.

To determine the environment tensor in the backward iteration, let us assume that the TRG iterations are terminated at the Nth step so that only four sites are left in the environment. In this case, M_N^{env} is determined by the formula

$$\left(M_N^{\mathrm{env}}\right)_{ijkl} = \sum_{\alpha\beta\gamma\delta} T_{i\gamma\delta}^b T_{j\alpha\beta}^b T_{k\gamma\delta}^a T_{l\alpha\beta}^a, \tag{23.24}$$

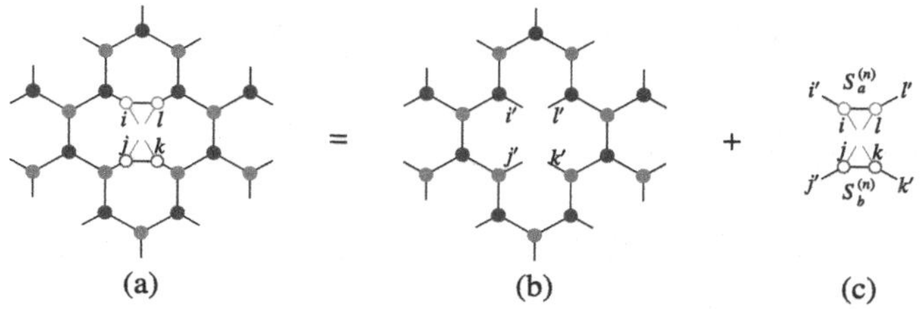

Figure 23.4 The relationship between two successive environment tensors: (a) M_n^{env} is a product of (b) M_{n+1}^{env} and four local tensors $S_a^{(n)}$ and $S_b^{(n)}$ shown in (c).

where T^a and T^b are the local tensors for the final environment lattice. The center-symmetric boundary conditions are assumed. In practical calculations, there is no need to use a large N because M^{env} generally converges fast with the lattice size.

This discussion indicates that both forward and backward iterations should be taken in order to determine the environment tensor, similar to the finite-lattice sweeping in DMRG. To improve the accuracy of the environment tensor, one can also repeat the forward and backward iteration steps a few times.

23.4.3 By Automatic Differentiation

The environment tensor can be also determined by automatic differentiation [120, 337, 338, 432]. To understand this, let us use $T_i = \left(T_i^a, T_i^b\right)$ and $W_i = (U_i, V_i, \lambda_i)$ to denote the local tensors and transformation matrices generated at the ith TRG iteration, respectively. Graphically, the forward iterations of TRG can be represented as

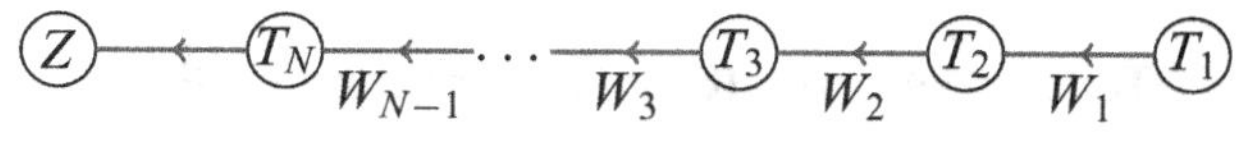

$$\textcircled{Z} \longleftarrow \textcircled{T_N} \underset{W_{N-1}}{\longleftarrow} \cdots \longleftarrow \textcircled{T_3} \underset{W_2}{\longleftarrow} \textcircled{T_2} \underset{W_1}{\longleftarrow} \textcircled{T_1} . \tag{23.25}$$

W_i depends on T_i. Initially, the local tensor is T_1. The initial M is obtained by contracting two initial local tensors, hence is a function of T_1.

The environment tensor is nothing but the derivative of the partition function Z with respect to M. This derivative, on the other hand, can be determined from the derivative of Z with respect to T_1:

$$M_{\text{env}} = \frac{\delta Z}{\delta M} = \frac{\delta Z}{\delta T_1} \left(\frac{\delta M}{\delta T_1}\right)^{-1} . \tag{23.26}$$

To evaluate $\delta Z/\delta T_1$, let us introduce the adjoint tensor of T_i:

$$\bar{T}_i = \frac{\delta Z}{\delta T_i} = \frac{\delta Z}{\delta T_{i+1}} \frac{\delta T_{i+1}}{\delta T_i} = \bar{T}_{i+1} \frac{\delta T_{i+1}}{\delta T_i} . \tag{23.27}$$

Taking the backward propagation for the adjoint variables [339] starting with
$\bar{Z} = 1$:

$$\textcircled{$\bar{Z}$}\longrightarrow\textcircled{$\bar{T}_N$}\longrightarrow\cdots\longrightarrow\textcircled{$\bar{T}_3$}\longrightarrow\textcircled{$\bar{T}_2$}\longrightarrow\textcircled{$\bar{T}_1$}\;, \tag{23.28}$$

we immediately obtain $\bar{T}_1$ and then

$$M_{\text{env}} = \bar{T}_1 \left(\frac{\delta M}{\delta T_1}\right)^{-1}. \tag{23.29}$$

The above discussion indicates that the backward iteration of TRG introduced in
§23.4.2 can be achieved by the back propagation of automatic differentiation. There
is, however, a subtle difference. In the forward iteration (23.25), the local tensor
obtained at the $(i + 1)$th iteration, T_{i+1}, is determined by the local tensor at the previous step T_i, as well as the transformation matrix W_{i+1}; that is, $T_{i+1} = T_{i+1}(T_i, W_i)$.
Since W_i is also a function of T_i, this means that $T_{i+1} = T_{i+1}(T_i, W_i(T_i))$ and

$$\frac{\delta T_{i+1}}{\delta T_i} = \frac{\partial T_{i+1}}{\partial T_i} + \frac{\partial T_{i+1}}{\partial W_i}\frac{\partial W_i}{\partial T_i}. \tag{23.30}$$

In the backward iteration introduced in §23.4.2, however, W_i is treated as a variable independent of T_i. This implies that the second term on the right-hand side
of the above equation is not included in the backward iteration. Thus if the same
number of iterations is used for calculating the environment tensor, the automatic
differentiation method is more accurate than the TRG backward iteration method.
However, the cost of backpropagation of automatic differentiation, in both computational time and memory space, is much higher than the TRG approach. It limits
N to be a relatively small integer compared to the backward TRG method.

23.5 Tensor Network Renormalization (TNR)

Both TRG and SRG previously introduced do not remove all the local or short-
range entanglement at each coarse-graining step. As a result, the coarse-grained
system still contains short-range entanglement, which is supposed to be irrelevant to long-range physics but can reduce the accuracy of results. In principle,
the coarse-graining tensor renormalization can be more efficient if the short-range
entanglement is effectively reduced.

The short-range entanglement can be filtered out by introducing the disentangler
(a unitary matrix), as in MERA, at each RG step [51]. Gu and Wen discussed
the reduction of short-range entanglement in the framework of TRG [132].
In 2015, Evenbly and Vidal introduced a novel coarse-graining RG approach,
dubbed the tensor network renormalization (TNR), to perform the TRG calculation by removing the short-range entanglement [133]. Their approach solves the

slow-convergence problem of TRG at a critical point and yields a reliable scheme to find a scale-invariant local tensor.

TNR introduces two kinds of transformation matrices to renormalize the tensors. One is a unitary matrix, called a disentangler, u, which is used to reduce the entanglement between two neighboring tensors. The other is an isometric matrix, denoted as v or w, used to decimate a local tensor. They form a set of projection operators:

$$P = \boxed{u} \cdots \boxed{u^\dagger} = \boxed{u} \cdots \boxed{u^\dagger} . \qquad (23.31)$$

Both the disentangler u and the isometries v and w are determined by minimizing the cost function:

$$f(u, v, w) = \|M - PM\|, \qquad (23.32)$$

where

$$M = \quad , \qquad PM = \boxed{u} \cdots \boxed{u^\dagger} . \qquad (23.33)$$

Using the property of projection operator, $P^\dagger = P$ and $P^2 = P$, the cost function can be also expressed as

$$f(u, v, w) = \mathrm{Tr}\left(M^\dagger M\right) - \mathrm{Tr}\left(M^\dagger P M\right). \qquad (23.34)$$

The first term on the right-hand side is independent of u, v, and w. Thus minimizing the cost function (23.32) is equivalent to maximizing $\mathrm{Tr}\left(M^\dagger P M\right)$.

We determine u, v, and w using a self-consistent iterative method. As an example, let us show how u is determined. The isometric matrices v and w can be similarly determined.

To find u, we first fix $u^\dagger$ and all v and w matrices, and write $\mathrm{Tr}\left(M^\dagger P M\right)$ as

$$\mathrm{Tr}\left(M^\dagger P M\right) = \mathrm{Tr}\left(T_u u\right), \qquad (23.35)$$

where T_u is the derivative of $M^\dagger P M$ with respect to u:

$$T_u = \boxed{\quad \cdots u^\dagger \quad} , \qquad (23.36)$$

which is obtained by removing the u tensor from $M^\dagger PM$. Factorizing T_u with SVD,

$$T_u = U\lambda V^\dagger, \tag{23.37}$$

a provisional solution of u is found to be

$$u = VU^\dagger. \tag{23.38}$$

$\mathrm{Tr}\left(M^\dagger PM\right)$ is now simply a sum of all singular values of T_u:

$$\mathrm{Tr}\left(M^\dagger PM\right) = \mathrm{Tr}\lambda. \tag{23.39}$$

Replacing $u^\dagger$ using the newly obtained u and repeating the above procedure until u is converged, we are able to find the optimized disentangler.

In this scheme, a product of four tensors on a square is approximated by inserting two projectors between the left and right two-column tensors

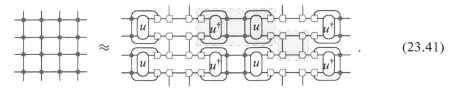

$$\tag{23.40}$$

For a translation-invariant system, the coarse-graining transformation starts by converting a square lattice tensor network into a new tensor network model using the above approximation:

$$\tag{23.41}$$

This newly generated tensor network model comprises two kinds of tensor clusters, as indicated by the shaded areas enclosed by the dashed lines. By contracting all the internal bonds in these clusters, we obtain two auxiliary tensors:

$$\tag{23.42}$$

and

$$\tag{23.43}$$

This changes the tensor network model to

$$(23.44)$$

Now let us factorize these two auxiliary tensors by SVD:

$$\times = \bigvee_{\wedge}, \qquad \times = \succ\!\!\prec. \tag{23.45}$$

In this step, a further approximation is taken to truncate the basis states by keeping the first D singular values for each factorization. It converts Eq. (23.44) into a tensor network model defined on a square-octagonal lattice, shown in panel (a) in the following equation:

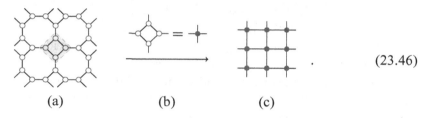

$$(23.46)$$

(a)   (b)   (c)

By reducing the four local tensors inside the shaded square to one tensor, as in panel (b), we obtain a renormalized tensor network model defined on the square lattice, shown in panel (c). However, the lattice size of this renormalized model is reduced by a factor of 4 compared with the original. This completes a cycle of TNR iteration. An accurate calculation for the free energy can be obtained by repeating these steps until the entire lattice is contracted.

23.6 Loop Tensor Network Renormalization (Loop-TNR)

Loop-TNR [134] is another way to perform the TRG calculation on the square lattice. It follows precisely the same coarse-graining steps as in TRG. Figure 23.2 shows a schematic representation of these steps.

Like TRG, Loop-TNR starts by rewiring a square lattice into a square-octagon one. Instead of factorizing each local tensor by SVD, it determines the eight tensors at the eight vertexes of an octagon used for approximately representing the four local tensors on a minimal square in the original square lattice:

$$
\begin{array}{c}
\begin{array}{cc}
T_1 & T_4 \\
 & \\
T_2 & T_3
\end{array}
\end{array}
\approx
\begin{array}{c}
\text{(octagon with } S_1, S_8, S_2, S_7, S_3, S_6, S_4, S_5)
\end{array}
, \tag{23.47}
$$

by variationally minimizing the cost function:

$$
f(S_1 \cdots S_8) =
\left\|
\begin{array}{cc}
T_1 & T_4 \\
 & \\
T_2 & T_3
\end{array}
-
\begin{array}{c}
\text{(octagon with } S_1, S_8, S_2, S_7, S_3, S_6, S_4, S_5)
\end{array}
\right\|^2 . \tag{23.48}
$$

Loop-TNR can be regarded as a cluster update approach of TRG. It works better than TRG because it provides an approximate scheme to optimize a four-tensor network simultaneously. In contrast, TRG optimizes each local tensor independently, which ignores the interplay between different local tensors.

Loop-TNR provides an intelligent scheme to implement the idea of tensor entanglement filtering renormalization (TEFR) initially proposed by Gu and Wen [132]. The local entanglement inside a shaded square in the rewired square-octagon lattice, shown in Fig. 23.2(b), is removed after the four local tensors on this square are contracted. The short-range entanglement inside an octagon is also removed in the minimization of the cost function. In Ref. [134], the authors introduced a pseudo-canonical step to remove the short-range entanglement inside a square that is rewired to an octagon. However, that step is unnecessary and may even introduce extra numerical errors. This method removes the short-range entanglement just in the step of variational minimization.

Similar to TRG, both TNR and Loop-TNR are local optimization approaches, not considering the renormalization effect of the environment. Nevertheless, they partially remove the short-range entanglement during the rewiring and work reasonably well at the critical point. However, as both approaches involve variational minimization, their computational cost is higher than TRG if the same bond dimension is adopted.

23.7 HOTRG

TRG and other coarse-graining tensor renormalization methods introduced in the preceding sections provide a powerful tool for studying two-dimensional classical models based on SVD. However, extending these methods to three dimensions is difficult due to the increase in the order of local tensors and the change of lattice topology in the coarse-graining process.

The coarse-graining tensor renormalization using HOSVD [334], abbreviated as HOTRG, nevertheless, provides a solution to this problem [131]. HOTRG is a coarse-graining approach that rescales the lattice size by performing tensor contractions. Unlike TRG, it contracts the lattice along the two principal axes, x- and y-axis, on the square lattice.

For example, a contraction along the y-axis is done in the following two steps:

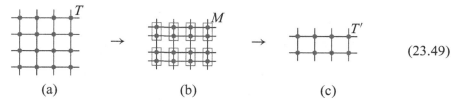

$$(23.49)$$

(a)   (b)   (c)

At the first step (a→b), two neighboring local tensors T along the y-axis are merged to form a new tensor M in the coarse-grained lattice:

$$x \text{—}\boxed{M}\text{—} x' \; = \; \begin{matrix} x_2 \text{—}\bullet\text{—} x'_2 \\ x_1 \text{—}\bullet\text{—} x'_1 \end{matrix} \quad , \qquad (23.50)$$

where $x = x_1 \otimes x_2$, $x' = x'_1 \otimes x'_2$. The bond dimension of M along the x-axis equals the square of the bond dimension of T. To truncate the horizontal bond dimension, we need to find a pair of basis transformation matrices, P and Q. P and Q are isometric or, more generally, oblique matrices.

Suppose we have already found the transformation matrices P and Q. The step following, (b→c), is to update the local tensor according to the formula

$$\boxed{T'} \; = \; \text{—}\!\!\left\langle Q \right.\!\! \boxed{M} \!\!\left. P \right\rangle\!\!\text{—} \; . \qquad (23.51)$$

This reduces the partition function to a tensor network model spanned by local tensor T'. Furthermore, the lattice size is reduced by a factor of 2 compared with the original lattice.

The coarse-graining contraction along the horizontal axis can be similarly done. Again an approximation is taken to reduce the lattice size by a factor of 2. This HOTRG calculation can be repeated iteratively until the free energy and other physical quantities calculated are converged. The cost of the calculation scales as D^7 in the computer time and D^4 in the memory space.

Two approaches can be used to find the transformation matrices. Here follows an introduction to these approaches.

23.7.1 HOSVD Approach

To find the transformation matrices, one can simply use HOSVD [334], introduced in §2.6, to pseudo-diagonalize tensor M:

$$x—\boxed{M}—x' \;=\; —\boxed{U_l}—\boxed{S}—\boxed{U_r}— , \tag{23.52}$$

where U's are unitary matrices. S is the core tensor, which is pseudo-diagonal and satisfies the properties of all orthogonality, as described in §2.6. The norm of this core tensor plays a similar role to the singular value of a matrix.

In M, the two vertical bonds (i.e. y and y'), are not renormalized. Thus in a practical calculation, only U_l and U_r need to be evaluated. Since the right bond of M is linked directly to the left bond of an identical tensor on its right-neighboring site (assuming the tensor network model is translation invariant), thus truncating any one of the horizontal bonds of M will automatically truncate the other horizontal bond. The truncation matrix U is determined by comparing the errors for the left and right truncations:

$$\varepsilon_1 = \sum_{i>D} |S_{i,:,:,:}|^2 \tag{23.53}$$

and

$$\varepsilon_2 = \sum_{j>D} |S_{:,j,:,:}|^2. \tag{23.54}$$

If $\varepsilon_1 < \varepsilon_2$, then $U = U_l$ and the first dimension of S is truncated from D^2 to D. Otherwise, $U = U_r^\dagger$ and the second dimension of S is truncated from D^2 to D. After truncation, U and $U^\dagger$ become isometric. They are just the transformation matrices to be found, $Q = U$ and $P = U^\dagger$.

HOSVD provides a simple and locally optimal approximation to minimize the truncation error [335, 336]. However, it should be pointed out that in multilinear algebra, there does not exist a general decomposition method for tensors with all the properties of a matrix SVD. A more general scheme for decomposing a tensor into a set of matrices and a reduced core tensor is called Tucker decomposition. An optimal decomposition can be achieved by variationally minimizing the difference between the tensors before and after the decomposition. However, the computational cost of this variational scheme is generally high.

23.7.2 Bond-Matrix Decomposition Approach

A more accurate approach for finding the oblique matrices, P and Q, is to insert them at each bond linking two neighboring M-tensors [433], and determine them by minimizing the difference:

$$\left\| \begin{array}{c} M \quad M \\ \end{array} - \begin{array}{c} M \quad M \\ P - Q \end{array} \right\| . \tag{23.55}$$

This minimization problem can be solved by decomposing the bond matrix between two M tensors with SVD. For doing this, we first take a QR decomposition for the left M-tensor:

$$x - \boxed{M} - x' = x - Q_l - R - x' , \tag{23.56}$$

and a LQ-decomposition for the right M-tensor:

$$x - \boxed{M} - x' = x - L - Q_r - x' . \tag{23.57}$$

As both Q_l and Q_r are unitary matrices, to truncate the bond dimension linking the two M-tensors, we only need to consider the product of R and L. Taking SVD for this product:

$$RL = U\lambda V^\dagger, \tag{23.58}$$

we then truncate the diagonal matrix λ by retaining its largest D singular values. This truncation can also be achieved by inserting the oblique pair, P and Q, between R and L. Assuming $\bar{U}$, $\bar{\lambda}$, and $\bar{V}$ to be the corresponding matrices after the truncation, we have

$$RPQL = \bar{U}\bar{\lambda}\bar{V}^\dagger; \tag{23.59}$$

hence

$$RP = \bar{U}\bar{\lambda}^{1/2}, \qquad QL = \bar{\lambda}^{1/2}\bar{V}^\dagger. \tag{23.60}$$

Using Eq. (23.58), we find P and Q to be

$$P = R^{-1}\bar{U}\bar{\lambda}^{1/2} = L\bar{V}\bar{\lambda}^{-1/2}, \tag{23.61}$$

$$Q = \bar{\lambda}^{1/2}\bar{V}^{\dagger}L^{-1} = \bar{\lambda}^{-1/2}\bar{U}^{\dagger}R. \tag{23.62}$$

R^{-1} and L^{-1} are the left and right inverses of R and L, respectively.

In this scheme, the oblique pair is found by taking QR and LQ for the left and right local tensors, respectively. It does not require that these two local tensors be the same. Therefore, this scheme can be readily extended to systems without translation symmetry.

23.8 Second Renormalized HOTRG

The idea of the second renormalization introduced in §23.3 can also be used to improve HOTRG. Again, it leads to global minimization of the truncation error, and the corresponding method is referred to as HOSRG [131].

Like in SRG, the key step in HOSRG is to evaluate the environment tensor M^{env} for a given system tensor M. The environment tensor is, in fact, an adjoint of the system tensor, defined by the derivative of the partition function Z with respect to M:

$$M^{\mathrm{env}} = \bar{M} = \frac{\delta Z}{\delta M}. \tag{23.63}$$

The partition function, on the other hand, is the scalar product of these two tensors:

$$Z = \mathrm{Tr}\left(MM^{\mathrm{env}}\right) = \mathrm{Tr}\left(M\bar{M}\right). \tag{23.64}$$

To calculate $\bar{M}$, it is more convenient first to determine the environment tensor corresponding to a single local tensor T, defined by

$$\bar{T} = \frac{\delta Z}{\delta T}. \tag{23.65}$$

$\bar{M}$ is related to $\bar{T}$ by the equation

$$\bar{M} = \bar{T}\left(\frac{\delta M}{\delta T}\right)^{-1}. \tag{23.66}$$

23.8.1 Determination of the Environment Tensors

Similar to SRG, there is more than one way to determine the environment tensor $\bar{T}$. For example, one can evaluate $\bar{T}$ by taking automatic differentiation, similar to the approach introduced in §23.4.3. One can also determine $\bar{T}$ by taking a forward-backward iteration with HOTRG. The former approach is simple to implement in programming; the latter, on the other hand, is physically more intuitive. Here we give an introduction to the latter approach.

To determine the environment tensor, again we first perform a HOTRG calculation to determine iteratively all the transformation matrices $U^{(n)}$ and local tensors $T^{(n)}$, starting with $T^{(1)} = T$. This iteration terminates when the system reaches the desired size, say 2^N lattice points with $N = 30 \sim 80$. We then carry out a backward iteration to evaluate the environment tensor $\bar{T}^{(n)}$ iteratively, starting from $\bar{T}^{(N+1)}$, which is assumed to be a unit tensor.

On the square lattice, the environment tensor simply has the structure

$$(23.67)$$

From the iteration steps of HOTRG, it is straightforward to show that $\bar{T}^{(n)}$ is recursively connected to $\bar{T}^{(n+1)}$ by the local tensor as well as the transformation matrices:

$$(23.68)$$

It can also be expressed as

$$\bar{T}^{(n)}_{y'\bar{y}x_1x_1'} = \sum_{xx'yx_2x_2'\bar{y}} \bar{T}^{(n+1)}_{xx'yy'} T^{(n)}_{x_2x_2'y\bar{y}} Q^{(n)}_{x_1x_2,x} P^{(n)}_{x_1'x_2',x'}. \tag{23.69}$$

This backward iteration is terminated once $\bar{T}$ is determined.

23.8.2 HOSRG Sweep

The backward iteration determines all coarse-grained environment tensors needed for carrying out the second renormalization calculation. We can repeat the forward and backward iterations from these tensors to further reduce the truncation error, similar to the DMRG sweep. Starting from the second cycle of iterations, the coarse-graining transformation matrices, $P^{(n)}$ and $Q^{(n)}$, are determined by evaluating the bond density matrix $\rho^{(n)}$ with full consideration of the environment contribution from $\bar{T}^{(n+2)}$ ($n = 1 \cdots N$):

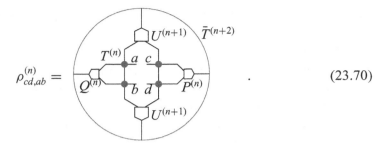

$$\rho^{(n)}_{cd,ab} = \qquad . \qquad (23.70)$$

This density matrix is diagonalized by a pair of canonical matrices, $P^{(n)}$ and its inverse $Q^{(n)}$:

$$\rho^{(n)} = P^{(n)} \Lambda Q^{(n)}, \qquad (23.71)$$

where Λ is the eigenspectrum of $\rho^{(n)}$. As in DMRG [18], the eigenvalues of this density matrix determine the probabilities of the corresponding eigenvectors in the virtual bond basis space. By keeping the largest D eigenvalues of Λ in their absolute values and the corresponding eigenvectors, one can update the local tensor $T^{(n+1)}$ using Eq. (23.51). In case $\rho^{(n)}$ is not semi-positive definite, the truncation schemes discussed in Chapter 15 should be invoked to determine $P^{(n)}$ and $Q^{(n)}$.

The forward-backward sweep of HOSRG establishes a global optimization scheme of coarse-graining tensor renormalization. It provides a self-consistent solution to the local and environment tensors and can significantly improve the accuracy of HOTRG.

23.9 Comparison of Different Methods

The RG methods introduced in this book for treating a two-dimensional classical lattice model, or equivalently a one-dimensional quantum lattice model, can be grouped into two classes according to the schemes in contracting the lattice. The first includes the coarse-graining methods introduced in this chapter. The second includes, for example, DMRG, TMRG, CTMRG, and iTEBD, introduced in the preceding chapters. These methods work based on the diagonalization of a Hamiltonian, or more generally a transfer matrix. Below we make a quantitative comparison between these methods by applying them to the two-dimensional Ising model.

Figure 23.5 compares the relative errors of the free energy:

$$\delta F(T) = 1 - \frac{F(T)}{F_{ex}(T)}, \qquad (23.72)$$

obtained with different RG methods for the Ising model on the square lattice. $F(T)$ is the free energy calculated with an RG method in the thermodynamic limit. The

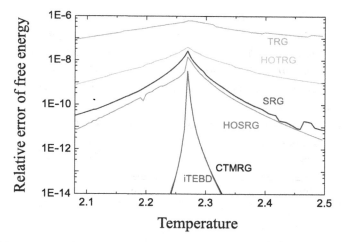

Figure 23.5 Temperature dependences of the relative errors of the free energy obtained by different RG methods with $D = 30$ for the two-dimensional Ising model on the square lattice.

maximal bond dimension used is $D = 30$. $F_{ex}(T)$ is the exact result [342]. For this simple model, all the RG methods produce very accurate results. In particular, the transfer matrix-based RG methods, CTMRG and iTEBD, show smaller errors than the coarse-graining ones. The relative error is less than 10^{-6} even at the worst case at the critical point $T_c = 2/\ln(1 + \sqrt{2}) \approx 2.26918531421$. It drops very fast with the temperature moving away from the critical point. The error becomes larger at T_c because the correlation length diverges at that point. Away from the critical point, the correlation length becomes finite, eliminating the logarithmic correction to the entanglement area law with the increase of the system size.

Figure 23.6 shows how the relative errors of the free energy at the critical temperature T_c converge with the bond dimension D for the two-dimensional Ising model on the square lattice. As expected, the errors drop with D. The free energy converges polynomially with D in most RG methods, except TNR and Loop-TNR, where the short-range entanglement is effectively suppressed at the step of RG decimation so that the errors can decay exponentially with D. It demonstrates the importance of reducing short-range entanglement in the RG procedure [133]. By comparing the results of TRG/HOTRG with those of HOTRG/HOSRG, it is clear that the coarse-graining RG results can also be significantly improved by considering the renormalization effect of the environment on the basis truncation. Furthermore, the improvement of SRG/HOSRG over TRG/HOTRG becomes more and more pronounced with increasing D.

Table 23.1 compares the scaling powers of the computational cost needed for the typical RG methods developed for studying one-dimensional quantum or

Table 23.1 *Minimal orders of computational resources needed for different methods. d is the physical degrees of freedom at each lattice site. D is the number of states retained at each RG decimation. For one-dimensional quantum lattice models, N is the lattice size for DMRG and TEBD and the Trotter number for QTMRG. If a translation-invariant MPS is used, the cost of TEBD is N-independent. For two-dimensional classical systems, N is the lattice length along one of the principal axes.*

Method	CPU Time	Memory
DMRG [18]	$d^3 D^3 N$	$d^2 D^3$
TMRG [68]	$d^3 D^3 N$	$d^2 D^3$
CTMRG [70]	$d^3 D^3 N$	$d^2 D^3$
QTMRG [72, 75]	$d^3 D^3 N$	$d^2 D^3$
TEBD [38]	$d^3 D^3 N$	$d^2 D^3$
TRG [128]	$D^6 \ln N$	D^4
SRG [129]	$D^6 \ln N$	D^4
HOTRG [131]	$D^7 \ln N$	D^4
HOSRG [131]	$D^8 \ln N$	D^6
TNR [133]	$D^7 \ln N$	D^5
Loop-TNR [134]	$D^6 \ln N$	D^4

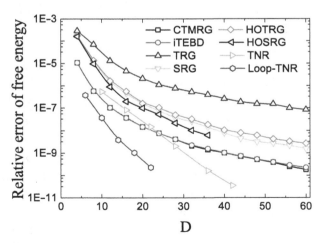

Figure 23.6 Relative errors of the free energy versus the bond dimension D at the critical temperature T_c obtained with different methods for the two-dimensional Ising model on the square lattice. The TNR and Loop-TNR data are taken from Refs. [133] and [134], respectively.

two-dimensional classical lattice models. Generally, the transfer matrix–based RG methods run more efficiently than the coarse-graining ones. The computational time of TNR scales as D^7, but it could be reduced to D^6 [133]. Besides the bond-dimension dependence, the computational time depends on how the local tensor is updated in the RG iteration. For example, Loop-TNR has the same scaling power as TRG in the computational time. However, the computational cost needed in a Loop-TNR calculation is much higher than TRG because the former relies on a variational determination of local tensors, which is generally time-consuming.

23.10 Three-Dimensional Classical Models

An advantage of HOTRG compared with TRG or other coarse-graining RG methods is that it can be readily extended to three dimensions. On the cubic lattice, a complete cycle of lattice contraction needs to be done in three steps: along the x-axis, y-axis, and z-axis, respectively. At each step, two neighboring tensors are merged to form a single coarse-grained tensor, and the lattice size is reduced by a factor of 2, like in two dimensions. A contraction along the z-axis, for example, is done in the following two steps:

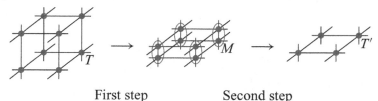

First step   Second step

The coarse-grained tensor M is determined by contracting two local tensors T. Nevertheless, only the bond dimensions on the xy-plane need to be renormalized.

Using the methods introduced in §23.7.1 or §23.7.2, we can determine from M the basis transformation matrices along both x and y directions. After that, the dominant D singular vectors along these two bond directions are retained, and M is renormalized and set as the updated local tensor T' for the next iteration:

$$\text{diagram} \qquad (23.73)$$

The coarse-graining and renormalization of local tensors along x- and y-directions can be similarly done. This three-step iteration is then repeated until the results converge.

In the HOTRG calculation, the computational time scales with D^{11} and the memory scales with D^6. This cost in the computational resource is significantly smaller than in other three-dimensional RG calculations [110, 318, 319, 320, 392, 434, 435, 436].

After each HOTRG (forward) iteration, one can also do a backward iteration to evaluate the environment tensors and carry out the HOSRG calculation in three dimensions. Again, the bond density matrix is evaluated from the environment tensors. Diagonalizing the bond density matrix allows us to determine the local transformation matrices at each step of the forward iteration.

23.10.1 Bond-Matrix Decomposition

As in two dimensions, the transformation matrices can be determined more accurately by decomposing the bond matrix. This decomposition is easy to implement, especially when the local tensors are not reflection and translation symmetric. Below we take a translation-invariant system as an example to show how this approach works.

To determine the bond transformation matrices along the x-axis, we consider a system of two M tensors linked by a horizontal bond. We first take a QR decomposition to separate the left bonds from the rest of the bonds for the left M-tensor:

$$ \text{(diagram)} = \text{(diagram)} \; , \tag{23.74} $$

and an LQ-decomposition to separate the right bonds from the rest bonds for the right M-tensor:

$$ \text{(diagram)} = \text{(diagram)} \; . \tag{23.75} $$

Here the thick bond has a dimension of D^2 if the thinner one has a dimension of D. Both Q_l and Q_r are isometric. The upper and lower triangular matrices, R and L, have the same structures as the quantities obtained in Eqs. (23.56) and (23.57), respectively. We can then follow all the steps below Eq. (23.57) in §23.7.2 to find the oblique transformation matrices along the x-axis using R and L. Similarly, one can also find the oblique transformation matrices along the y-axis.

In this calculation, the cost for obtaining R using Eq. (23.74) is of the order $O(D^{12})$. This cost can be dramatically reduced by taking the following two QR decompositions for the two T tensors in M before contraction, instead of one QR decomposition for the contracted local tensor M:

$$\text{(figure)} \qquad = \qquad \text{(figure)} \;, \qquad\qquad (23.76)$$

$$\text{(figure)} \qquad = \qquad \text{(figure)} \;. \qquad\qquad (23.77)$$

Equations (23.76) and (23.77) are applied to the upper and lower local tensors in M, respectively. Both Q_1 and Q_2 are isometric. Using R_1 and R_2, we obtain a new matrix A:

$$\text{(figure)} \qquad = \qquad \text{(figure)} \;. \qquad\qquad (23.78)$$

By further taking a QR decomposition for A, we then obtain the R-matrix:

$$\text{(figure)} \qquad = \qquad \text{(figure)} \;. \qquad\qquad (23.79)$$

Q_l is a product of Q_1, Q_2, and Q_3, and does not need to be explicitly evaluated. The cost, in this case, scales as D^8. The L-matrix in Eq. (23.75) can be similarly determined.

23.10.2 Example: Three-Dimensional Ising Model

As an example of application, we take the Ising model on the cubic lattice to demonstrate how the three-dimensional HOTRG works.

Figures 23.7 and 23.8 show the temperature dependence of the internal energy, the specific heat, and the magnetization for the three-dimensional Ising model obtained from HOTRG with $D = 14$, respectively. For comparison, the Monte Carlo results for the specific heat [437] and the magnetization [438] are also shown in the corresponding figures. The HOTRG result of the specific heat is less accurate than the internal energy around the critical point because the former is obtained from the numerical derivative of the internal energy. Nevertheless, the HOTRG results agree with the Monte Carlo results in the whole temperature range.

The critical transition temperature T_c can be determined from the singular behavior of the specific heat and that of the magnetization. A comparison for the value of T_c obtained by different methods is shown in Table 23.2. Furthermore, by fitting the data of the magnetization M in the critical regime with the formula

$$M \sim \left(1 - \frac{T}{T_c}\right)^{\gamma}, \qquad\qquad (23.80)$$

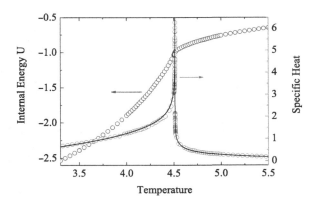

Figure 23.7 Temperature dependence of the internal energy (left) and the specific heat (right) for the three-dimensional Ising model obtained by the HOTRG with $D = 14$. The Monte Carlo result (solid curve) is obtained from an empirical formula in Ref. [437]. (Adopted from Ref. [131]. Copyright by the American Physical Society.)

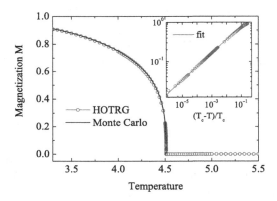

Figure 23.8 Temperature dependence of the magnetization for the three dimensional Ising model obtained from HOTRG with $D = 14$. The Monte Carlo result obtained using the formula from Ref. [438] is shown for comparison. Inset: Logarithmic plot of the magnetization around the critical point. (Adopted from Ref. [131]. Copyright by the American Physical Society.)

the exponent γ is estimated to be 0.3295, which agrees with the results obtained by the Monte Carlo [444] (0.3262) and series expansion [445] (0.3265) results.

23.11 Two-Dimensional Quantum Lattice Models

The three-dimensional HOTRG and HOSRG introduced in the previous section are applicable to two-dimensional quantum systems. In this case, one needs first to

Table 23.2 *Comparison of critical temperatures T_c obtained by different methods for the Ising model on the simple cubic lattice.*

Method (year)	T_c
CTMRG (2001) [110]	4.5393
TPVA (2005) [320]	4.554
Algebraic variation (2006) [439]	4.547
Series expansion (2000) [440]	4.511536(21)
Monte Carlo RG (1996) [441]	4.5115(2)
Monte Carlo (2003) [442]	4.5115248(6)
Monte Carlo (2010) [443]	4.5115232(17)
HOTRG ($D = 23$) [135]	4.51152469(1)

represent the partition function of a two-dimensional quantum lattice model as a three-dimensional tensor network model.

Below we take the two-dimensional quantum Ising model as an example to show how the method works. The Hamiltonian of the quantum Ising model reads

$$H = - \sum_{\langle ij \rangle} \sigma_z^i \sigma_z^j - h \sum_i \sigma_x^i. \tag{23.81}$$

Using the Trotter–Suzuki decompositions, the partition function of this model can be expressed as [392]:

$$Z = \mathrm{Tr} e^{-\beta H} \approx \mathrm{Tr} \left(e^{-\tau H_z} e^{-\tau H_x} \right)^M + O(\tau^2), \tag{23.82}$$

where

$$H_z = - \sum_{\langle ij \rangle} \sigma_z^i \sigma_z^j, \qquad H_x = -h \sum_i \sigma_x^i. \tag{23.83}$$

$\beta = M\tau$ is the inverse temperature, and τ is the Trotter parameter. After inserting the complete basis states of all σ_z^i between the evolution operators and simplification, we can rewrite the partition function as a three-dimensional tensor product:

$$Z \approx \mathrm{Tr} \prod_i T_{l_i r_i f_i b_i \mu_i \nu_i} + O(\tau^2), \tag{23.84}$$

where the local tensor is defined by

$$T_{lrfb\mu\nu} = \sum_\sigma W_{\sigma l} W_{\sigma r} W_{\sigma f} W_{\sigma b} P_{\sigma\mu} P_{\sigma\nu}, \tag{23.85}$$

$$W = \begin{pmatrix} \sqrt{\cosh \tau}, & \sqrt{\sinh \tau} \\ \sqrt{\cosh \tau}, & -\sqrt{\sinh \tau} \end{pmatrix}, \tag{23.86}$$

$$P = \frac{1}{\sqrt{2}} \begin{pmatrix} e^{\tau h/2}, & e^{-\tau h/2} \\ e^{\tau h/2}, & -e^{-\tau h/2} \end{pmatrix}. \tag{23.87}$$

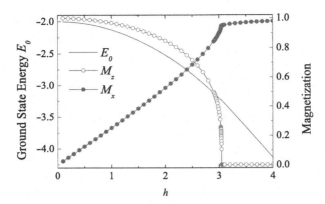

Figure 23.9 The ground state energy E_0, the magnetization $M_x \equiv \langle \sigma_x \rangle$ and $M_z \equiv \langle \sigma_z \rangle$ versus the applied field h for the two-dimensional quantum Ising model obtained by HOTRG with $D = 14$. (Adopted from Ref. [131]. Copyright by the American Physical Society.)

W results from the Ising interaction term. It is simple to show that

$$e^{\tau \sigma_1 \sigma_2} = \sum_\alpha W_{\sigma_1,\alpha} W_{\sigma_2,\alpha}. \tag{23.88}$$

Here α serves the role of the bond index. On the other hand, P results from the basis transformation imposed by $\exp(\tau h \sigma_x^i)$ at each lattice site:

$$\langle \sigma_1 | e^{\tau h \sigma_x^i} | \sigma_2 \rangle = \sum_\alpha P_{\sigma_1,\alpha} P_{\sigma_2,\alpha}. \tag{23.89}$$

The transverse field h appears only in the P-matrix. When $h = 0$, it is simple to show that $PP^\dagger = I$. In this case, the Hamiltonian returns to the two-dimensional classical Ising model, and Eq. (23.84) is reduced to a product of M two-dimensional tensor network states for the classical Ising model with an inverse temperature τ.

Unlike the three-dimensional Ising model, the tensor network model defined here is highly anisotropic along the imaginary time direction and the two spatial directions. If we directly apply HOTRG to this tensor network model, the truncation error would become large if we contract the lattice along any one of the spatial directions because the singular values of the local evolution operator in $\exp(-\tau H)$ are all very close to 1 for small τ. To avoid this singular problem, we first perform the imaginary time evolution without contracting the tensors along the two spatial directions. At each step of evolution, one more Trotter layer is added to the system, and the truncation for the bond dimensions along the two spatial directions is taken using HOTRG. The inverse temperature increases linearly with the evolution number. After performing this imaginary time evolution to a temperature T at which both τ and τh are of order 1, we then apply HOTRG to contract the whole tensor network model along all three directions successively.

At finite temperature, the lattice dimension of the tensor network model is finite along the imaginary time direction. The contraction along the Trotter direction is terminated if all bond variables along that direction are contracted. The iteration is then carried out along the two spatial directions for a pure two-dimensional classical model. This approach is accurate and efficient for evaluating physical quantities. However, the number of temperature points that can be studied with this approach is quite limited for a given τ since the temperature is reduced by a factor of 2 at each contraction along the Trotter direction.

This approach allows us to study the physical properties of the ground state without evaluating the ground-state wave function. At zero temperature, $M \rightarrow \infty$, the tensor network of the partition function becomes infinite along all three directions.

Figure 23.9 shows the field dependence of the ground-state energy, the transverse magnetization M_x and the longitudinal magnetization M_z for the transverse-field Ising model obtained by HOTRG with $D = 14$ and $\tau = 0.01$. This model exhibits a phase transition at a finite field. The critical field obtained from this calculation is $h_c = 3.0439$, consistent with other published results [71, 117, 392, 431, 435, 446, 447, 448].

Appendix

Other Numerical Methods

A.1 Power Method

The power method is a simple but powerful numerical method for finding the leading eigenvalue, the largest eigenvalue in its absolute value, and the corresponding eigenvector of a matrix. It works by applying the matrix successively to an arbitrary initial vector to generate a sequent vector. In the end, the vector will converge to the leading eigenvector.

The power method is easy to implement. Let us consider a matrix A (not necessary to be Hermitian) and an arbitrary initial vector $|x_1\rangle$ that is not orthogonal to the leading eigenvector. Formally, one can expand $|x_1\rangle$ using all right eigenvectors of A, $\{|\lambda_k\rangle\}$:

$$|x_1\rangle = \sum_k a_k |\lambda_k\rangle, \tag{A.1}$$

where a_k is the expansion coefficient of the eigenvector $|\lambda_k\rangle$ in $|x_1\rangle$. By applying A to these vectors, one obtains a new vector $|x_2\rangle$;

$$|x_2\rangle = A|x_1\rangle = \sum_k a_k \lambda_k |\lambda_k\rangle. \tag{A.2}$$

In comparison with $|x_1\rangle$, the probability of eigenvector $|\lambda_k\rangle$ in this new vector is amplified or diminished by the eigenvalue λ_k. Clearly, the probability of the leading eigenvector is always amplified with respect to all other eigenvectors. One can repeat the above projection iteratively to generate a sequent of vectors

$$|x_n\rangle = A|x_{n-1}\rangle, \tag{A.3}$$

then in the limit $n \to \infty$, $|x_n\rangle$ will converge to the leading eigenvector $|\lambda_{\max}\rangle$:

$$A|\lambda_{\max}\rangle = \lambda_{\max}|\lambda_{\max}\rangle, \tag{A.4}$$

$$|\lambda_{\max}\rangle = \lim_{n \to \infty} |x_n\rangle, \tag{A.5}$$

where $\lambda_{\max}$ is the largest eigenvalue.

If the largest eigenvalue is degenerate, then $|x_n\rangle$ will converge to a vector in the subspace spanned by the eigenvectors associated with this largest eigenvalue. After finding the first eigenvector, one can successively restrict the algorithm to the null space of the known eigenvectors to get other eigenvectors. In practice, this simple algorithm does not work for computing many of the eigenvectors because any round-off error will tend to introduce some small components of the more significant eigenvectors back into the computation, degrading the accuracy of the computation. Pure power methods also can converge slowly, even for the first eigenvector.

In a physical system, the absolute value of the ground-state energy may not be the largest eigenvalue of the Hamiltonian. However, one can always add a constant term to the Hamiltonian to make it the largest eigenvalue. Thus, the power method can also be applied to find the eigenvalue and eigenvector of the ground state.

The power method can be applied to an arbitrary square matrix. For a non-symmetric matrix, its left and right eigenvectors are not conjugate to each other. In this case, the left and right eigenvectors should be evaluated independently.

The power method does not work as efficiently as the Lanczos or other numerical methods, especially for a large-dimensional matrix. However, the idea of the power method is useful. Many approximate methods, for example, the Green's function Monte Carlo [449], are developed based on this simple idea.

A.2 Lanczos Method

Applying the power method, while getting the ultimate eigenvector, also produces a series of vectors $(|x_1\rangle, |x_2\rangle, \ldots, |x_n\rangle, \ldots,)$, which are eventually discarded, resulting in a considerable waste of computational resources since these vectors implicitly incorporate the information of matrix elements. Some advanced algorithms, such as the Lanczos algorithm, save this information and use the Gram–Schmidt process or Householder algorithm to reorthogonalize them into a tridiagonal matrix. This reorthogonalization can significantly improve the efficiency of the matrix diagonalization.

Cornelius Lanczos invented an iterative algorithm that is an adaptation of power methods to find eigenvalues and eigenvectors of a square matrix, or the singular value decomposition of a rectangular matrix. It transforms the original matrix into a tridiagonal matrix which is real and symmetric. It is particularly useful for finding the largest or smallest eigenvalue and the corresponding eigenvector of a sparse symmetric (or Hermitian) matrix. In the study of strongly correlated problems, this method is commonly used to find the ground state energy and eigenfunction of a Hamiltonian H.

The Lanczos iteration starts by applying the Hamiltonian H to an arbitrary initial state $|e_1\rangle$, that is normalized (i.e. $\langle e_1|e_1\rangle = 1$) and not orthogonal to the true ground state. By decomposing $H|e_1\rangle$ using $|e_1\rangle$ and a vector in a direction perpendicular, this defines a new basis vector $|e_2\rangle$ which is orthogonal to $|e_1\rangle$,

$$H|e_1\rangle = \alpha_1|e_1\rangle + \beta_2|e_2\rangle. \tag{A.6}$$

By multiplying both sides by $\langle e_1|$ and using the orthogonality condition $\langle e_1|e_2\rangle = 0$, α_1 is found to be

$$\alpha_1 = \langle e_1|H|e_1\rangle. \tag{A.7}$$

The vector $|\tilde{e}_2\rangle = \beta_2|e_2\rangle$ is then determined by moving the α_1 term from the right-to the left-hand side of Eq. (A.6)

$$|\tilde{e}_2\rangle = H|e_1\rangle - \alpha_1|e_1\rangle. \tag{A.8}$$

By further normalizing this vector, the value of β_2 is then determined to be

$$\beta_2 = \langle \tilde{e}_2|\tilde{e}_2\rangle^{1/2}. \tag{A.9}$$

β_2 can contain an arbitrary phase factor. However, it is more convenient to take β_2 as a real positive number in practical calculations.

By acting H onto $|e_2\rangle$, a new basis vector $|e_3\rangle$, which is orthogonal to $|e_1\rangle$ and $|e_2\rangle$, can be defined by expanding the resulting state using these three basis vectors:

$$H|e_2\rangle = \beta_2|e_1\rangle + \alpha_2|e_2\rangle + \beta_3|e_3\rangle. \tag{A.10}$$

The coefficient of the $|e_1\rangle$ term is equal to β_2 because the Hamiltonian is symmetric, $\langle e_1|H|e_2\rangle = \langle e_2|H|e_1\rangle$. The value of α_2 can be determined by multiplying both sides of the equation by $\langle e_2|$ and using the orthogonal conditions

$$\alpha_2 = \langle e_2|H|e_2\rangle. \tag{A.11}$$

The vector $|\tilde{e}_3\rangle = \beta_3|e_3\rangle$ is then given by

$$|\tilde{e}_3\rangle = H|e_2\rangle - \beta_2|e_1\rangle - \alpha_2|e_2\rangle. \tag{A.12}$$

β_3 is determined by normalizing this vector:

$$\beta_3 = \langle \tilde{e}_3|\tilde{e}_3\rangle^{1/2}. \tag{A.13}$$

Like β_2, β_3 can always be set real and positive.

The next step is to apply H to $|e_3\rangle$ and expand the resulting state using $(|e_1\rangle, |e_2\rangle, |e_3\rangle))$ and a normalized basis state $|e_4\rangle$ that is orthogonal to $(|e_1\rangle, |e_2\rangle, |e_3\rangle))$:

$$H|e_3\rangle = \beta_3|e_2\rangle + \alpha_3|e_3\rangle + \beta_4|e_4\rangle. \tag{A.14}$$

From the symmetry of the Hamiltonian, this expansion shows there is no term that is proportional to $|e_1\rangle$ and the coefficient of $|e_2\rangle$ is equal to β_3. α_3 can be determined from the expectation value of $\langle e_3|H|e_3\rangle$. As for Eq. (A.13), β_4 is determined by the normalization condition of $|e_4\rangle$.

Repeating this step generates a set of orthonormal basis states. In general, in the nth iteration, the Hamiltonian can be tridiagonalized as

$$H|e_n\rangle = \beta_n|e_{n-1}\rangle + \alpha_n|e_n\rangle + \beta_{n+1}|e_{n+1}\rangle, \tag{A.15}$$

where

$$\alpha_n = \langle e_n|H|e_n\rangle. \tag{A.16}$$

Substituting it into Eq. (A.15), one obtains a vector that is orthogonal to $(|e_1\rangle, \cdots, |e_n\rangle)$:

$$|\tilde{e}_{n+1}\rangle = \beta_{n+1}|e_{n+1}\rangle = H|e_n\rangle - \beta_n|e_{n-1}\rangle - \alpha_n|e_n\rangle. \tag{A.17}$$

β_{n+1} is then obtained by normalizing this vector:

$$\beta_{n+1} = \langle \tilde{e}_{n+1}|\tilde{e}_{n+1}\rangle^{1/2}. \tag{A.18}$$

In principle, this iteration can be repeated until all orthonormal basis vectors are found. The ground state can then be determined by diagonalizing the Hamiltonian in this basis representation. However, in practical calculations, this iteration has to be terminated to the limit of computer memory space. After the iteration, we get α_j and β_j, which constructs a tridiagonal matrix

$$H = \begin{pmatrix} \alpha_1 & \beta_2 & 0 & \cdots & 0 & 0 \\ \beta_2 & \alpha_2 & \beta_3 & \cdots & 0 & 0 \\ 0 & \beta_3 & \alpha_3 & \cdots & 0 & 0 \\ \vdots & \vdots & \vdots & \vdots & \vdots & \vdots \\ 0 & 0 & 0 & \cdots & \alpha_{M-1} & \beta_M \\ 0 & 0 & 0 & \cdots & \beta_M & \alpha_M \end{pmatrix}, \tag{A.19}$$

which is a real and symmetric matrix with positive nondiagonal matrix elements.

After the matrix H is calculated, one can solve its eigenvalues and their corresponding eigenvectors. This process is pretty simple because H is a tridiagonal matrix. The Lanczos vectors, $(|e_1\rangle, \cdots, |e_M\rangle)$, generated on the fly, construct the transformation matrix, which is useful for calculating the eigenvector. In practice, it could be saved after generations. For large M, storing all these vectors is difficult if the Hamiltonian dimension is high.

Usually, the Lanczos algorithm restarts after a certain number of iterations. One of the most significant is the implicitly restarted Lanczos method. The restarting vector is the approximate ground-state wave function obtained in the previous iteration. From the variational principle, it can be shown that the lowest eigenvalue obtained in each iteration is always lower than that obtained in its previous iteration. Thus, the lowest eigenvalue always converges towards the true ground state energy. For example, for $M = 2$, the Hamiltonian is a 2×2 matrix in the subspace spanned by $|e_1\rangle$ and $|e_2\rangle$:

$$H = \begin{pmatrix} \alpha_1 & \beta_2 \\ \beta_2 & \alpha_2 \end{pmatrix},$$
(A.20)

where α_1 is the value of ground-state energy obtained in the previous iteration. By diagonalizing this matrix, the ground state energy is found to be

$$E_0 = \frac{1}{2}\left(\alpha_1 + \alpha_2 - \sqrt{(\alpha_2 - \alpha_1)^2 + 4\beta_2^2}\right).$$
(A.21)

It is simple to show that E_0 is always less than or equal to α_1.

For the Lanczos algorithm, it can be shown that with exact arithmetic, the vectors $(|e_1\rangle, \cdots, |e_M\rangle)$ constitute an orthonormal basis set. However, in practical calculations where small computer round-off errors are inevitable, the orthogonality is quickly lost, and, in some cases, the new vector could even be linearly dependent on the basis set already constructed. As a result, some of the eigenvalues of the resultant tridiagonal matrix may not be good approximations of the original Hamiltonian. This nonorthogonality often happens when the ground state is degenerate or nearly degenerate. In this case, the ground state energy may oscillate with the iteration and not converge. As a result, the Lanczos algorithm is often not stable.

The Lanczos algorithm is attractive because the multiplication between H and basis vectors is the only large-scale linear operation. If H is a sparse matrix whose average number of nonzero matrix elements in each row is independent of the matrix dimension D, then the cost of Lanczos calculation scales just linearly with D.

The Lanczos algorithm can be extended to evaluate more than one eigenvector and eigenvalue. A commonly adopted approach is to start from a block of basis vectors instead of just one basis vector. Following similar steps, one can generate a block tridiagonal matrix as defined by Eq. (A.19), but α_i and β_i are now block matrices. By diagonalizing this block tridiagonal Hamiltonian, one can find a few lowest-energy states. The number of block vectors should be larger than or equal to the number of states to find. These are called "block" Lanczos algorithms.

A.3 Conjugate Gradient Method

Instead of diagonalizing the Hamiltonian using the Lanczos method, one can also find the ground-state energy and its corresponding eigenfunction by minimizing the expectation value of the Hamiltonian:

$$E_x = \frac{\langle x|H|x\rangle}{\langle x|x\rangle}. \tag{A.22}$$

To solve this minimization problem, one can use the steepest descent method. In this method, the minimum of E_x is searched along a path on which each local value of E_x decreases most quickly, namely along the direction opposite to the gradient of E_x:

$$|r_x\rangle = -\nabla_x E_x = -\frac{2}{\langle x|x\rangle}\left(H|x\rangle - E_x|x\rangle\right). \tag{A.23}$$

Here, $|r_x\rangle$ determines the descent direction. The descent stops when the gradient direction curves are about to ascend. One then computes another gradient vector and performs another descent in that direction. In the Lanczos method introduced in §A.2, $|e_2\rangle$ is a unit vector along the gradient of E_x at $|x\rangle = |e_1\rangle$. Thus, when $m = 2$, the implicit restarted Lanczos method is a steepest descent method.

The steepest descent method is stable and straightforward to implement. If minimum points exist, the method is guaranteed to locate them after sufficiently many iterations. However, despite all these positive characteristics, the method has a crucial drawback. If the descent directions deviate strongly from the minimum, it generally converges slowly.

The conjugate gradient is more efficient than the steepest descent method [450]. Like the steepest descent, this method proceeds by generating a set of vectors to search for the minimum of E_x. These vectors are generated along a conjugate path, $(p_1, p_2, \ldots, p_k, \ldots)$, rather than along the steepest descent path. In particular, they satisfy the following conjugate conditions of H:

$$\langle p_i| H|p_j\rangle = 0, \qquad (i \neq j). \tag{A.24}$$

Along this path, the new searching direction is always orthogonal to all previous searching directions. This improves efficiency since there is no overlap between any two searching directions.

The conjugate gradient algorithm selects the successive direction vectors as a conjugate version of the successive gradients. Thus, the directions are not specified beforehand but determined sequentially at each iteration step by evaluating the negative gradient vector and obtaining a new conjugate direction vector in combination with the previous direction vectors. This set of conjugate vectors can be found as follows.

Starting from an arbitrary initial state $|x_0\rangle$, which has a finite overlap with the true ground state but is not necessary to be normalized, first calculate the steepest descent vector

$$|p_1\rangle = |r_{x_0}\rangle. \tag{A.25}$$

Along this direction, one can find a new vector,

$$|x_1\rangle = |x_0\rangle + \alpha_1|p_1\rangle, \tag{A.26}$$

at which E_x is minimized. α_1 is a coefficient determined by minimizing E_x.

From $|p_1\rangle$ and $|x_1\rangle$, one can iteratively find new conjugate vectors. At each itera-tion, the conjugate vector should be chosen closely parallel to the steepest descent direction, but at the same time conjugate orthogonal to all previous descent vectors.

Assuming that $(|p_1\rangle, \ldots, |p_k\rangle)$ are the conjugate vectors determined from the first k iterations and the corresponding minima of E_x are located at $(|x_1\rangle, \ldots, |x_k\rangle)$, now let us consider how to find the next conjugate vector $|p_{k+1}\rangle$. Mathematically, it can be shown that this vector can be determined simply by searching for a vector on the plane spanned by $|r_{x_k}\rangle$ and $|p_k\rangle$, which is conjugate orthogonal to $|p_k\rangle$. The vector thus obtained is conjugate orthogonal to all other conjugate vectors obtained in the previous iterations. Thus, $|p_{k+1}\rangle$ is given by

$$|p_{k+1}\rangle = |r_{x_k}\rangle + \beta_k|p_k\rangle, \tag{A.27}$$

where β_k can be determined from the conjugate condition between $|p_k\rangle$ and $|p_{k+1}\rangle$ as

$$\beta_k = -\frac{\langle p_k|H|r_{x_k}\rangle}{\langle p_k|H|p_k\rangle}. \tag{A.28}$$

From this formula, it can be further shown that

$$\beta_k = \frac{\langle r_{x_k}|r_{x_k}\rangle}{\langle r_{x_{k-1}}|r_{x_{k-1}}\rangle}. \tag{A.29}$$

From $|x_k\rangle$, one can also find the minimum of E_x along the direction of $|p_k\rangle$:

$$|x_{k+1}\rangle = |x_k\rangle + \alpha_k|p_k\rangle, \tag{A.30}$$

where α_k (including α_1 in Eq. (A.25)) can be found by solving the minimal equation

$$\partial E_{x_{k+1}}/\partial\alpha_k = 0. \tag{A.31}$$

It is straightforward to show that

$$\alpha_k = \frac{\sqrt{B_k^2 - 4A_kC_k} - B_k}{2A_k}, \tag{A.32}$$

where

$$A_k = \langle p_k|H|p_k\rangle\langle x_k|p_k\rangle - \langle x_k|H|p_k\rangle\langle p_k|p_k\rangle, \tag{A.33}$$

$$B_k = \langle p_k|H|p_k\rangle\langle x_k|x_k\rangle - \langle x_k|H|x_k\rangle\langle p_k|p_k\rangle, \tag{A.34}$$

$$C_k = \langle x_k|H|p_k\rangle\langle x_k|x_k\rangle - \langle x_k|H|x_k\rangle\langle x_k|p_k\rangle. \tag{A.35}$$

From this iteration, E_x and $|x_k\rangle$ will gradually converge to the true ground-state energy and the corresponding eigenvector. This iteration is terminated if the energy deviation is smaller than an error bar ε:

$$\sqrt{\frac{\langle x_k|H^2|x_k\rangle\langle x_k|x_k\rangle - \langle x_k|H|x_k\rangle^2}{\langle x_k|H|x_k\rangle^2}} < \varepsilon. \tag{A.36}$$

From Eq. (A.23), one can simplify this convergence criterion as

$$\sqrt{\frac{\langle r_k|r_k\rangle\langle x_k|x_k\rangle}{E_{x_k}^2}} < 2\varepsilon. \tag{A.37}$$

Implementing the conjugate gradient method requires only a single matrix multiplication per iteration but needs memory for storing three to four vectors, including $(|x_k\rangle, |r_{x_k}\rangle, |p_k\rangle)$ and an auxiliary vector, simultaneously.

The conjugate gradient method has three primary advantages: 1. Unless the solution is attained, the gradient is always nonzero and linearly independent of all previous direction vectors. 2. The formula for determining the new conjugate vector is simple, allowing the method to be not much more complicated than the steepest descent method. 3. Because the directions are determined based on the gradients, the process makes good uniform progress towards the solution at every iteration.

A.4 Arnoldi Method

The Lanczos method introduced in Sec. A.2 and the conjugate gradient method introduced in Sec. A.3 can be used for diagonalizing symmetric (or Hermitian) matrices. However, we often need to diagonalize a sparse nonsymmetric matrix in the study of quantum many-body problems. For example, the quantum transfer matrix in TMRG is often a sparse nonsymmetric matrix. To diagonalize this kind of matrix, the Arnoldi algorithm, invented by W. E. Arnoldi in 1951 [451], is generally used. This algorithm is an iterative eigensolver of a general matrix by constructing an orthonormal basis of the Krylov subspace. It reduces to the Lanczos algorithm when the matrix is symmetric.

Starting from an initial normalized basis vector $|\psi_1\rangle$, one can generate n Krylov vectors by applying A to this initial state $(n-1)$ times.

$$|\psi_1\rangle \quad A|\psi_1\rangle \quad A^2|\psi_1\rangle \quad A^3|\psi_1\rangle \quad \dots \quad A^{n-1}|\psi_1\rangle . \tag{A.38}$$

This sequence of vectors converges to the eigenvector corresponding to the eigenvalue with the largest absolute value in the large n limit. These vectors are not orthogonal to each other. They are also ill-conditioned, because $A^{n-1}|\psi_1\rangle$ is dominated by the largest eigenvector of A when n becomes large.

The Arnoldi method, instead, generates a sequence of orthonormal basis vectors. It starts by applying A to vector $|\psi_1\rangle$. This generates a new vector $|\psi_2'\rangle$, which is orthogonal to $|\psi_1\rangle$:

$$|\psi_2'\rangle = A|\psi_1\rangle - |\psi_1\rangle\langle\psi_1|A|\psi_1\rangle. \tag{A.39}$$

After normalization, it becomes the second orthonormal basis vector:

$$|\psi_2\rangle = \frac{1}{\sqrt{\langle\psi_2'|\psi_2'\rangle}}|\psi_2'\rangle. \tag{A.40}$$

We then apply A to $|\psi_2\rangle$ to generate another vector $|\psi_3'\rangle$, which is orthogonal to both $|\psi_1\rangle$ and $|\psi_2\rangle$:

$$|\psi_3'\rangle = A|\psi_2\rangle - |\psi_1\rangle\langle\psi_1|A|\psi_2\rangle - |\psi_2\rangle\langle\psi_2|A|\psi_2\rangle. \tag{A.41}$$

Again this vector is normalized to become the third orthonormal basis vector:

$$|\psi_3\rangle = \frac{1}{\sqrt{\langle\psi_3'|\psi_3'\rangle}}|\psi_3'\rangle. \tag{A.42}$$

Repeating the above process recursively, we obtain a set of orthonormal basis vectors. Each time, A is applied to the last basis vector, say $|\psi_k\rangle$, to generate a new vector that is orthogonal to the basis vectors previously generated:

$$|\psi_{k+1}'\rangle = A|\psi_k\rangle - \sum_{i=1}^{k}|\psi_i\rangle\langle\psi_i|A|\psi_k\rangle. \tag{A.43}$$

After normalization, it becomes the $(k+1)$th orthonormal basis vector:

$$|\psi_{k+1}\rangle = \frac{1}{\sqrt{\langle\psi_{k+1}'|\psi_{k+1}'\rangle}}|\psi_{k+1}'\rangle. \tag{A.44}$$

By carrying out the Arnoldi iteration $(n-1)$ times, we obtain n orthonormal basis states. Within the subspace spanned by these basis states, A is represented as an upper Hessenberg matrix:

$$A_n = \begin{pmatrix} a_{1,1} & a_{1,2} & a_{1,3} & \cdots & a_{1,n} \\ a_{2,1} & a_{2,2} & a_{2,3} & \cdots & a_{2,n} \\ 0 & a_{3,2} & a_{3,3} & \cdots & a_{3,n} \\ \vdots & \ddots & \ddots & \ddots & \vdots \\ 0 & \cdots & 0 & a_{n,n-1} & a_{n,n} \end{pmatrix}, \tag{A.45}$$

where

$$a_{i,j} = \langle\psi_i|A|\psi_j\rangle. \tag{A.46}$$

By diagonalizing this upper Hessenberg matrix, we obtain an approximate solution for the largest eigenvalue and the corresponding eigenvector. This eigensolution should be sufficiently accurate by taking a large enough n. In practical calculations, however, due to storage considerations, the Arnoldi iteration is typically truncated at a small n and restarted by taking the largest eigenvector as the starting vector for the next iteration. Repeating the Arnoldi iterations recursively, the eigensolution would converge. Lehoucq and Sorensen proposed this implicitly restarted Arnoldi method [452].

In the diagonalization of a nonsymmetric method using the Arnoldi method, one needs to independently determine the left and right eigenvectors because they are not conjugate pairs, and the Arnoldi algorithm cannot find them in one run.

A.5 Quantum Monte Carlo Simulation

In the calculation of thermodynamic partition function or other physical quantities, we often need to evaluate the following type of high-dimensional integrations:

$$\langle A \rangle_p = \frac{\int \mathrm{d}x A(x) p(x)}{\int \mathrm{d}x p(x)}, \tag{A.47}$$

where x represents a point in the phase space, $p(x)$ is a probability distribution function, and $\langle A \rangle_p$ is the expectation value of physical quantity A under this distribution. In a classical system, each particle has six degrees of freedom, three for coordinates and three for momenta. The total dimension of phase space for N_p particles is $D = 6N_p$. However, in a quantum system, the phase space dimension grows exponentially with N_p, and is much larger than in the classical limit.

The high dimensional integration can be efficiently done with the Monte Carlo simulation, in which a Markov process is constructed to generate a random walk in the phase space, $\{x_0, x_1, \ldots\}$, according to the distribution function $p(x)$. The average of A is given by the sum of $A(x)$ at these sampling points. However, in the real calculation, the ratio of $p(x)$ in the whole phase space is unknown, and direct sampling is infeasible. The Monte Carlo is instead to generate a set of samples through a Markov random walk. To do this, a transition rule $W(x_i \rightarrow x_{i+1})$ from x_i to x_{i+1} is exploited in such a way that the probability distribution converges to an equilibrium state, satisfying

$$\sum_x p(x) W(x \rightarrow x') = p(x'). \tag{A.48}$$

This random walk must satisfy the following two conditions:

(i) Ergodicity: Markov random walks can go through all points in the phase space for a finite time.

(ii) Detailed balance: this is the requirement of probability conservation, which states that the probability of going from x to y is equal to the probability of going back from y to x:

$$p(x)W(x \to y) = p(y)W(y \to x). \tag{A.49}$$

The Markov process satisfying the above conditions is not unique. Different algorithms of Monte Carlo methods take different transition rules. A simple but commonly used Monte Carlo method is the Metropolis algorithm [453]. In this algorithm, the transition probability $W(x_i \to x_{x+1})$ from x_i to a neighboring point

$$x_{i+1} = x_i + \Delta x_i, \tag{A.50}$$

is determined by the following steps:

(i) Calculate the acceptance ratio:

$$P = \frac{p(x_{i+1})}{p(x_i)}. \tag{A.51}$$

(iii) If $P > 1$, then state x_{i+1} has a higher probability than x_i and is automatically accepted.

(iii) If $P < 1$, then x_{i+1} is accepted with probability P. This can be achieved by generating a random number r within the interval $[0, 1]$. If $r < P$, the random walk is accepted; otherwise it is rejected and the state stays at x_i; that is, $x_{i+1} = x_i$.

The acceptance ratio depends strongly on the random walk step size Δx_i. In order to improve the efficiency of calculation, the average step size should be chosen so that the average acceptance ratio is roughly equal to $1/2$. The expectation value of physical quantity A defined by Eq. (A.47) is then determined by the sum of A on the sampling points:

$$\langle A \rangle_p \approx \frac{1}{N} \sum_{i=1}^{N} A(x_i). \tag{A.52}$$

This algorithm proceeds by randomly attempting to move in the sample space. The acceptance ratio indicates the probability that the new proposed sample is accepted with respect to the current sample, according to the distribution $p(x)$. However, this algorithm has some disadvantages since the samples are correlated: even though they do correctly follow $p(x)$ over the long term, a set of nearby samples will be correlated with each other. This implies that if we want a set of independent samples, we have to throw away the majority of samples and only take every nth sample for some value of n (typically determined by examining the autocorrelation between adjacent samples). Autocorrelation can be reduced by increasing the jumping width (the average size of a jump, which is related to the variance of the jumping distribution), but this will also increase the likelihood of rejection of the proposed jump. Too large or too small a jumping size will lead to a slow-mixing Markov chain; that is, a highly correlated set of samples so that a vast number of samples will be needed to get a reasonable estimate of any desired property of the distribution.

In a classical statistical system, $p(x)$ in Eq. (A.47) is the Boltzmann distribution function, which is proportional to the thermal density matrix

$$p \propto e^{-\beta H}. \tag{A.53}$$

However, in a quantum system, H is an operator (or matrix), and the integration in Eq. (A.47) can be reexpressed as

$$\langle A \rangle_p = \frac{\mathrm{Tr} A e^{-\beta H}}{\mathrm{Tr} e^{-\beta H}}, \tag{A.54}$$

where the denominator is the partition function

$$Z = \mathrm{Tr} e^{-\beta H}. \tag{A.55}$$

Unlike in a classical system, H contains not only diagonal matrix elements but also off-diagonal ones. It is not easy to directly evaluate the value of $e^{-\beta H}$. To overcome this difficulty, one needs to first map a d-dimensional quantum model onto a $d + 1$-dimensional classical one by taking β as a coordinate in an imaginary time axis and expressing $\exp(-\beta H)$ as a path integral along the imaginary time direction. This is achieved by dividing β into M pieces and expressing $\exp(-\beta H)$ into a product of M short-time evolution operator $\exp(-\tau H)$:

$$e^{-\beta H} = \prod_{i=1}^{M} e^{-\tau H}, \tag{A.56}$$

where $\tau = \beta/M$. For sufficiently small τ, $\exp(-\tau H)$ can be evaluated with reliable approximation. A simple approximation is to take τ as a small parameter and expand $\exp(-\tau H)$ with respect to τ up to the leading order term:

$$e^{-\tau H} = 1 - \tau H + O(\tau^2). \tag{A.57}$$

With this approximation, the matrix elements of this exponential operator under a convenient basis representation $\{|\alpha\rangle\}$,

$$\langle \alpha | e^{-\tau H} | \alpha' \rangle, \tag{A.58}$$

can be readily determined once the matrix elements of H are known. However, this approximation is poor since the error is of order τ^2. In practical calculations, a higher-order Trotter–Suzuki decomposition formula is generally used to reduce the error to the order of τ^3 or even lower.

Under this approximation, the partition function can be written as

$$Z = \sum_{\alpha_1 \cdots \alpha_M} p(\alpha_1, \cdots \alpha_M), \tag{A.59}$$

where

$$p(\alpha_1, \cdots \alpha_M) = \langle \alpha_1 | e^{-\tau H} | \alpha_M \rangle \langle \alpha_M | e^{-\tau H} | \alpha_{M-1} \rangle \cdots \langle \alpha_2 | e^{-\tau H} | \alpha_1 \rangle. \tag{A.60}$$

Similarly, one can find the expression for $\text{Tr} A \exp(-\beta H)$. The expectation value of A is then given by

$$\langle A \rangle_p = \frac{\displaystyle\sum_{\alpha_1 \cdots \alpha_M} A_{\alpha_1, \alpha_M} p(\alpha_1, \cdots \alpha_M)}{\displaystyle\sum_{\alpha_1 \cdots \alpha_M} p(\alpha_1, \cdots \alpha_M)}, \tag{A.61}$$

where

$$A_{\alpha_1, \alpha_M} = \frac{\langle \alpha_1 | A e^{-\tau H} | \alpha_M \rangle}{\langle \alpha_1 | e^{-\tau H} | \alpha_M \rangle}. \tag{A.62}$$

Comparing Eq. (A.61) with Eq. (A.47), we find that $(\alpha_1, \ldots, \alpha_M)$ corresponds to a configuration in the phase space and $p(\alpha_1, \ldots, \alpha_M)$ is the distribution function. If $p(\alpha_1, \ldots, \alpha_M)$ is larger than or equal to 0, then the expectation value of A can be evaluated using the method previously introduced for the classical system. However, for a system of interacting fermions or frustrated spin models, $p(\alpha_1, \ldots, \alpha_M)$ can be less than 0. In this case, the Monte Carlo simulation can still be done. But the phase and the amplitude of $p(\alpha_1, \ldots, \alpha_M)$ must be separated. Only the amplitude $|p(\alpha_1, \ldots, \alpha_M)|$ is taken as the probability distribution function and the phase factor is taken as part of the measured physical quantity. The numerator and the denominator in Eq. (A.61) are then determined separately using the Monte Carlo. Finally we can find the expectation value of A using the formula

$$\langle A \rangle_p = \frac{\langle A \, \text{sign}(p) \rangle_{|p|}}{\langle \text{sign}(p) \rangle_{|p|}}, \tag{A.63}$$

where

$$\langle A \, \text{sign}(p) \rangle_{|p|} = \frac{\displaystyle\sum_{\alpha_1 \cdots \alpha_M} A_{\alpha_1, \alpha_M} \text{sign}[(\alpha_1, \ldots, \alpha_M)] |p(\alpha_1, \ldots, \alpha_M)|}{\displaystyle\sum_{\alpha_1 \cdots \alpha_M} |p(\alpha_1, \ldots, \alpha_M)|}, \tag{A.64}$$

$$\langle \text{sign}(p) \rangle_{|p|} = \frac{\displaystyle\sum_{\alpha_1 \cdots \alpha_M} \text{sign}[\, p(\alpha_1, \ldots, \alpha_M)]\, |\, p(\alpha_1, \ldots, \alpha_M)|}{\displaystyle\sum_{\alpha_1 \cdots \alpha_M} |\, p(\alpha_1, \ldots, \alpha_M)|}. \tag{A.65}$$

Equation (A.63) is always valid. But in an interacting fermion or frustrated spin system, it is found that the average sign of $p(\alpha_1, \ldots, \alpha_M)$ drops exponentially with the increase of the particle number N_p or the decrease of temperature

$$\langle \text{sign}(p) \rangle_{|p|} \propto e^{-c\beta N_p}, \tag{A.66}$$

where c is a constant independent of N_p and the temperature. In this case, the average value of $A\text{sign}(p)$ also approaches zero exponentially, and the expectation value of A is determined by the ratio of two nearly zero numbers, resulting in a considerable error in the final result, which is the so-called minus sign problem.

The minus sign problem originates from the off-diagonal matrix elements of H, $\langle \alpha'|H|\alpha \rangle$ ($\alpha' \neq \alpha$). The diagonal matrix element of H does not lead to any minus sign problem because

$$\langle \alpha|e^{-\tau H}|\alpha \rangle \approx \langle \alpha|1 - \tau H|\alpha \rangle \tag{A.67}$$

is always positive for sufficiently small τ. If the off-diagonal matrix element $\langle \alpha'|H|\alpha \rangle \leq 0$,

$$\langle \alpha'|e^{-\tau H}|\alpha \rangle \approx -\tau \langle \alpha'|H|\alpha \rangle \geq 0, \qquad (\alpha' \neq \alpha), \tag{A.68}$$

there is also no minus sign problem. However, if $\langle \alpha'|H|\alpha \rangle > 0$, then

$$\langle \alpha'|e^{-\tau H}|\alpha \rangle \approx -\tau \langle \alpha'|H|\alpha \rangle < 0, \qquad (\alpha' \neq \alpha), \tag{A.69}$$

and $p(\alpha_1, \ldots, \alpha_M)$ may become negative.

The minus sign problem depends on the choice of the basis states $\{|\alpha\rangle\}$. In some specific systems, for example, the negative U Hubbard model, the positive U Hubbard model on a bipartite lattice at half filling, or the antiferromagnetic Heisenberg model without frustration, the minus sign problem can be gauged out. The variational Monte Carlo is commonly used for studying the ground-state properties. This method also does not have the minus sign problem since the probability distribution used in the Monte Carlo sampling is the absolute value of the trial wave function, which is always positive.

References

[1] W. Kohn. Nobel Lecture: Electronic structure of matter-wave functions and density functionals. *Rev. Mod. Phys.*, 71:1253–1266, 1999.

[2] P. W. Anderson. More is different: Broken symmetry and nature of hierarchical structure of science. *Science*, 177:393–396, 1972.

[3] E. C. G. Stueckelberg and A. Peterman. The normalization group in quantum theory. *Helv. Phys. Acta*, 24:317–319, 1951.

[4] E. C. G. Stueckelberg and A. Peterman. La renormalisation des constants dans la theorie de quanta. *Helv. Phys. Acta*, 26:499–520, 1953.

[5] M. Gell-Mann and F. E. Low. Quantum electrodynamics at small distances. *Phys. Rev.*, 95:1300–1312, 1954.

[6] C. G. Callan. Broken scale invariance in scalar field theory. *Phys. Rev. D*, 2:1541–1547, 1970.

[7] K. Symanzik. Small distance behaviour in field theory and power counting. *Commun. Math. Phys.*, 18:227–246, 1970.

[8] L. P. Kadanoff. Scaling laws for Ising models near T_c. *Physics Physique Fizika*, 2:263–272, 1966.

[9] K. G. Wilson. The renormalization group: Critical phenomena and the Kondo problem. *Rev. Mod. Phys.*, 47:773–840, 1975.

[10] J. W. Bray and S. T. Chui. Computer renormalization-group calculations of $2k_F$ and $4k_F$ correlation functions of the one-dimensional Hubbard model. *Phys. Rev. B*, 19:4876–4882, 1979.

[11] C. Y. Pan and X. Y. Chen. Renormalization-group study of high-spin Heisenberg antiferromagnets. *Phys. Rev. B*, 36:8600–8606, 1987.

[12] M. D. Kovarik. Numerical solution of large S=1/2 and S=1 Heisenberg anti-ferromagnetic spin chains using a truncated basis expansion. *Phys. Rev. B*, 41:6889–6898, 1990.

[13] T. Xiang and G. A. Gehring. Real space renormalization group study of Heisenberg spin chain. *J. Mag. Mag. Mat.*, 104–107:861–862, 1992.

[14] J. Hubbard. Electron correlations in narrow energy bands. *Proc. R. Soc. Lond. A. Math. Phys. Sci.*, 276:238–257, 1963.

[15] S. R. White. Numerical renormalization group for finite Hubbard lattices. *Phys. Rev. B*, 45:5752–5755, 1992.

[16] S. R. White and R. M. Noack. Real-space quantum renormalization groups. *Phys. Rev. Lett.*, 68:3487–3490, 1992.

[17] T. Xiang and G. A. Gehring. Numerical solution of S=1 antiferromagnetic spin chains using a truncated basis expansion. *Phys. Rev. B*, 48:303–310, 1993.

[18] S. R. White. Density matrix formulation for quantum renormalization groups. *Phys. Rev. Lett.*, 69:2863–2866, 1992.

[19] S. R. White and D. A. Huse. Numerical renormalization-group study of low-lying eigenstates of the antiferromagnetic S=1 Heisenberg chain. *Phys. Rev. B*, 48:3844–3852, 1993.

[20] S. Östlund and S. Rommer. Thermodynamic limit of density matrix renormalization. *Phys. Rev. Lett.*, 75:3537–3540, 1995.

[21] J. Dukelsky, M. A. Martin-Delgado, T. Nishino, and G. Sierra. Equivalence of the variational matrix product method and the density matrix renormalization group applied to spin chains. *Europhys. Lett.*, 43:457–462, 1998.

[22] M. B. Hastings. Solving gapped Hamiltonians locally. *Phys. Rev. B*, 73:085115, 2006.

[23] F. Verstraete and J. I. Cirac. Matrix product states represent ground states faithfully. *Phys. Rev. B*, 73:094423, 2006.

[24] R. J. Baxter. Dimers on a rectangular lattice. *J. Math. Phys.*, 9:650–654, 1968.

[25] L. Accardi. Topics in quantum probability. *Phys. Rep.*, 77:169–192, 1981.

[26] I. Affleck, T. Kennedy, E. H. Lieb, and H. Tasaki. Rigorous results on valence-bond ground states in antiferromagnets. *Phys. Rev. Lett.*, 59:799–802, 1987.

[27] I. Affleck, T. Kennedy, E. H. Lieb, and H. Tasaki. Valence bond ground states in isotropic quantum antiferromagnets. *Commun. Math. Phys.*, 115:477–528, 1988.

[28] F. D. M. Haldane. Continuum dynamics of the 1-D Heisenberg anti-ferromagnet: Identification with the O(3) non-linear sigma-model. *Phys. Lett. A*, 93:464–468, 1993

[29] F. D. M. Haldane. Non-linear field-theory of large-spin Heisenberg anti-ferromagnets - semi-classically quantized solitons of the one-dimensional easy-axis Neel state. *Phys. Rev. Lett.*, 50:1153–1156, 1983.

[30] F. D. M. Haldane. Theta-physics and quantum spin chains. *J. Appl. Phys.*, 57:3359, 1985.

[31] A. Klümper, A. Schadschneider, and J. Zittartz. Equivalence and solution of anisotropic spin-1 models and generalized t-J fermion models in one dimension. *J. Phys. A*, 24:L955–L959, 1991.

[32] A. Klümper, A. Schadschneider, and J. Zittartz. Groundstate properties of a generalized VBS-model. *Z. Phys. B*, 87:281–287, 1992.

[33] M. Fannes, B. Nachtergaele, and R. F. Werner. Finitely correlated states on quantum spin chains. *Commun. Math. Phys.*, 144:443–490, 1992.

[34] T. Xiang, J. Lou, and Z. Su. Two-dimensional algorithm of the density-matrix renormalization group. *Phys. Rev. B*, 64:104414, 2001.

[35] G. Vidal. Efficient classical simulation of slightly entangled quantum computations. *Phys. Rev. Lett.*, 91:147902, 2003.

[36] M. B. Hastings. An area law for one-dimensional quantum systems. *J. Stat. Mech.: Theo. Exp.*, 2007:P08024, 2007.

[37] J. Eisert, M. Cramer, and M. B. Plenio. Colloquium: Area laws for the entanglement entropy. *Rev. Mod. Phys.*, 82:277–306, 2010.

[38] G. Vidal. Efficient simulation of one-dimensional quantum many-body systems. *Phys. Rev. Lett.*, 93:040502, 2004.

[39] F. Verstraete, D. Porras, and J. I. Cirac. Density matrix renormalization group and periodic boundary conditions: A quantum information perspective. *Phys. Rev. Lett.*, 93:227205, 2004.

[40] A. Bijl, J. De Boer, and A. Michels. Properties of liquid helium II. *Physica*, 8:655–675, 1941.

[41] R. P. Feynman. Atomic theory of the λ transition in helium. *Phys. Rev.*, 91:1291–1301, 1953.

[42] R. P. Feynman. Atomic theory of the two-fluid model of liquid helium. *Phys. Rev.*, 94:262–277, 1954.

[43] S. G. Chung and L. Wang. Entanglement perturbation theory for the elementary excitation in one dimension. *Phys. Lett. A*, 373:2277–2280, 2009.

[44] B. Pirvu, J. Haegeman, and F. Verstraete. Matrix product state based algorithm for determining dispersion relations of quantum spin chains with periodic boundary conditions. *Phys. Rev. B*, 85:035130, 2012.

[45] J. Haegeman, B. Pirvu, D. J. Weir et al. Variational matrix product ansatz for dispersion relations. *Phys. Rev. B*, 85:100408, 2012.

[46] C. Holzhey, F. Larsen, and F. Wilczek. Geometric and renormalized entropy in conformal field theory. *Nucl. Phys. B*, 424:443–467, 1994.

[47] G. Vidal, J. I. Latorre, E. Rico, and A. Kitaev. Entanglement in quantum critical phenomena. *Phys. Rev. Lett.*, 90:227902, 2003.

[48] P. Calabrese and J. Cardy. Entanglement entropy and quantum field theory. *J. Stat. Mech.: Theo. Exp.*, 2004:P06002, 2004.

[49] J. I. Cirac and G. Sierra. Infinite matrix product states, conformal field theory, and the Haldane–Shastry model. *Phys. Rev. B*, 81:104431, 2010.

[50] H. H. Tu and G. Sierra. Infinite matrix product states, boundary conformal field theory, and the open Haldane–Shastry model. *Phys. Rev. B*, 92:041119, 2015.

[51] G. Vidal. Entanglement renormalization. *Phys. Rev. Lett.*, 99:220405, 2007.

[52] F. Verstraete and J. I. Cirac. Continuous matrix product states for quantum fields. *Phys. Rev. Lett.*, 104:190405, 2010.

[53] J. Haegeman, T. J. Osborne, H. Verschelde, and F. Verstraete. Entanglement renormalization for quantum fields in real space. *Phys. Rev. Lett.*, 110:100402, 2013.

[54] D. Jennings, C. Brockt, J. Haegeman, T. J. Osborne, and F. Verstraete. Continuum tensor network field states, path integral representations and spatial symmetries. *New J. Phys.*, 17:063039, 2015.

[55] A. Tilloy and J. I. Cirac. Continuous tensor network states for quantum fields. *Phys. Rev. X*, 9:021040, 2019.

[56] T. Xiang. Density-matrix renormalization-group method in momentum space. *Phys. Rev. B*, 53:R10445–R10448, 1996.

[57] S. R. White and R. L Martin. Ab initio quantum chemistry using the density matrix renormalization group. *J. Chem. Phys.*, 110:4127–4130, 1999.

[58] J. Dukelsky and S. Pittel. New approach to large-scale nuclear structure calculations. *Phys. Rev. C*, 63:061303, 2001.

[59] N. Shibata and D. Yoshioka. Ground-state phase diagram of 2D electrons in a high Landau level: A density-matrix renormalization group study. *Phys. Rev. Lett.*, 86:5755–5758, 2001.

[60] A. E. Feiguin, E. Rezayi, C. Nayak, and S. Das Sarma. Density matrix renormalization group study of incompressible fractional quantum Hall states. *Phys. Rev. Lett.*, 100:166803, 2008.

[61] J. Zhao, D. N. Sheng, and F. D. M. Haldane. Fractional quantum Hall states at 1/3 and 5/2 filling: Density-matrix renormalization group calculations. *Phys. Rev. B*, 83:195135, 2011.

[62] Y. Anusooya, S. K. Pati, and S. Ramasesha. Symmetrized density matrix renormalization group studies of the properties of low-lying states of the poly-para-phenylene system. *J. Chem. Phys.*, 106:10230–10237, 1997.

[63] Z. Shuai, J. L. Brdas, A. Saxena, and A. R. Bishop. Linear and nonlinear optical response of polyenes: A density matrix renormalization group study. *J. Chem. Phys.*, 109:2549–2555, 1998.

[64] W. Barford, R. J. Bursill, and M. Y. Lavrentiev. Density matrix renormalization group calculations of the low-lying excitations and non-linear optical properties of poly(para-phenylene). *J. Phys.: Conden. Matt.*, 10:6429–6446, 1998.

[65] G. Fano, F. Ortolani, and L. Ziosi. The density matrix renormalization group method: Application to the PPP model of a cyclic polyene chain. *J. Chem. Phys.*, 108:9246–9252, 1998.

[66] G. K. L. Chan and M. Head-Gordon. Exact solution (within a triple-zeta, double polarization basis set) of the electronic Schrödinger equation for water. *J. Chem. Phys.*, 118:8551–8554, 2003.

[67] H. G. Luo, M. P. Qin, and T. Xiang. Optimizing Hartree–Fock orbitals by the density-matrix renormalization group. *Phys. Rev. B*, 81:235129, 2010.

[68] T. Nishino. Density-matrix renormalization-group method for 2D classical-models. *J. Phys. Soc. Jpn.*, 64:3598–3601, 1995.

[69] Baxter R. J. Corner transfer matrices. *Physica A*, 106:18–27, 1981.

[70] T. Nishino and K. Okunishi. Corner transfer matrix renormalization group method. *J. Phys. Soc. Jpn.*, 65:891–894, 1996.

[71] R. Orús and G. Vidal. Simulation of two-dimensional quantum systems on an infinite lattice revisited: Corner transfer matrix for tensor contraction. *Phys. Rev. B*, 80:094403, 2009.

[72] R. J. Bursill, T. Xiang, and G. A. Gehring. The density matrix renormalization group for a quantum spin chain at non-zero temperature. *J. Phys.: Conden. Matt.*, 8:L583–590, 1996.

[73] H. F. Trotter. On the product of semi-groups of operators. *Proc. Am . Math. Soc.*, 10:545–551, 1959.

[74] M. Suzuki. Generalized Trotter's formula and systematic approximants of exponential operators and inner derivations with applications to many-body problems. *Commun. Math. Phys.*, 51:183–190, 1976.

[75] X. Wang and T. Xiang. Transfer-matrix density-matrix renormalization-group theory for thermodynamics of one-dimensional quantum systems. *Phys. Rev. B*, 56:5061–5064, 1997.

[76] N. Shibata. Thermodynamics of the anisotropic Heisenberg chain calculated by the density matrix renormalization group method. *J. Phys. Soc. Jpn.*, 66:2221–2223, 1997.

[77] Y. K. Huang. Biorthonormal transfer-matrix renormalization-group method for non-Hermitian matrices. *Phys. Rev. E*, 83:036702, 2011.

[78] F. Verstraete, J. J. Garcia-Ripoll, and J. I. Cirac. Matrix product density operators: Simulation of finite-temperature and dissipative systems. *Phys. Rev. Lett.*, 93:207204, 2004.

[79] A. E. Feiguin and S. R. White. Finite-temperature density matrix renormalization using an enlarged Hilbert space. *Phys. Rev. B*, 72:220401, 2005.

[80] E. R. Gagliano and C. A. Balseiro. Dynamical properties of quantum many-body systems at zero temperature. *Phys. Rev. Lett.*, 59:2999–3002, 1987.

[81] K. A. Hallberg. Density-matrix algorithm for the calculation of dynamical properties of low-dimensional systems. *Phys. Rev. B*, 52:R9827–R9830, 1995.

[82] Z. G. Soos and S. Ramasesha. Valence bond approach to exact nonlinear optical-properties of conjugated systems. *J. Chem. Phys.*, 90:1067–1076, 1989.

[83]   S. K. Pati, S. Ramasesha, and D. Sen. Low-lying excited states and low-temperature properties of an alternating spin-1–spin-1/2 chain: A density-matrix renormalization-group study. *Phys. Rev. B*, 55:8894–8904, 1997.

[84]   T. D. Kühner and S. R. White. Dynamical correlation functions using the density matrix renormalization group. *Phys. Rev. B*, 60:335–343, 1999.

[85]   E. Jeckelmann. Dynamical density-matrix renormalization-group method. *Phys. Rev. B*, 66:045114, 2002.

[86]   P. E. Dargel, A. Honecker, R. Peters, R. M. Noack, and T. Pruschke. Adaptive Lanczos-vector method for dynamic properties within the density matrix renormalization group. *Phys. Rev. B*, 83:161104, 2011.

[87]   A. Nocera and G. Alvarez. Spectral functions with the density matrix renormalization group: Krylov-space approach for correction vectors. *Phys. Rev. E*, 94:053308, 2016.

[88]   A. Holzner, A. Weichselbaum, I. P. McCulloch, U. Schollwöck, and J. von Delft. Chebyshev matrix product state approach for spectral functions. *Phys. Rev. B*, 83:195115, 2011.

[89]   H. D. Xie, R. Z. Huang, X. J. Han et al. Reorthonormalization of Chebyshev matrix product states for dynamical correlation functions. *Phys. Rev. B*, 97:075111, 2018.

[90]   P. E. Dargel, A. Wöllert, A. Honecker et al. Lanczos algorithm with matrix product states for dynamical correlation functions. *Phys. Rev. B*, 85:205119, 2012.

[91]   M. A. Cazalilla and J. B. Marston. Time-dependent density-matrix renormalization group: A systematic method for the study of quantum many-body out-of-equilibrium systems. *Phys. Rev. Lett.*, 88:256403, 2002.

[92]   H. G. Luo, T. Xiang, and X. Q. Wang. Comment on "Time-dependent density-matrix renormalization group: A systematic method for the study of quantum many-body out-of-equilibrium systems." *Phys. Rev. Lett.*, 91:049701, 2003.

[93]   A. J. Daley, C. Kollath, U. Schollwöck, and G. Vidal. Time-dependent density-matrix renormalization-group using adaptive effective Hilbert spaces. *J. Stat. Mech.: Theo. Exp.*, 2004:P04005, 2004.

[94]   S. R. White and A. E. Feiguin. Real-time evolution using the density matrix renormalization group. *Phys. Rev. Lett.*, 93:076401, 2004.

[95]   J. Haegeman, J. I. Cirac, T. J. Osborne et al. Time-dependent variational principle for quantum lattices. *Phys. Rev. Lett.*, 107:070601, 2011.

[96]   P. Calabrese and J. Cardy. Evolution of entanglement entropy in one-dimensional systems. *J. Stat. Mech.: Theo. Exp.*, 2005:P04010, 2005.

[97]   M. C. Bañuls, M. B. Hastings, F. Verstraete, and J. I. Cirac. Matrix product states for dynamical simulation of infinite chains. *Phys. Rev. Lett.*, 102:240603, 2009.

[98]   Y. K. Huang, P. Chen, Y. J. Kao, and T. Xiang. Long-time dynamics of quantum chains: Transfer-matrix renormalization group and entanglement of the maximal eigenvector. *Phys. Rev. B*, 89:201102, 2014.

[99]   T. Barthel, U. Schollwöck, and S. R. White. Spectral functions in one-dimensional quantum systems at finite temperature using the density matrix renormalization group. *Phys. Rev. B*, 79:245101, 2009.

[100] C. Karrasch, J. H. Bardarson, and J. E. Moore. Finite-temperature dynamical density matrix renormalization group and the Drude weight of spin-1/2 chains. *Phys. Rev. Lett.*, 108:227206, 2012.

[101] S. Liang and H. Pang. Approximate diagonalization using the density matrix renormalization-group method: A two-dimensional-systems perspective. *Phys. Rev. B*, 49:9214–9217, 1994.

[102] M. C. Chung and I. Peschel. Density-matrix spectra for two-dimensional quantum systems. *Phys. Rev. B*, 62:4191–4193, 2000.

[103] S. R. White and D. J. Scalapino. Ground states of the doped four-leg t-J ladder. *Phys. Rev. B*, 55:R14701–R14704, 1997.

[104] J. P. F. LeBlanc, A. E. Antipov, F. Becca et al. Solutions of the two-dimensional Hubbard model: Benchmarks and results from a wide range of numerical algorithms. *Phys. Rev. X*, 5:041041, 2015.

[105] S. Yan, D. A. Huse, and S. R. White. Spin-liquid ground state of the S=1/2 Kagome Heisenberg antiferromagnet. *Science*, 332:1173–1176, 2011.

[106] S. Depenbrock, I. P. McCulloch, and U. Schollwöck. Nature of the spin-liquid ground state of the $S = 1/2$ Heisenberg model on the kagome lattice. *Phys. Rev. Lett.*, 109:067201, 2012.

[107] H. J. Liao, Z. Y. Xie, J. Chen et al. Gapless spin-liquid ground state in the $S = 1/2$ Kagome antiferromagnet. *Phys. Rev. Lett.*, 118:137202, 2017.

[108] H. Niggemann, A. Klumper, and J. Zittartz. Quantum phase transition in spin-3/2 systems on the hexagonal lattice: Optimum ground state approach. *Z. Phys. B*, 104:103–110, 1997.

[109] G. Sierra and M. A. Martin-Delgado. The density matrix renormalization group, quantum groups and conformal field theory. In A. Krasnitz, Y. A. Kubyshin, R. Potting, and P. Sa (eds.), *Proceedings of the Workshop of the Exact Renormalization Group*, pages 126–41. World Scientific, 1999.

[110] T. Nishino, Y. Hieida, K. Okunishi et al. Two-dimensional tensor product variational formulation. *Prog. Theo. Phys.*, 105:409–417, 2001.

[111] F. Verstraete and J. I. Cirac. Renormalization algorithms for quantum-many body systems in two and higher dimensions. *cond-mat/0407066*.

[112] M. M. Wolf, F. Verstraete, M. B. Hastings, and J. I. Cirac. Area laws in quantum systems: Mutual information and correlations. *Phys. Rev. Lett.*, 100:070502, 2008.

[113] D. P. Arovas. Simplex solid states of SU(*N*) quantum antiferromagnets. *Phys. Rev. B*, 77:104404, 2008.

[114] Z. Y. Xie, J. Chen, J. F. Yu et al. Tensor renormalization of quantum many-body systems using projected entangled simplex states. *Phys. Rev. X*, 4:011025, 2014.

[115] M. Suzuki. Relationship between d-dimensional quantal spin systems and (d+1)-dimensional Ising systems: Equivalence, critical exponents and systematic approximants of partition-function and spin correlations. *Prog. Theo. Phys.*, 56:1454–1469, 1976.

[116] H. C. Jiang, Z. Y. Weng, and T. Xiang. Accurate determination of tensor network state of quantum lattice models in two dimensions. *Phys. Rev. Lett.*, 101:090603, 2008.

[117] J. Jordan, R. Orús, G. Vidal, F. Verstraete, and J. I. Cirac. Classical simulation of infinite-size quantum lattice systems in two spatial dimensions. *Phys. Rev. Lett.*, 101:250602, 2008.

[118] W. Li, J. von Delft, and T. Xiang. Efficient simulation of infinite tree tensor network states on the Bethe lattice. *Phys. Rev. B*, 86:195137, 2012.

[119] L. Wang and F. Verstraete. Cluster update for tensor network states. arXiv:1110.4362.

[120] H. J. Liao, J. G. Liu, L. Wang, and T. Xiang. Differentiable programming tensor networks. *Phys. Rev. X*, 9:031041, 2019.

[121] L. Vanderstraeten, M. Mariën, F. Verstraete, and J. Haegeman. Excitations and the tangent space of projected entangled-pair states. *Phys. Rev. B*, 92:201111, 2015.

[122] B. Ponsioen, F. F. Assaad, and P. Corboz. Automatic differentiation applied to excitations with projected entangled pair states. *SciPost Phys.*, 12:006, 2022.

[123] R. Z. Chi, Y. Liu, Y. Wan, H. J. Liao, and T. Xiang. Spin excitation spectra of anisotropic spin-1/2 triangular lattice Heisenberg antiferromagnets. *Phys. Rev. Lett.*, 129:227201, 2022.

[124] Y. Y. Shi, L. M. Duan, and G. Vidal. Classical simulation of quantum many-body systems with a tree tensor network. *Phys. Rev. A*, 74:022320, 2006.

[125] N. Schuch, M. M. Wolf, F. Verstraete, and J. I. Cirac. Simulation of quantum many-body systems with strings of operators and Monte Carlo tensor contractions. *Phys. Rev. Lett.*, 100:040501, 2008.

[126] F. Mezzacapo, N. Schuch, M. Boninsegni, and J. I. Cirac. Ground-state properties of quantum many-body systems: Entangled-plaquette states and variational Monte Carlo. *New J. Phys.*, 11:083026, 2009.

[127] G. Evenbly and G. Vidal. Class of highly entangled many-body states that can be efficiently simulated. *Phys. Rev. Lett.*, 112:240502, 2014.

[128] M. Levin and C. P. Nave. Tensor renormalization group approach to two-dimensional classical lattice models. *Phys. Rev. Lett.*, 99:120601, 2007.

[129] Z. Y. Xie, H. C. Jiang, Q. N. Chen, Z. Y. Weng, and T. Xiang. Second renormalization of tensor-network states. *Phys. Rev. Lett.*, 103:160601, 2009.

[130] H. H. Zhao, Z. Y. Xie, Q. N. Chen et al. Renormalization of tensor-network states. *Phys. Rev. B*, 81:174411, 2010.

[131] Z. Y. Xie, J. Chen, M. P. Qin et al. Coarse-graining renormalization by higher-order singular value decomposition. *Phys. Rev. B*, 86:045139, 2012.

[132] Z. C. Gu and X. G. Wen. Tensor-entanglement-filtering renormalization approach and symmetry-protected topological order. *Phys. Rev. B*, 80:155131, 2009.

[133] G. Evenbly and G. Vidal. Tensor network renormalization. *Phys. Rev. Lett.*, 115:180405, 2015.

[134] S. Yang, Z. C. Gu, and X. G. Wen. Loop optimization for tensor network renormalization. *Phys. Rev. Lett.*, 118:110504, 2017.

[135] S. Wang, Z. Y. Xie, J. Chen, B. Normand, and T. Xiang. Phase transitions of ferromagnetic Potts models on the simple cubic lattice. *Chin. Phys. Lett.*, 31:070503, 2014.

[136] I. Peschel, M. Kaulke, X. Wang, and K. Hallberg (eds.). *Density-Matrix Renormalization: A New Numerical Method in Physics*. Lecture Notes in Physics 528, Springer, 1999.

[137] S. R. White. Strongly correlated electron systems and the density matrix renormalization group. *Phys. Rep.*, 301:187 – 204, 1998.

[138] U. Schollwöck. The density-matrix renormalization group. *Rev. Mod. Phys.*, 77:259–315, 2005.

[139] K. A. Hallberg. New trends in density matrix renormalization. *Adv. Phys.*, 55:477–526, 2006.

[140] F. Verstraete, V. Murg, and J. I. Cirac. Matrix product states, projected entangled pair states, and variational renormalization group methods for quantum spin systems. *Adv. Phys.*, 57:143–224, 2008.

[141] U. Schollwöck. The density-matrix renormalization group in the age of matrix product states. *Ann. Phys.*, 326:96 – 192, 2011.

[142] E. M. Stoudenmire and S. R. White. Studying two-dimensional systems with the density matrix renormalization group. *Ann. Rev. Conden. Matt. Phys.*, 3:111–128, 2012.

[143] R. Orús. A practical introduction to tensor networks: Matrix product states and projected entangled pair states. *Ann. Phys.*, 349:117–158, 2014.

[144] S. Paeckel, T. Köhler, A. Swoboda et al. Time-evolution methods for matrix-product states. *Ann. Phys.*, 411:167998, 2019.

[145] R. Orús. Tensor networks for complex quantum systems. *Nat. Rev. Phys.*, 1:538–550, 2019.

[146] S. J. Ran, E. Tirrito, C. Peng et al. *Tensor Network Contractions: Methods and Applications to Quantum Many-Body Systems*. Springer, 2020.

[147] A. Baiardi and M. Reiher. The density matrix renormalization group in chemistry and molecular physics: Recent developments and new challenges. *J. Chem. Phys.*, 152:040903, 2020.

[148] M. C. Bañuls and K. Cichy. Review on novel methods for lattice gauge theories. *Rep. Prog. Phys.*, 83:024401, 2020.

[149] J. I. Cirac, D. Pérez-García, N. Schuch, and F. Verstraete. Matrix product states and projected entangled pair states: Concepts, symmetries, theorems. *Rev. Mod. Phys.*, 93:045003, 2021.

[150] H. Weimer, A. Kshetrimayum, and R. Orús. Simulation methods for open quantum many-body systems. *Rev. Mod. Phys.*, 93:015008, 2021.

[151] K. Okunishi, T. Nishino, and H. Ueda. Developments in the tensor network - from statistical mechanics to quantum entanglement. *J. Phys. Soc. Jpn.*, 91:062001, 2022.

[152] E. Efrati, Z. Wang, A. Kolan, and L. P. Kadanoff. Real-space renormalization in statistical mechanics. *Rev. Mod. Phys.*, 86:647–667, 2014.

[153] E. S. Sørensen and I. Affleck. Equal-time correlations in Haldane-gap antiferromagnets. *Phys. Rev. B*, 49:15771–15788, 1994.

[154] U. Schollwock and T. Jolicoeur. Haldane-gap and hidden order in the S = 2 antiferromagnetic quantum spin chain. *Europhys. Lett.*, 30:493–498, 1995.

[155] K. A. Hallberg, P. Horsch, and G. Martínez. Numerical renormalization-group study of the correlation functions of the antiferromagnetic spin-1/2 Heisenberg chain. *Phys. Rev. B*, 52:R719–R722, 1995.

[156] S. Qin, X. Wang, and L. Yu. Universality class of integer quantum spin chains: S=2 case study. *Phys. Rev. B*, 56:R14251–R14254, 1997.

[157] T. Xiang. Thermodynamics of quantum Heisenberg spin chains. *Phys. Rev. B*, 58:9142–9149, 1998.

[158] J. Lou, X. Dai, S. Qin, Z. Su, and L. Yu. Heisenberg spin-1 chain in a staggered magnetic field: A density-matrix-renormalization-group study. *Phys. Rev. B*, 60:52–55, 1999.

[159] D. Coombes, T. Xiang, and G. A. Gehring. Calculation of the susceptibility of the S = 1 antiferromagnetic Heisenberg chain with single-ion anisotropy using the transfer matrix renormalization group method. *J. Phys.: Conden. Matt.*, 10:L159–L165, 1998.

[160] J. Z. Zhao, X. Q. Wang, T. Xiang, Z. B. Su, and L. Yu. Effects of the Dzyaloshinskii–Moriya interaction on low-energy magnetic excitations in copper benzoate. *Phys. Rev. Lett.*, 90:207204, 2003.

[161] R. Chitra, S. Pati, H. R. Krishnamurthy, D. Sen, and S. Ramasesha. Density-matrix renormalization-group studies of the spin-1/2 Heisenberg system with dimerization and frustration. *Phys. Rev. B*, 52:6581–6587, 1995.

[162] R. J. Bursill, T. Xiang, and G. A. Gehring. Density matrix renormalization group study of the correlation function of the bilinear-biquadratic spin-1 chain. *J. Phys. A: Math. Gen.*, 28:2109–2119, 1995.

[163] R. Bursill, G. A. Gehring, D. J. J. Farnell et al. Numerical and approximate analytical results for the frustrated spin-1/2 quantum spin chain. *J. Phys.: Conden. Matt.*, 7:8605–8618, 1995.

[164] U. Schollwöck, T. Jolicœur, and T. Garel. Onset of incommensurability at the valence-bond-solid point in the S=1 quantum spin chain. *Phys. Rev. B*, 53:3304–3311, 1996.

[165] X. Wang and S. Mallwitz. Impurity state in the Haldane gap for an $S = 1$ Heisenberg antiferromagnetic chain with bond doping. *Phys. Rev. B*, 53:R492–R495, 1996.

[166] C. D. Batista, K. Hallberg, and A. A. Aligia. Specific heat of defects in the Haldane system Y_2BaNiO_5. *Phys. Rev. B*, 58:9248–9251, 1998.

[167] S. Rommer and S. Eggert. Impurity corrections to the thermodynamics in spin chains using a transfer-matrix DMRG method. *Phys. Rev. B*, 59:6301–6308, 1999.

[168] K. Hida. Low energy properties of the random spin-1/2 ferromagnetic-antiferromagnetic Heisenberg chain. *J. Phys. Soc. Jpn.*, 66:330–333, 1997.

[169] J. Lou, T. Xiang, and Z. Su. Thermodynamics of the bilinear-biquadratic spin-one Heisenberg chain. *Phys. Rev. Lett.*, 85:2380–2383, 2000.

[170] J. Lou, C. Chen, J. Zhao et al. Midgap states in antiferromagnetic Heisenberg chains with a staggered field. *Phys. Rev. Lett.*, 94:217207, 2005.

[171] Y. Yamashita, N. Shibata, and K. Ueda. Crossover phenomena in the one-dimensional SU(4) spin-orbit model under magnetic fields. *Phys. Rev. B*, 61:4012–4018, 2000.

[172] M. Führinger, S. Rachel, R. Thomale, M. Greiter, and P. Schmitteckert. DMRG studies of critical SU(N) spin chains. *Annalen der Physik*, 17:922–936, 2008.

[173] R. M. Noack, S. R. White, and D. J. Scalapino. Correlations in a two-chain Hubbard model. *Phys. Rev. Lett.*, 73:882–885, 1994.

[174] K. Penc, K. Hallberg, F. Mila, and H. Shiba. Shadow band in the one-dimensional infinite-U Hubbard model. *Phys. Rev. Lett.*, 77:1390–1393, 1996.

[175] H. Benthien, F. Gebhard, and E. Jeckelmann. Spectral function of the one-dimensional Hubbard model away from half filling. *Phys. Rev. Lett.*, 92:256401, 2004.

[176] E. Jeckelmann. Mott–Peierls transition in the extended Peierls–Hubbard model. *Phys. Rev. B*, 57:11838–11841, 1998.

[177] J. Lou, S. Qin, T. Xiang et al. Transition from band insulator to Mott insulator in one dimension: Critical behavior and phase diagram. *Phys. Rev. B*, 68:045110, 2003.

[178] S. Nishimoto, E. Jeckelmann, and D. J. Scalapino. Differences between hole and electron doping of a two-leg CuO ladder. *Phys. Rev. B*, 66:245109, 2002.

[179] K. Buchta, Ö. Legeza, E. Szirmai, and J. Sólyom. Mott transition and dimerization in the one-dimensional SU(n) Hubbard model. *Phys. Rev. B*, 75:155108, 2007.

[180] C. A. Hayward, D. Poilblanc, R. M. Noack, D. J. Scalapino, and W. Hanke. Evidence for a superfluid density in t-J ladders. *Phys. Rev. Lett.*, 75:926–929, 1995.

[181] L. Chen and S. Moukouri. Numerical renormalization-group study of the one-dimensional t-J model. *Phys. Rev. B*, 53:1866–1870, 1996.

[182] M. Guerrero and C. C. Yu. Kondo insulators modeled by the one-dimensional Anderson lattice: A numerical-renormalization-group study. *Phys. Rev. B*, 51:10301–10312, 1995.

[183] M. Guerrero and R. M. Noack. Phase diagram of the one-dimensional Anderson lattice. *Phys. Rev. B*, 53:3707–3712, 1996.

[184] C. C. Yu and S. R. White. Numerical renormalization group study of the one-dimensional Kondo insulator. *Phys. Rev. Lett.*, 71:3866–3869, 1993.

[185] S. Moukouri and L. G. Caron. Ground-state properties of the one-dimensional Kondo lattice at partial band filling. *Phys. Rev. B*, 52:R15723–R15726, 1995.

[186] H. Pang, S. Liang, and J. F. Annett. Effect of randomness on the Mott state. *Phys. Rev. Lett.*, 71:4377–4380, 1993.

[187] P. Schmitteckert, T. Schulze, C. Schuster, P. Schwab, and U. Eckern. Anderson localization versus delocalization of interacting fermions in one dimension. *Phys. Rev. Lett.*, 80:560–563, 1998.

[188] D. J. Garcia, K. Hallberg, B. Alascio, and M. Avignon. Spin order in one-dimensional Kondo and Hund lattices. *Phys. Rev. Lett.*, 93:177204, 2004.

[189] P. W. Anderson. Resonating valence bond state in La_2CuO_4 and superconductivity. *Science*, 235:1196–1198, 1987.

[190] F. C. Zhang and T. M. Rice. Effective Hamiltonian for the superconducting Cu oxides. *Phys. Rev. B*, 37:3759–3761, 1988.

[191] P. Sinjukow and W. Nolting. Exact mapping of periodic Anderson model to Kondo lattice model. *Phys. Rev. B*, 65:212303, 2002.

[192] L. G. Caron and S. Moukouri. Density-matrix renormalization-group study of one-dimensional acoustic phonons. *Phys. Rev. B*, 56:R8471–R8474, 1997.

[193] E. Jeckelmann and S. R. White. Density-matrix renormalization-group study of the polaron problem in the Holstein model. *Phys. Rev. B*, 57:6376–6385, 1998.

[194] C. Zhang, E. Jeckelmann, and S. R. White. Density matrix approach to local Hilbert space reduction. *Phys. Rev. Lett.*, 80:2661–2664, 1998.

[195] R. J. Bursill. Density-matrix renormalization-group algorithm for quantum lattice systems with a large number of states per site. *Phys. Rev. B*, 60:1643–1649, 1999.

[196] T. D. Kühner, S. R. White, and H. Monien. One-dimensional Bose–Hubbard model with nearest-neighbor interaction. *Phys. Rev. B*, 61:12474–12489, 2000.

[197] E. G. Dalla Torre, E. Berg, and E. Altman. Hidden order in 1D Bose insulators. *Phys. Rev. Lett.*, 97:260401, 2006.

[198] M. Greiner, O. Mandel, T. Esslinger, T. Haensch, and I. Bloch. Quantum phase transition from a superfluid to a Mott insulator in a gas of ultracold atoms. *Nature*, 415:39–44, 2002.

[199] C. Kollath, U. Schollwöck, J. von Delft, and W. Zwerger. Spatial correlations of trapped one-dimensional bosons in an optical lattice. *Phys. Rev. A*, 69:031601, 2004.

[200] J. J. Garcia-Ripoll, J. I. Cirac, P. Zoller et al. Variational ansatz for the super-fluid Mott-insulator transition in optical lattices. *Optics Express*, 12:42–54, 2004.

[201] R. J. Bursill, R. H. McKenzie, and C. J. Hamer. Phase diagram of a Heisenberg spin-Peierls model with quantum phonons. *Phys. Rev. Lett.*, 83:408–411, 1999.

[202] J. Xue, K. Seo, L. Tian, and T. Xiang. Quantum phase transition in a multiconnected Jaynes–Cummings lattice. *Phys. Rev. B*, 96:174502, 2017.

[203] Y. Liu, Y. Meurice, M. P. Qin et al. Exact blocking formulas for spin and gauge models. *Phys. Rev. D*, 88:056005, 2013.

[204] A. Denbleyker, Y. Liu, Y. Meurice et al. Controlling sign problems in spin models using tensor renormalization. *Phys. Rev. D*, 89:016008, 2014.

[205] T. M. R. Byrnes, P. Sriganesh, R. J. Bursill, and C. J. Hamer. Density matrix renormalization group approach to the massive Schwinger model. *Phys. Rev. D*, 66:013002, 2002.

[206] M. C. Bañuls, K. Cichy, J. I. Cirac, and K. Jansen. The mass spectrum of the Schwinger model with matrix product states. *J. High Energy Phys.*, 11:158, 2013.

[207] B. Buyens, J. Haegeman, K. Van Acoleyen, H. Verschelde, and F. Verstraete. Matrix product states for gauge field theories. *Phys. Rev. Lett.*, 113:091601, 2014.

[208] T. Sugihara. Density matrix renormalization group in a two-dimensional Hamiltonian lattice model. *J. High Energy Phys.*, 2004:007, 2004.

[209] A. Milsted, J. Haegeman, and T. J. Osborne. Matrix product states and variational methods applied to critical quantum field theory. *Phys. Rev. D*, 88:085030, 2013.

[210] T. Sugihara. Matrix product representation of gauge invariant states in a Z_2 lattice gauge theory. *J. High Energy Phys.*, 2005:022, 2005.

[211] S. Kühn, E. Zohar, J. I. Cirac, and M. C. Bañuls. Non-Abelian string breaking phenomena with matrix product states. *J. High Energy Phys.*, 2015:130, 2015.

[212] P. Silvi, Y. Sauer, F. Tschirsich, and S. Montangero. Tensor network simulation of an SU(3) lattice gauge theory in 1D. *Phys. Rev. D*, 100:074512, 2019.

[213] L. P. Yang, Y. Liu, H. Zou, Z. Y. Xie, and Y. Meurice. Fine structure of the entanglement entropy in the O(2) model. *Phys. Rev. E*, 93:012138, 2016.

[214] A. Milsted. Matrix product states and the non-Abelian rotor model. *Phys. Rev. D*, 93:085012, 2016.

[215] F. Bruckmann, K. Jansen, and S. Kühn. O(3) nonlinear sigma model in $1 + 1$ dimensions with matrix product states. *Phys. Rev. D*, 99:074501, 2019.

[216] W. Tang, X. C. Xie, L. Wang, and H. H. Tu. Tensor network simulation of the $(1+1)$-dimensional $O(3)$ nonlinear σ-model with $\theta = \pi$ term. *Phys. Rev. D*, 104:114513, 2021.

[217] A. Bermudez, E. Tirrito, M. Rizzi, M. Lewenstein, and S. Hands. Gross–Neveu–Wilson model and correlated symmetry-protected topological phases. *Ann. Phys.*, 399:149–180, 2018.

[218] J. Unmuth-Yockey, J. Zhang, A. Bazavov, Y. Meurice, and S. W. Tsai. Universal features of the Abelian Polyakov loop in $1+1$ dimensions. *Phys. Rev. D*, 98:094511, 2018.

[219] M. C. Bañuls, K. Cichy, Y. J. Kao et al. Phase structure of the $(1 + 1)$-dimensional massive Thirring model from matrix product states. *Phys. Rev. D*, 100:094504, 2019.

[220] M. Ganahl, J. Rincón, and G. Vidal. Continuous matrix product states for quantum fields: An energy minimization algorithm. *Phys. Rev. Lett.*, 118:220402, 2017.

[221] D. Draxler, J. Haegeman, T. J. Osborne et al. Particles, holes, and solitons: A matrix product state approach. *Phys. Rev. Lett.*, 111:020402, 2013.

[222] J. Haegeman, J. I. Cirac, T. J. Osborne, H. Verschelde, and F. Verstraete. Applying the variational principle to $(1 + 1)$-dimensional quantum field theories. *Phys. Rev. Lett.*, 105:251601, 2010.

[223] F. Quijandría, J. J. García-Ripoll, and D. Zueco. Continuous matrix product states for coupled fields: Application to Luttinger liquids and quantum simulators. *Phys. Rev. B*, 90:235142, 2014.

[224] S. S. Chung, K. Sun, and C. J. Bolech. Matrix product ansatz for fermi fields in one dimension. *Phys. Rev. B*, 91:121108, 2015.

[225] W. Tang, H. H. Tu, and L. Wang. Continuous matrix product operator approach to finite temperature quantum states. *Phys. Rev. Lett.*, 125:170604, 2020.

[226] B. Swingle. Entanglement renormalization and holography. *Phys. Rev. D*, 86:065007, 2012.

[227] A. Milsted and G. Vidal. Geometric interpretation of the multi-scale entanglement renormalization ansatz. arXiv:1812.00529.

[228] S. R. White. Spin gaps in a frustrated Heisenberg model for CaV_4O_9. *Phys. Rev. Lett.*, 77:3633–3636, 1996.

[229] S. R. White and A. L. Chernyshev. Neél order in square and triangular lattice Heisenberg models. *Phys. Rev. Lett.*, 99:127004, 2007.

[230] P. Corboz. Variational optimization with infinite projected entangled-pair states. *Phys. Rev. B*, 94:035133, 2016.

[231] P. C. G. Vlaar and P. Corboz. Simulation of three-dimensional quantum systems with projected entangled-pair states. *Phys. Rev. B*, 103:205137, 2021.

[232] A. W. Sandvik. Finite-size scaling of the ground-state parameters of the two-dimensional Heisenberg model. *Phys. Rev. B*, 56:11678–11690, 1997.

[233] P. Corboz, M. Lajkó, K. Penc, F. Mila, and A. M. Läuchli. Competing states in the SU(3) Heisenberg model on the honeycomb lattice: Plaquette valence-bond crystal versus dimerized color-ordered state. *Phys. Rev. B*, 87:195113, 2013.

[234] P. Corboz and F. Mila. Crystals of bound states in the magnetization plateaus of the Shastry–Sutherland model. *Phys. Rev. Lett.*, 112:147203, 2014.

[235] Z. Zhu and S. R. White. Spin liquid phase of the S=1/2 J_1-J_2 Heisenberg model on the triangular lattice. *Phys. Rev. B*, 92:041105, 2015.

[236] L. Wang and A. W. Sandvik. Critical level crossings and gapless spin liquid in the square-lattice spin-1/2 J_1-J_2 Heisenberg antiferromagnet. *Phys. Rev. Lett.*, 121:107202, 2018.

[237] J. Hasik, D. Poilblanc, and F. Becca. Investigation of the Neel phase of the frustrated Heisenberg antiferromagnet by differentiable symmetric tensor networks. *SciPost Phys.*, 10:012, 2021.

[238] L. Capriotti, D. J. Scalapino, and S. R. White. Spin-liquid versus dimerized ground states in a frustrated Heisenberg antiferromagnet. *Phys. Rev. Lett.*, 93:177004, 2004.

[239] M. S. Block, D. N. Sheng, O. I. Motrunich, and M. P. A. Fisher. Spin Bose-metal and valence bond solid phases in a spin-1/2 model with ring exchanges on a four-leg triangular ladder. *Phys. Rev. Lett.*, 106:157202, 2011.

[240] S. R. White and D. J. Scalapino. Density matrix renormalization group study of the striped phase in the 2D t-J model. *Phys. Rev. Lett.*, 80:1272–1275, 1998.

[241] S. R. White and D. J. Scalapino. Checkerboard patterns in the t-J model. *Phys. Rev. B*, 70:220506, 2004.

[242] P. Corboz, T. M. Rice, and M. Troyer. Competing states in the *t-J* model: Uniform *d*-wave state versus stripe state. *Phys. Rev. Lett.*, 113:046402, 2014.

[243] S. R. White and D. J. Scalapino. Stripes on a 6-leg Hubbard ladder. *Phys. Rev. Lett.*, 91:136403, 2003.

[244] B. X. Zheng, C. M. Chung, P. Corboz et al. Stripe order in the underdoped region of the two-dimensional Hubbard model. *Science*, 358:1155–1160, 2017.

[245] J. C. Xavier and E. Dagotto. Robust *d*-wave pairing correlations in the Heisenberg Kondo lattice model. *Phys. Rev. Lett.*, 100:146403, 2008.

[246] J. M. Tranquada, B. J. Sternlieb, J. D. Axe, Y. Nakamura, and S. Uchida. Evidence for stripe correlations of spins and holes in copper oxide superconductors. *Nature*, 375:561–563, 1995.

[247] V. Murg, F. Verstraete, and J. I. Cirac. Variational study of hard-core bosons in a two-dimensional optical lattice using projected entangled pair states. *Phys. Rev. A*, 75:033605, 2007.

[248] J. Jordan, R. Orús, and G. Vidal. Numerical study of the hard-core Bose–Hubbard model on an infinite square lattice. *Phys. Rev. B*, 79:174515, 2009.

[249] W. L. Tu, H. K. Wu, and T. Suzuki. Frustration-induced supersolid phases of extended Bose–Hubbard model in the hard-core limit. *J. Phys.: Conden. Matt.*, 32:455401, 2020.

[250] R. V. Mishmash, M. S. Block, R. K. Kaul et al. Bose metals and insulators on multileg ladders with ring exchange. *Phys. Rev. B*, 84:245127, 2011.

[251] H. Otsuka. Density-matrix renormalization-group study of the spin-1/2 XXZ antiferromagnet on the Bethe lattice. *Phys. Rev. B*, 53:14004–14007, 1996.

[252] B. Friedman. A density matrix renormalization group approach to interacting quantum systems on Cayley trees. *J. Phys.: Conden. Matt.*, 9:9021–9029, 1997.

[253] H. J. Changlani, S. Ghosh, S. Pujari, and C. L. Henley. Emergent spin excitations in a Bethe lattice at percolation. *Phys. Rev. Lett.*, 111:157201, 2013.

[254] H. J. Liao, Z. Y. Xie, J. Chen et al. Heisenberg antiferromagnet on the Husimi lattice. *Phys. Rev. B*, 93:075154, 2016.

[255] L. Marie-Bernadette, M. Cousy, and G. Pastor. Density-matrix renormalization study of the Hubbard model on a Bethe lattice. *Euro. Phys. J. B*, 13:421–427, 2000.

[256] P. Lunts, A. Georges, E. M. Stoudenmire, and M. Fishman. Hubbard model on the Bethe lattice via variational uniform tree states: Metal-insulator transition and a Fermi liquid. *Phys. Rev. Research*, 3:023054, 2021.

[257] D. W. Qu, W. Li, and T. Xiang. Thermal tensor network simulations of the Heisenberg model on the Bethe lattice. *Phys. Rev. B*, 100:125121, 2019.

[258] Ö. Legeza and J. Sólyom. Optimizing the density-matrix renormalization group method using quantum information entropy. *Phys. Rev. B*, 68:195116, 2003.

[259] C. Krumnow, L. Veis, Ö. Legeza, and J. Eisert. Fermionic orbital optimization in tensor network states. *Phys. Rev. Lett.*, 117:210402, 2016.

[260] S. Nishimoto, E. Jeckelmann, F. Gebhard, and R. M. Noack. Application of the density matrix renormalization group in momentum space. *Phys. Rev. B*, 65:165114, 2002.

[261] G. Ehlers, J. Sólyom, Ö. Legeza, and R. M. Noack. Entanglement structure of the Hubbard model in momentum space. *Phys. Rev. B*, 92:235116, 2015.

[262] J. Motruk, M. P. Zaletel, R. S. K. Mong, and F. Pollmann. Density matrix renormalization group on a cylinder in mixed real and momentum space. *Phys. Rev. B*, 93:155139, 2016.

[263] G. Ehlers, S. R. White, and R. M. Noack. Hybrid-space density matrix renormalization group study of the doped two-dimensional Hubbard model. *Phys. Rev. B*, 95:125125, 2017.

[264] E. H. Rezayi and F. D. M. Haldane. Laughlin state on stretched and squeezed cylinders and edge excitations in the quantum Hall effect. *Phys. Rev. B*, 50:17199–17207, 1994.

[265] R. B. Laughlin. Anomalous quantum Hall effect: An incompressible quantum fluid with fractionally charged excitations. *Phys. Rev. Lett.*, 50:1395–1398, 1983.

[266] G. Moore and N. Read. Nonabelions in the fractional quantum Hall effect. *Nucl. Phys. B*, 360:362–396, 1991.

[267] M. P. Zaletel and R. S. K. Mong. Exact matrix product states for quantum Hall wave functions. *Phys. Rev. B*, 86:245305, 2012.

[268] B. Estienne, N. Regnault, and B. A. Bernevig. Correlation lengths and topological entanglement entropies of unitary and nonunitary fractional quantum Hall wave functions. *Phys. Rev. Lett.*, 114:186801, 2015.

[269] H. Li and F. D. M. Haldane. Entanglement spectrum as a generalization of entanglement entropy: Identification of topological order in non-Abelian fractional quantum Hall effect states. *Phys. Rev. Lett.*, 101:010504, 2008.

[270] Z. X. Hu, Z. Papić, S. Johri, R. N. Bhatt, and P. Schmitteckert. Comparison of the density-matrix renormalization group method applied to fractional quantum Hall systems in different geometries. *Phys. Lett. A*, 376:2157–2161, 2012.

[271] M. P. Zaletel, R. S. K. Mong, and F. Pollmann. Topological characterization of fractional quantum Hall ground states from microscopic Hamiltonians. *Phys. Rev. Lett.*, 110:236801, 2013.

[272] D. L. Kovrizhin. Density matrix renormalization group for bosonic quantum Hall effect. *Phys. Rev. B*, 81:125130, 2010.

[273] S. S. Gong, W. Zhu, and D. N. Sheng. Emergent chiral spin liquid: Fractional quantum Hall effect in a Kagome Heisenberg model. *Sci. Rep.*, 4:6317, 2014.

[274] S. Daul, I. Ciofini, C. Daul, and S. R. White. Full-CI quantum chemistry using the density matrix renormalization group. *Int. J. Quan. Chem.*, 79:331–342, 2000.

[275] A. O. Mitrushenkov, G. Fano, F. Ortolani, R. Linguerri, and P. Palmieri. Quantum chemistry using the density matrix renormalization group. *J. Chem. Phys.*, 115:6815–6821, 2001.

[276] G. K. L. Chan and M. Head-Gordon. Highly correlated calculations with a polynomial cost algorithm: A study of the density matrix renormalization group. *J. Chem. Phys.*, 116:4462–4476, 2002.

[277] Ö. Legeza, J. Röder, and B. A. Hess. Controlling the accuracy of the density-matrix renormalization-group method: The dynamical block state selection approach. *Phys. Rev. B*, 67:125114, 2003.

[278] J. Hachmann, W. Cardoen, and G. K. L. Chan. Multireference correlation in long molecules with the quadratic scaling density matrix renormalization group. *J. Chem. Phys.*, 125:144101, 2006.

[279] E. Ronca, Z. Li, C. A. Jimenez-Hoyos, and G. K. L. Chan. Time-step targeting time-dependent and dynamical density matrix renormalization group algorithms with ab initio Hamiltonians. *J. Chem. Theo. Comput.*, 13:5560–5571, 2017.

[280] S. Wouters, V. Van Speybroeck, and D. Van Neck. DMRG-CASPT2 study of the longitudinal static second hyperpolarizability of all-trans polyenes. *J. Chem. Phys.*, 145:054120, 2016.

[281] W. Mizukami, Y. Kurashige, and T. Yanai. Communication: Novel quantum states of electron spins in polycarbenes from ab initio density matrix renormalization group calculations. *J. Chem. Phys.*, 133:091101, 2010.

[282] K. H. Marti, I. M. Ondík, G. Moritz, and M. Reiher. Density matrix renormalization group calculations on relative energies of transition metal complexes and clusters. *J. Chem. Phys.*, 128:014104, 2008.

[283] K. H. Marti and M. Reiher. New electron correlation theories for transition metal chemistry. *Phys. Chem. Chem. Phys.*, 13:6750–6759, 2011.

[284] D. Zgid and M. Nooijen. The density matrix renormalization group self-consistent field method: Orbital optimization with the density matrix renormalization group method in the active space. *J. Chem. Phys.*, 128:144116, 2008.

[285] D. Ghosh, J. Hachmann, T. Yanai, and G. K. L. Chan. Orbital optimization in the density matrix renormalization group, with applications to polyenes and β-carotene. *J. Chem. Phys.*, 128:144117, 2008.

[286] G. K. L. Chan and S. Sharma. The density matrix renormalization group in quantum chemistry. *Ann. Rev. Phys. Chem.*, 62:465–481, 2011.

[287] V. Murg, F. Verstraete, Ö. Legeza, and R. M. Noack. Simulating strongly correlated quantum systems with tree tensor networks. *Phys. Rev. B*, 82:205105, 2010.

[288] N. Nakatani and G. K. L. Chan. Efficient tree tensor network states (TTNS) for quantum chemistry: Generalizations of the density matrix renormalization group algorithm. *J. Chem. Phys.*, 138:134113, 2013.

[289] G. Moritz, A. Wolf, and M. Reiher. Relativistic DMRG calculations on the curve crossing of cesium hydride. *J. Chem. Phys.*, 123:184105, 2005.

[290] S. Knecht, Ö. Legeza, and M. Reiher. Communication: Four-component density matrix renormalization group. *J. Chem. Phys.*, 140:041101, 2014.

[291] S. Battaglia, S. Keller, and S. Knecht. Efficient relativistic density-matrix renormalization group implementation in a matrix-product formulation. *J. Chem. Theo. Comp.*, 14:2353–2369, 2018.

[292] Y. Kurashige and T. Yanai. Second-order perturbation theory with a density matrix renormalization group self-consistent field reference function: Theory and application to the study of chromium dimer. *J. Chem. Phys.*, 135:094104, 2011.

[293] Y. Kurashige, J. Chalupsky, T. N. Lan, and T. Yanai. Complete active space second-order perturbation theory with cumulant approximation for extended active-space wavefunction from density matrix renormalization group. *J. Chem. Phys.*, 141:174111, 2014.

[294] L. Veis, A. Antalík, J. Brabec et al. Coupled cluster method with single and double excitations tailored by matrix product state wave functions. *J. Phys. Chem. Lett.*, 7:4072–4078, 2016.

[295] T. Yanai and G. K. L. Chan. Canonical transformation theory for multireference problems. *J. Chem. Phys.*, 124:194106, 2006.

[296] E. D. Hedegard, S. Knecht, J. S. Kielberg, H. J. A. Jensen, and M. Reiher. Density matrix renormalization group with efficient dynamical electron correlation through range separation. *J. Chem. Phys.*, 142:224108, 2015.

[297] J. Dukelsky, S. Pittel, S. S. Dimitrova, and M. V. Stoitsov. Density matrix renormalization group method and large-scale nuclear shell-model calculations. *Phys. Rev. C*, 65:054319, 2002.

[298] T. Papenbrock and D. J. Dean. Density matrix renormalization group and wavefunction factorization for nuclei. *J. Phys. G*, 31:S1377–S1383, 2005.

[299] S. Pittel and N. Sandulescu. Density matrix renormalization group and the nuclear shell model. *Phys. Rev. C*, 73:014301, 2006.

[300] J. Rotureau, N. Michel, W. Nazarewicz, M. Płoszajczak, and J. Dukelsky. Density matrix renormalization group approach for many-body open quantum systems. *Phys. Rev. Lett.*, 97:110603, 2006.

[301] B. Thakur, S. Pittel, and N. Sandulescu. Density matrix renormalization group study of ^{48}Cr and ^{56}Ni. *Phys. Rev. C*, 78:041303, 2008.

[302] Ö. Legeza, L. Veis, A. Poves, and J. Dukelsky. Advanced density matrix renormalization group method for nuclear structure calculations. *Phys. Rev. C*, 92:051303, 2015.

[303] T. Nishino and K. Okunishi. Corner transfer matrix algorithm for classical renormalization group. *J. Phys. Soc. Jpn.*, 66:3040–3047, 1997.

[304] W. Lay and J. Rudnick. Analysis of a continuous field theory in two dimensions with use of the density matrix renormalization group. *Phys. Rev. Lett.*, 88:057203, 2002.

[305] T. Nishino and K. Okunishi. Numerical latent heat observation of the q = 5 Potts model. *J. Phys. Soc. Jpn.*, 67:1492–1493, 1998.

[306] E. Carlon, C. Chatelain, and B. Berche. Critical behavior of the random Potts chain. *Phys. Rev. B*, 60:12974–12981, 1999.

[307] Q. N. Chen, M. P. Qin, J. Chen et al. Partial order and finite-temperature phase transitions in Potts models on irregular lattices. *Phys. Rev. Lett.*, 107:165701, 2011.

[308] M. P. Qin, Q. N. Chen, Z. Y. Xie et al. Partial long-range order in antiferromagnetic Potts models. *Phys. Rev. B*, 90:144424, 2014.

[309] H. Sato and K. Sasaki. Numerical study of the two-dimensional three-state chiral clock model by the density matrix renormalization group method. *J. Phys. Soc. Jpn.*, 69:1050–1054, 2000.

[310] A. Gendiar, R. Krcmar, K. Ueda, and T. Nishino. Phase transition of clock models on a hyperbolic lattice studied by corner transfer matrix renormalization group method. *Phys. Rev. E*, 77:041123, 2008.

[311] Z. Q. Li, L. P. Yang, Z. Y. Xie et al. Critical properties of the two-dimensional q-state clock model. *Phys. Rev. E*, 101:060105, 2020.

[312] Y. Honda and T. Horiguchi. Density-matrix renormalization group for the Berezinskii–Kosterlitz–Thouless transition of the 19-vertex model. *Phys. Rev. E*, 56:3920–3926, 1997.

[313] H. Takasaki, T. Nishino, and Y. Hieida. Phase diagram of a 2D vertex model. *J. Phys. Soc. Jpn.*, 70:1429–1430, 2001.

[314] D. P. Foster and C. Pinettes. Corner-transfer-matrix renormalization-group method for two-dimensional self-avoiding walks and other $O(n)$ models. *Phys. Rev. E*, 67:045105, 2003.

[315] J. F. Yu, Z. Y. Xie, Y. Meurice et al. Tensor renormalization group study of classical XY model on the square lattice. *Phys. Rev. E*, 89:013308, 2014.

[316] L. Vanderstraeten, B. Vanhecke, A. M. Läuchli, and F. Verstraete. Approaching the Kosterlitz–Thouless transition for the classical XY model with tensor networks. *Phys. Rev. E*, 100:062136, 2019.

[317] P. Schmoll, A. Kshetrimayum, J. Eisert, R. Orús, and M. Rizzi. The classical two-dimensional Heisenberg model revisited: An SU(2)-symmetric tensor network study. *SciPost Phys.*, 11:098, 2021.

[318] T. Nishino and K. Okunishi. A density matrix algorithm for 3D classical models. *J. Phys. Soc. Jpn.*, 67:3066–3072, 1998.

[319] T. Nishino, K. Okunishi, Y. Hieida, N. Maeshima, and Y. Akutsu. Self-consistent tensor product variational approximation for 3D classical models. *Nucl. Phys. B*, 575:504 − 512, 2000.

[320] A. Gendiar and T. Nishino. Phase diagram of the three-dimensional axial next-nearest-neighbor Ising model. *Phys. Rev. B*, 71:024404, 2005.

[321] J. I. Cirac, D. Poilblanc, N. Schuch, and F. Verstraete. Entanglement spectrum and boundary theories with projected entangled-pair states. *Phys. Rev. B*, 83:245134, 2011.

[322] X. G. Wen. Topological orders and edge excitations in fractional quantum Hall states. *Adv. Phys.*, 44:405–473, 1995.

[323] X. Chen, Z. C. Gu, and X. G. Wen. Classification of gapped symmetric phases in one-dimensional spin systems. *Phys. Rev. B*, 83:035107, 2011.

[324] Y. Levine, O. Sharir, N. Cohen, and A. Shashua. Quantum entanglement in deep learning architectures. *Phys. Rev. Lett.*, 122:065301, 2019.

[325] C. Giuseppe and T. Matthias Solving the quantum many-body problem with artificial neural networks. *Science*, 355:602–606, 2017.

[326] J. Chen, S. Cheng, H. Xie, L. Wang, and T. Xiang. Equivalence of restricted Boltzmann machines and tensor network states. *Phys. Rev. B*, 97:085104, 2018.

[327] E. M. Stoudenmire and D. J. Schwab. Supervised learning with tensor networks. In D. D. Lee, M. Sugiyama, U. V. Luxburg, I. Guyon, and R. Garnett (eds.), *Advances in Neural Information Processing Systems, vol. 29 (NIPS 2016)*.

[328] S. Cheng, L. Wang, and P. Zhang. Supervised learning with projected entangled pair states. *Phys. Rev. B*, 103:125117, 2021.

[329] E. M. Stoudenmire. Learning relevant features of data with multi-scale tensor networks. *Quantum Sci. Tech.*, 3:034003, 2018.

[330] S. Cheng, L. Wang, T. Xiang, and P. Zhang. Tree tensor networks for generative modeling. *Phys. Rev. B*, 99:155131, 2019.

[331] Y. Liu, W. J. Li, X. Zhang, M. Lewenstein, G. Su, and S. J. Ran. Entanglement-based feature extraction by tensor network machine learning. *Front. App. Math. Stat.*, 7:716044, 2021.

[332] Z. F. Gao, S. Cheng, R. Q. He et al. Compressing deep neural networks by matrix product operators. *Phys. Rev. Research*, 2:023300, 2020.

[333] J. Martyn, G. Vidal, C. Roberts, and S. Leichenauer. Entanglement and tensor networks for supervised image classification. arXiv:2007.06082.

[334] L. de Lathauwer, B. de Moor, and J. Vandewalle. A multilinear singular value decomposition. *SIAM J. Matrix Anal. Appl*, 21:1253–1278, 2000.

[335] L. de Lathauwer, B. de Moor, and J. Vandewalle. On the best rank-1 and rank-(R1, R2,...,Rn) approximation of higher-order tensors. *SIAM J. Matrix Anal. Appl.*, 21:1324–1342, 2000.

[336] V. de. Silva and L. H. Lim. Tensor rank and the ill-posedness of the best low-rank approximation problem. *SIAM. J. Matrix Anal. Appl.*, 30:1084–1127, 2008.

[337] W. Baur and V. Strassen The complexity of partial derivatives. *Theo. Comp. Sci.*, 22:317–330, 1983.

[338] A. Griewank. *On Automatic Differentiation*. Math. Program. Recent Dev. Appl. Kluwer Academic Publishers, Dordrecht, The Netherlands, 1989.

[339] S. Linnainmaa. Taylor expansion of the accumulated rounding error. *BIT Numerical Mathematics*, 16:146–160, 1976.

[340] M. Suzuki. General-theory of fractal path-integrals with applications to many-body theories and statistical physics. *J. Math. Phys.*, 32:400–407, 1991.

[341] W. Janke and T. Sauer. Properties of higher-order Trotter formulas. *Phys. Lett. A*, 165:199–205, 1992.

[342] L. Onsager. Crystal statistics. I. A two-dimensional model with an order-disorder transition. *Phys. Rev.*, 65:117–149, 1994.

[343] G. H. Wannier. Antiferromagnetism: The triangular Ising net. *Phys. Rev.*, 79:357–364, 1950.

[344] G. H. Wannier. Antiferromagnetism: The triangular Ising net. *Phys. Rev. B*, 7:5017–5017, 1973.

[345] J. Chen, H. J. Liao, H. D. Xie et al. Phase transition of the q-state clock model: Duality and tensor renormalization. *Chin. Phys. Lett.*, 34:050503, 2017.

[346] F. Y. Wu. The Potts model. *Rev. Mod. Phys.*, 54:235–268, 1982.

[347] M. Zwolak and G. Vidal. Mixed-state dynamics in one-dimensional quantum lattice systems: A time-dependent superoperator renormalization algorithm. *Phys. Rev. Lett.*, 93:207205, 2004.

[348] M. P. Zaletel, R. S. K. Mong, C. Karrasch, J. E. Moore, and F. Pollmann. Time-evolving a matrix product state with long-ranged interactions. *Phys. Rev. B*, 91:165112, 2015.

[349] C. Hubig, I. P. McCulloch, and U. Schollwöck. Generic construction of efficient matrix product operators. *Phys. Rev. B*, 95:035129, 2017.

[350] M. B. Hastings. Locality in quantum and Markov dynamics on lattices and networks. *Phys. Rev. Lett.*, 93:140402, 2004.

[351] L. Masanes. Area law for the entropy of low-energy states. *Phys. Rev. A*, 80:052104, 2009.

[352] M. B. Hastings. Entropy and entanglement in quantum ground states. *Phys. Rev. B*, 76:035114, 2007.

[353] S. Iblisdir, D. Perez-Garcia, M. Aguado, and J. Pachos. Scaling law for topologically ordered systems at finite temperature. *Phys. Rev. B*, 79:134303, 2009.

[354] M. M. Wolf. Violation of the entropic area law for fermions. *Phys. Rev. Lett.*, 96:010404, 2006.

[355] D. Gioev and I. Klich. Entanglement entropy of fermions in any dimension and the Widom conjecture. *Phys. Rev. Lett.*, 96:100503, 2006.

[356] F. Verstraete, M. M. Wolf, D. Perez-Garcia, and J. I. Cirac. Criticality, the area law, and the computational power of projected entangled pair states. *Phys. Rev. Lett.*, 96:220601, 2006.

[357] M. Cramer, J. Eisert, M. B. Plenio, and J. Dreißig. Entanglement-area law for general bosonic harmonic lattice systems. *Phys. Rev. A*, 73:012309, 2006.

[358] A. Kitaev and J. Preskill. Topological entanglement entropy. *Phys. Rev. Lett.*, 96:110404, 2006.

[359] M. Levin and X. G. Wen. Detecting topological order in a ground state wave function. *Phys. Rev. Lett.*, 96:110405, 2006.

[360] S. T. Flammia, A. Hamma, T. L. Hughes, and X. G. Wen. Topological entanglement Rényi entropy and reduced density matrix structure. *Phys. Rev. Lett.*, 103:261601, 2009.

[361] Y. Zhang, T. Grover, and A. Vishwanath. Entanglement entropy of critical spin liquids. *Phys. Rev. Lett.*, 107:067202, 2011.

[362] Y. Zhang, T. Grover, and A. Vishwanath. Topological entanglement entropy of Z_2 spin liquids and lattice Laughlin states. *Phys. Rev. B*, 84:075128, 2011.

[363] M. den Nijs and K. Rommelse. Preroughening transitions in crystal surfaces and valence-bond phases in quantum spin chains. *Phys. Rev. B*, 40:4709–4734, 1989.

[364] T. Kennedy and H. Tasaki. Hidden $Z_2 \times Z_2$ symmetry breaking in Haldane-gap antiferromagnets. *Phys. Rev. B*, 45:304–307, 1992.

[365] M. Oshikawa. Hidden $Z_2 * Z_2$ symmetry in quantum spin chains with arbitrary integer spin. *J. Phys.: Conden. Matt.*, 4:7469–7488, 1992.

[366] N. Schuch, D. Poilblanc, J. I. Cirac, and D. Perez-Garcia. Topological order in the projected entangled-pair states formalism: Transfer operator and boundary Hamiltonians. *Phys. Rev. Lett.*, 111:090501, 2013.

[367] S. Hu, B. Normand, X. Wang, and L. Yu. Accurate determination of the Gaussian transition in spin-1 chains with single-ion anisotropy. *Phys. Rev. B*, 84:220402, 2011.

[368] M. P. Qin, J. M. Leinaas, S. Ryu et al. Quantum torus chain. *Phys. Rev. B*, 86:134430, 2012.

[369] L. Hulthén. Über das Austauschproblem eines Kristalles (Ph.D thesis, monograph). *Arkiv för matematik, astronomi och fysik*, 26A:1–106, 1938.

[370] H. W. J. Blöte, J. L. Cardy, and M. P. Nightingale. Conformal invariance, the central charge, and universal finite-size amplitudes at criticality. *Phys. Rev. Lett.*, 56:742–745, 1986.

[371] I. Affleck. Universal term in the free energy at a critical point and the conformal anomaly. *Phys. Rev. Lett.*, 56:746–748, 1986.

[372] W. Tatsuaki. Interaction-round-a-face density-matrix renormalization-group method applied to rotational-invariant quantum spin chains. *Phys. Rev. E*, 61:3199–3206, 2000.

[373] I. P. McCulloch and M. Gulacsi. Total spin in the density matrix renormalization group algorithm. *Phil. Mag. Lett.*, 81:447–453, 2001.

[374] I. P. McCulloch and M. Gulácsi. The non-Abelian density matrix renormalization group algorithm. *Europhys. Lett.*, 57:852, 2002.

[375] A. Weichselbaum. Non-Abelian symmetries in tensor networks: A quantum symmetry space approach. *Ann. Phys.*, 327:2972–3047, 2012.

[376] G. Ehlers, J. Sólyom, Ö. Legeza, and R. M. Noack. Entanglement structure of the Hubbard model in momentum space. *Phys. Rev. B*, 92:235116, 2015.

[377] J. Rissler, R. M. Noack, and S. R. White. Measuring orbital interaction using quantum information theory. *Chem. Phys.*, 323:519–531, 2006.

[378] G. Barcza, Ö. Legeza, K. H. Marti, and M. Reiher. Quantum-information analysis of electronic states of different molecular structures. *Phys. Rev. A*, 83:012508, 2011.

[379] D. Pérez-García, M. M. Wolf, M. Sanz, F. Verstraete, and J. I. Cirac. String order and symmetries in quantum spin lattices. *Phys. Rev. Lett.*, 100:167202, 2008.

[380] E. K. U. Gross, L. N. Oliveira, and W. Kohn. Rayleigh–Ritz variational principle for ensembles of fractionally occupied states. *Phys. Rev. A*, 37:2805–2808, 1988.

[381] M. Hauru, M. Van Damme, and J. Haegeman. Riemannian optimization of isometric tensor networks. *SciPost Phys.*, 10:040, 2021.

[382] R. Orús and G. Vidal. Infinite time-evolving block decimation algorithm beyond unitary evolution. *Phys. Rev. B*, 78:155117, 2008.

[383] G. Vidal. Classical simulation of infinite-size quantum lattice systems in one spatial dimension. *Phys. Rev. Lett.*, 98:070201, 2007.

[384] B. Paredes, F. Verstraete, and J. I. Cirac. Exploiting quantum parallelism to simulate quantum random many-body systems. *Phys. Rev. Lett.*, 95:140501, 2005.

[385] C. Guo, A. Weichselbaum, S. Kehrein, T. Xiang, and J. von Delft. Density matrix renormalization group study of a quantum impurity model with Landau–Zener time-dependent Hamiltonian. *Phys. Rev. B*, 79:115137, 2009.

[386] E. H. Lieb and W. Liniger. Exact analysis of an interacting Bose gas. I: The general solution and the ground state. *Phys. Rev.*, 130:1605–1616, 1963.

[387] E. H. Lieb. Exact analysis of an interacting Bose gas. II: The excitation spectrum. *Phys. Rev.*, 130:1616–1624, 1963.

[388] J. Haegeman, J. I. Cirac, T. J. Osborne, and F. Verstraete. Calculus of continuous matrix product states. *Phys. Rev. B*, 88:085118, 2013.

[389] M. Ganahl and G. Vidal. Continuous matrix product states for nonrelativistic quantum fields: A lattice algorithm for inhomogeneous systems. *Phys. Rev. B*, 98:195105, 2018.

[390] M. T. Fishman, L. Vanderstraeten, V. Zauner-Stauber, J. Haegeman, and F. Verstraete. Faster methods for contracting infinite two-dimensional tensor networks. *Phys. Rev. B*, 98:235148, 2018.

[391] R. J. Baxter Corner transfer matrices of 8-vertex model. 1: Low-temperature expansions and conjectured properties. *J. Stat. Phys.*, 15:485–503, 1976.

[392] R. Orús. Exploring corner transfer matrices and corner tensors for the classical simulation of quantum lattice systems. *Phys. Rev. B*, 85:205117, 2012.

[393] Y. K. Huang, P. Chen, and Y. J. Kao. Accurate computation of low-temperature thermodynamicds for quantum spin chains. *Phys. Rev. B*, 86:235102, 2012.

[394] Y. H. Su, Q. H. Xiao, T. Xiang, X. Q. Wang, and Z. B. Su. Instability of the Fermi surface in the one-dimensional Kondo lattice. *J. Phys.: Conden. Matt.*, 16:5163–5169, 2004.

[395] J. C. Bonner and M. E. Fisher. Linear magnetic chains with anisotropic coupling. *Phys. Rev.*, 135:A640–A658, 1964.

[396] S. Eggert, I. Affleck, and M. Takahashi. Susceptibility of the spin-1/2 Heisenberg antiferromagnetic chain. *Phys. Rev. Lett.*, 73:332–335, 1994.

[397] R. B. Griffiths. Magnetization curve at zero temperature for the antiferromagnetic Heisenberg linear chain. *Phys. Rev.*, 133:A768–A775, 1964.

[398] K. Nomura and M. Yamada. Thermal Bethe-ansatz study of the correlation length of the one-dimensional S=1/2 Heisenberg antiferromagnet. *Phys. Rev. B*, 43:8217–8223, 1991.

[399] E. S. Sørensen and I. Affleck. Large-scale numerical evidence for Bose condensation in the S=1 antiferromagnetic chain in a strong field. *Phys. Rev. Lett.*, 71:1633–1636, 1993.

[400] T. Xiang and X. Q. Wang. Quantum transfer-matrix and momentum-space DMRG. In I. Peschel, X. Wang, M. Kaulke, and K. Hallberg (eds.), *Density-Matrix Renormalization: A New Numerical Method in Physics*, Lecture Notes in Physics, Vol. 528, pages 149–172. Springer, 1999.

[401] E. Dagotto. Correlated electrons in high-temperature superconductors. *Rev. Mod. Phys.*, 66:763–840, 1994.

[402] D. G. Pettifor and D. L. Weaire. *The Recursion Method and Its Applications*, volume 58 of *Springer Series in Solid State Sciences*. Springer, 1985.

[403] V. S. Viswanath and G. Muller. Recursion method in quantum spin dynamics - the art of terminating a continued-fraction. *J. Appl. Phys.*, 67:5486–5488, 1990.

[404] Y. Saad. *Iterative Methods for Sparse Linear Systems*. Siam, 2003.

[405] J. des Cloizeaux and J. J. Pearson. Spin-wave spectrum of the antiferromagnetic linear chain. *Phys. Rev.*, 128:2131–2135, 1962.

[406] T. Yamada. Fermi-liquid theory of linear antiferromagnetic chains. *Prog. Theo. Phys.*, 41:880–890, 1969.

[407] G. Müller, H. Thomas, H. Beck, and J. C. Bonner. Quantum spin dynamics of the antiferromagnetic linear chain in zero and nonzero magnetic field. *Phys. Rev. B*, 24:1429–1467, 1981.

[408] L. D. Faddeev and L. A. Takhtajan. What is the spin of a spin wave? *Phys. Lett. A*, 85:375–377, 1981.

[409] J. D. Johnson and B. M. McCoy. Low-temperature thermodynamics of the $|\Delta|\geq1$ Heisenberg–Ising ring. *Phys. Rev. A*, 6:1613–1626, 1972.

[410] M. Karbach, G. Müller, A. H. Bougourzi, A. Fledderjohann, and K. H. Mütter. Two-spinon dynamic structure factor of the one-dimensional S=1/2 Heisenberg antiferromagnet. *Phys. Rev. B*, 55:12510–12517, 1997.

[411] J. S. Caux and R. Hagemans. The four-spinon dynamical structure factor of the Heisenberg chain. *J. Stat. Mech.: Theo. Exp.*, 2006:P12013–P12013, 2006.

[412] A. E. Feiguin and S. R. White. Time-step targeting methods for real-time dynamics using the density matrix renormalization group. *Phys. Rev. B*, 72:020404, 2005.

[413] T. Dutta and S. Ramasesha. Double time window targeting technique: Real-time DMRG dynamics in the Pariser–Parr–Pople model. *Phys. Rev. B*, 82:035115, 2010.

[414] D. P. Arovas, A. Auerbach, and F. D. M. Haldane. Extended Heisenberg models of antiferromagnetism: Analogies to the fractional quantum Hall effect. *Phys. Rev. Lett.*, 60:531–534, 1988.

[415] J. Haegeman, T. J. Osborne, and F. Verstraete. Post-matrix product state methods: To tangent space and beyond. *Phys. Rev. B*, 88:075133, 2013.

[416] L. Vanderstraeten, J. Haegeman, and F. Verstraete. Simulating excitation spectra with projected entangled-pair states. *Phys. Rev. B*, 99:165121, 2019.

[417] B. Ponsioen and P. Corboz. Excitations with projected entangled pair states using the corner transfer matrix method. *Phys. Rev. B*, 101:195109, 2020.

[418] A. W. Sandvik and R. R. P. Singh. High-energy magnon dispersion and multi-magnon continuum in the two-dimensional Heisenberg antiferromagnet. *Phys. Rev. Lett.*, 86:528–531, 2001.

[419] A. Cayley. Desiderata and suggestions: No. 2. The theory of groups: Graphical representation. *Am. J. Math.*, 1:174–176, 1878.

[420] M. Ostilli. Cayley trees and Bethe lattices: A concise analysis for mathematicians and physicists. *Physica A*, 391:3417 – 3423, 2012.

[421] K. Husimi. Note on Mayers' theory of cluster integrals. *J. Chem. Phys.*, 18:682–684, 1950.

[422] R. J. Riddell and G. E. Uhlenbeck. On the theory of the virial development of the equation of state of monoatomic gases. *J. Chem. Phys.*, 21:2056–2064, 1953.

[423] F. Harary and G. E. Uhlenbeck. On the number of Husimi trees. *Proc. Nat. Acad. Sci.*, 39:315–322, 1953.

[424] H. A. Bethe. Statistical theory of superlattices. *Proc. R. Soc. Lond. A: Math. Phys. Eng. Sci.*, 150:552–575, 1935.

[425] M. Eckstein, M. Kollar, K. Byczuk, and D. Vollhardt. Hopping on the Bethe lattice: Exact results for densities of states and dynamical mean-field theory. *Phys. Rev. B*, 71:235119, 2005.

[426] K. Byczuk and D. Vollhardt. Correlated bosons on a lattice: Dynamical mean-field theory for Bose–Einstein condensed and normal phases. *Phys. Rev. B*, 77:235106, 2008.

[427] D. Nagaj, E. Farhi, J. Goldstone, P. Shor, and I. Sylvester. Quantum transverse-field Ising model on an infinite tree from matrix product states. *Phys. Rev. B*, 77:214431, 2008.

[428] A. Nagy. Simulating quantum systems on the Bethe lattice by translationally invariant infinite-tree tensor network. *Ann. Phys.*, 327:542 – 552, 2012.

[429] G. Semerjian, M. Tarzia, and F. Zamponi. Exact solution of the Bose–Hubbard model on the Bethe lattice. *Phys. Rev. B*, 80:014524, 2009.

[430] Z. Y. Xie, H. J. Liao, R. Z. Huang et al. Optimized contraction scheme for tensor-network states. *Phys. Rev. B*, 96:045128, 2017.

[431] Z. C. Gu, M. Levin, and X. G. Wen. Tensor-entanglement renormalization group approach as a unified method for symmetry breaking and topological phase transitions. *Phys. Rev. B*, 78:205116, 2008.

[432] B. B. Chen, Y. Gao, Y. B. Guo et al. Automatic differentiation for second renormalization of tensor networks. *Phys. Rev. B*, 101:220409, 2020.

[433] C. Wang, S. M. Qin, and H. J. Zhou. Topologically invariant tensor renormalization group method for the Edwards–Anderson spin glasses model. *Phys. Rev. B*, 90:174201, 2014.

[434] K. Okunishi and T. Nishino. Kramers–Wannier approximation for the 3D Ising model. *Prog. Theo. Phys.*, 103:541–548, 2000.

[435] N. Maeshima, Y. Hieida, Y. Akutsu, T. Nishino, and K. Okunishi. Vertical density matrix algorithm: A higher-dimensional numerical renormalization scheme based on the tensor product state ansatz. *Phys. Rev. E*, 64:016705, 2001.

[436] A. García-Sáez and J. I. Latorre. Renormalization group contraction of tensor networks in three dimensions. *Phys. Rev. B*, 87:085130, 2013.

[437] X. Feng and H. W. J. Blöte. Specific heat of the simple-cubic Ising model. *Phys. Rev. E*, 81:031103, 2010.

[438] A. L. Talapov and H. W. J. Blote. The magnetization of the 3D Ising model. *J. Phys. A*, 29:5727–5733, 1996.

[439] S. G. Chung. New method for the 3D Ising model. *Phys. Lett. A*, 359:707–711, 2006.

[440] P. Butera and M. Comi. Extension to order β^{23} of the high-temperature expansions for the spin-1/2 Ising model on simple cubic and body-centered cubic lattices. *Phys. Rev. B*, 62:14837–14843, 2000.

[441] R. Gupta and P. Tamayo. Critical exponents of the 3-D Ising model. *Int. J. Mod. Phys. C*, 7:305–319, 1996.

[442] Y. Deng and H. W. J. Blöte. Simultaneous analysis of several models in the three-dimensional Ising universality class. *Phys. Rev. E*, 68:036125, 2003.

[443] M. Hasenbusch. Finite size scaling study of lattice models in the three-dimensional Ising universality class. *Phys. Rev. B*, 82:174433, 2010.

[444] M. Hasenbusch. Monte Carlo studies of the three-dimensional Ising model in equilibrium. *Int. J. Mod. Phys. C*, 12:911–1009, 2001.

[445] M. Campostrini, A. Pelissetto, P. Rossi, and E. Vicari. 25th-order high-temperature expansion results for three-dimensional Ising-like systems on the simple-cubic lattice. *Phys. Rev. E*, 65:066127, 2002.

[446] H. X. He, C. J. Hamer, and J. Oitmaa. High-temperature series expansions for the (2+1)-dimensional Ising model. *J. Phys. A: Math. Gen.*, 23:1775, 1990.

[447] J. Oitmaa, C. J. Hamer, and W. Zheng. Low-temperature series expansions for the (2+1)-dimensional Ising model. *J. Phys. A: Math. Gen.*, 24:2863, 1991.

[448] H. W. J. Blöte and Y. Deng. Cluster Monte Carlo simulation of the transverse Ising model. *Phys. Rev. E*, 66:066110, 2002.

[449] M. H. Kalos. Monte Carlo calculations of the ground state of three- and four-body nuclei. *Phys. Rev.*, 128:1791–1795, 1962.

[450] M. P. Nightingale, V. S. Viswanath, and G. Müller. Computation of dominant eigenvalues and eigenvectors: A comparative study of algorithms. *Phys. Rev. B*, 48:7696–7699, 1993.

[451] W. E. Arnoldi. The principle of minimized iterations in the solution of the matrix eigenvalue problem. *Quarterly Appl. Math.*, 9:17–29, 1951.

[452] R. B. Lehoucq and D. C. Sorensen. Deflation techniques for an implicitly restarted Arnoldi iteration. *SIAM J. Matrix Anal. Appl.*, 17:789–821, 1996.

[453] N. Metropolis, A. W. Rosenbluth, M. N. Rosenbluth, A. H. Teller, and E. Teller. Equation of state calculations by fast computing machines. *J. Chem. Phys.*, 21:1087–1092, 1953.

Index